Friction-Induced Vibrations and Self-Organization

Mechanics and Non-Equilibrium Thermodynamics of Sliding Contact

Friction-Induced Vibrations and Self-Organization

Mechanics and Non-Equilibrium Thermodynamics of Sliding Contact

Michael Nosonovsky
Vahid Mortazavi

CRC Press
Taylor & Francis Group
Boca Raton London New York

CRC Press is an imprint of the
Taylor & Francis Group, an **informa** business

CRC Press
Taylor & Francis Group
6000 Broken Sound Parkway NW, Suite 300
Boca Raton, FL 33487-2742

First issued in paperback 2017

© 2014 by Taylor & Francis Group, LLC
CRC Press is an imprint of Taylor & Francis Group, an Informa business

No claim to original U.S. Government works

ISBN-13: 978-1-4665-0401-1 (hbk)
ISBN-13: 978-1-138-07432-3 (pbk)

Library of Congress Cataloging-in-Publication Data

Nosonovsky, Michael.
    Friction-induced vibrations and self-organization : mechanics and non-equilibrium thermodynamics of sliding contact / Michael Nosonovsky, Vahid Mortazavi.
    pages cm
    Includes bibliographical references and index.
    ISBN 978-1-4665-0401-1 (hardback)
    1. Nonequilibrium thermodynamics. 2. Sliding friction. 3. Vibration. I. Mortazavi, Vahid. II. Title.

QC318.I7.N67 2013
531'.4--dc23   2013005555

Visit the Taylor & Francis Web site at
http://www.taylorandfrancis.com

and the CRC Press Web site at
http://www.crcpress.com

Contents

Preface: Rediscovering Friction in the Twenty-First Century

This book is different from numerous other textbooks and monographs about friction. In this work we study various effects and manifestations of friction in order to identify those properties of friction that are *invariant for different friction mechanisms* and thus constitute the very essence of friction as a physical phenomenon. By doing that, we are making the case that friction is a fundamental force of nature whose properties can be deduced directly from the second law of thermodynamics, rather than treated as a collection of phenomenological unrelated effects and empirical facts.

When this book was conceived by the authors, we had several goals and ambitions in mind. First, there is a growing research field of friction-induced self-organization (pattern formation), which, however, still remains a somewhat marginal area of mainstream tribological studies. Friction is usually thought of as a process that leads to irreversible dissipation of energy, and wear is thought of as irreversible deterioration. Many scientists and engineers do not realize that, under certain conditions, friction can lead to the formation of new structures at the interface, including in situ tribofilms and various patterns at the interface. Friction-induced self-organization was studied mostly by scholars in Eastern Europe in the 1970s–1990s, although the field remains exotic to many tribologists in other countries.

Second, we wanted to emphasize the relationship between friction-induced instabilities and friction-induced self-organization. Friction is usually thought of as a stabilizing factor; however, sometimes friction leads to the instability of sliding, in particular when friction is coupled with another process. For example, thermoelastic instabilities were studied extensively by J. R. Barber. These instabilities arise from the fact that friction is coupled with heat generation, which, in turn, is coupled with material expansion, creating a positive feedback. If friction increases locally due to a random fluctuation, more heat is generated, leading to higher local normal pressures at the interface and, in turn, to increased friction. Thus, a random fluctuation has a tendency to grow, signifying the instability.

A similar situation occurs when the coefficient of friction decreases with the increasing sliding velocity: A small random increase of the sliding velocity leads to a decrease of the frictional resistance and to the further increasing of velocity. It was discovered in the 1990s that frictional elastodynamic instabilities (the Adams–Martins instabilities) occur even in the case when the coefficient of friction is constant. Instabilities are related to self-organization because they constitute the main mechanism for pattern formation. At first, a stationary structure loses its stability; after that, vibrations with increasing amplitude occur, leading to a limit cycle corresponding to a periodic pattern.

We formulate a general variational stability criterion of the stationary state of frictional sliding, $\delta^2 \dot{S} > 0$, where $\dot{S}$ is the rate of entropy production. The criterion is very powerful since it allows combining very diverse mechanisms of frictional instabilities within one general theory. The entropy S can include pure mechanical, thermodynamic, heat and mass transfer, chemical reactions, and other terms. Variations of the relevant parameters (physically corresponding to small random fluctuation) can be suppressed or expanded. The stability criterion captures this trend. For the case of the constant temperature T, sliding velocity V, and normal load W, the criterion can be significantly simplified.

Third, we wanted to combine the mechanical and thermodynamic methods in tribology and thus to extend the field of mechanical friction-induced vibrations to non-mechanical instabilities and self-organizing processes at the frictional interface. Thermodynamic analysis places it in a broader context. From the thermodynamic point of view, friction and wear are two sides of the same phenomenon: irreversible energy dissipation and material deterioration during sliding. As with any irreversibility, both friction and wear are the consequences of the second law of thermodynamics. It would therefore be logical to expect that the laws of friction and wear (such as the Coulomb and Archard laws) are deduced from the thermodynamic principles of irreversibility. However, in practice, this is difficult to achieve. We suggest a procedure on how this can be done using Onsager's linearized laws of irreversible thermodynamics combined with the asymptotic transition from the bulk to the interface. Non-equilibrium (irreversible) thermodynamics is also useful for the study of friction-induced self-organization. In general, the place of thermodynamic methods in tribology is growing.

Fourth, we wanted to relate the area of friction-induced self-organization to novel biomimetic materials, such as self-lubricating, self-cleaning, and self-healing materials. These "smart" materials have the embedded capacity for self-organization, leading to their unusual properties. Understanding the structure–property relationships leading to self-organization is the key to designing these novel materials. It is noted that these materials often have hierarchical organization, which makes them similar to biological materials.

All this background information caused us to look at friction, as a physical phenomenon, from a different angle. For most conventional textbooks on mechanics, Coulomb's dry friction is quite an external phenomenon, which is postulated in the form of laws of friction (usually, the Coulomb–Amontons laws) introduced in an arbitrary and ad hoc manner in addition to the constitutive laws of mechanics. Furthermore, the very compatibility of the Coulomb friction laws with the laws of mechanics is questionable due to the existence of the so-called frictional paradoxes or logical contradictions in the mechanical problems with friction. The Coulomb–Amontons law is not considered a fundamental law of nature, but an approximate empirical rule, whereas friction is perceived as a collective name for various unrelated effects of different natures and mechanisms—such as adhesion, fracture, and deformation—lacking any internal unity or universality.

Despite this artificial character of friction laws in mechanics, Coulomb friction is a fundamental and universal phenomenon that is observed for all classes of materials and for loads ranging from nanonewtons in nanotribology to millions of tons in seismology. There is a clear contradiction between the generality and universality of friction and the artificial manner of how the friction laws are postulated in mechanics and physics. Is it by chance that all these diverse conditions and mechanisms lead to the same (or at least similar) laws of friction? We think that if a thermodynamic approach is used consistently, the laws of friction and wear can be introduced in a much more consistent way (if not in an axiomatic manner in the sense of Hilbert's sixth problem).

These reflections about the fundamental nature of friction and its status in physics caused us also to examine with great attention the role of friction in physics throughout the history of science. In the days of Aristotle's *Physics,* friction was seen as a fundamental force having the same status as inertia force. The problem of inertia force was one of the central issues of physics throughout the Middle Ages, until it was finally resolved by Galileo, leading to the foundation of modern mechanics by Newton in the late seventeenth century. In contrast, friction force was not usually seen by the philosophers of science as a fundamental force of nature and became somewhat marginal in modern mechanics. Viewing friction as a fundamental force of nature to some extent restores the original Aristotelian approach. The introductory chapter concentrates on these historical matters and their applications to a modern philosophy of science in general and of mechanics in particular.

The organization of the rest of this book is the following: The second chapter introduces general concepts related to vibrations, instabilities, and self-organization in the bulk of materials and at the interface. The third chapter discusses the principles of non-equilibrium thermodynamics in application to the interface. On the basis of these principles, the laws of friction are formulated in the fourth chapter and various important implications of these laws are studied. Wear and lubrication are analyzed in the fifth chapter. The entire sixth chapter, the longest chapter of this book, is devoted to the theoretical study of different types of friction-induced vibration. The seventh chapter deals with practical situations and applications where friction-induced vibrations are important. The eighth chapter considers various types of friction-induced self-organization, and the ninth chapter is concerned with the use of these effects for novel self-lubricating, self-cleaning, and self-healing materials. The ninth chapter reviews novel materials with the capacity of self-organization (self-lubrication).

The book is based to a large degree on the original research of the authors: the study of friction-induced vibrations and waves in 1997–2001 by one author (M. N.) that led to his PhD dissertation at Northeastern University (advisor: Dr. George G. Adams), his further studies in nanotribology and biomimetic tribology since 2002, and a joint study of friction-induced self-organization by both authors, which led to another PhD dissertation (V. M.) at the University of Wisconsin-Milwaukee (UWM).

We leave to the readers to judge to what extent our goals of presenting a new, twenty-first century view on friction and tribology have been achieved.

We would like to thank our colleagues from the UWM, including Dr. Roshan Dsouza, Dr. W. Tysoe, and those who helped in preparation of this book—in particular, Dr. Marjorie P. Piechowski, who proofread the manuscript. The support of the University of Wisconsin-Milwaukee Research Growth Initiative is also appreciated.

Dr. Michael Nosonovsky and Vahid Mortazavi
University of Wisconsin-Milwaukee

Authors

Michael Nosonovsky is an assistant professor at the University of Wisconsin-Milwaukee. He received his MSc from St. Petersburg Polytechnic University (Russia) and PhD in mechanical engineering from Northeastern University (Boston). After that, he worked at Ohio State University and the National Institute of Standards and Technology. Michael's interests include biomimetic surfaces, capillary effects, nanotribology, and friction-induced self-organization.

Vahid Mortazavi is a doctoral student at the University of Wisconsin-Milwaukee. He received his MSc from Tarbiat Modares University in Tehran, Iran. Vahid's research interests include friction, tribology, and heat transfer.

1

Introduction: Friction as a Fundamental Force of Nature in the History of Mechanics

One of the objectives of this work is to present the principles of friction as a fundamental force of nature in a rigorous and deductive manner. As a first step toward this purpose, in this chapter we will review the logical foundations of mechanics in general. Therefore, this chapter is different from the rest of the book in the sense that it discusses many historical, methodological, and even philosophical problems of mechanics. However, we believe that without understanding these logical and historical foundations, it is impossible to understand properly many current developments. In particular, we will discuss the difference between the Newtonian and Eulerian approaches in mechanics and the attempts to create axiomatic descriptions of the classical (or "rational") mechanics, ranging from the formulation of Hilbert's sixth problem to modern work by Clifford Truesdell, Walter Noll, and others. This background will allow us to understand the role of the deductive approach in mechanics in general, the relationship between mechanics and thermodynamics, and the place that the contact interactions and interfaces occupy in mechanics.

1.1 Historical Background

Mechanics, as we know, has its roots in ancient Greece. In his *Physics*, Aristotle (384–322 BC, Figure 1.1) introduced the concept of force as a cause of motion. According to his teaching, bodies tend to move with the velocity V, which is proportional to the applied force

$$F = \eta V \tag{1.1}$$

where η is a coefficient of proportionality (the viscosity; Nosonovsky 2012).

When no force is applied, bodies move ideally toward the "natural place of the elements" (earth, water, air, and fire) of which they consist. The ideal terrestrial motion is straightforward (upward for bodies dominated by the "air" and "fire" elements and downward for bodies dominated by the "earth" and "water" elements) with a constant speed. The ideal celestial motion is

FIGURE 1.1
Aristotle (384–322 BC)

circular with a constant speed. Aristotle considered place (defined as a two-dimensional [2D] boundary, as opposed to the three-dimensional [3D] volume), time, and void as three preconditions of motion. He also introduced the concepts of potential and actual cause. Aristotle believed that matter is continuous and does not consist of discrete atoms, as some ancient scholars thought (Merkin 1994).

The most evident flaw of Aristotle's physics, which contradicted the everyday experience, was the problem of a projectile motion; that is, why an arrow or a stone thrown by a person would continue its motion after force was no longer applied. Apparently Aristotle himself believed that this phenomenon was due to rare air in front of the moving body. However, this could not be considered a satisfactory explanation. The answer was suggested by Jean Buridan (1300–1358), who believed that a special acting force called "impetus" is transferred to the body by the mover, so that the body continues its motion under the action of impetus. The quantitative value of impetus was equal to the mass of the body multiplied by its velocity, making it similar to the modern concept of momentum. Buridan also believed that celestial bodies possessed what he called the circular impetus (angular momentum), which makes them move with a constant speed along a circular trajectory.

The further development of mechanics was related mostly to astronomical observations. According to the Ptolemaic system, the Sun rotates around the Earth. However, it was clear that the trajectories of the Sun and the planets are not perfectly circular, as Aristotle's physics would suggest for the celestial motion. In his treatise "Almagest," Claudius Ptolemy (90–168) brought forward the idea of epicycles. The planets were assumed to move in a small circle called an epicycle, which in turn moves along a larger circle called a deferent. With the increasing amount of astronomical observations and data, the geocentric system became quite complex, involving a large number of

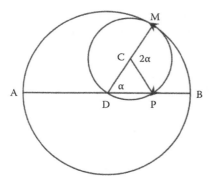

FIGURE 1.2
The "Tusi couple" theorem: a small sphere rolls inside a large sphere of doubled diameter, so that the point P remains always on the straight line AB.

epicycles and an even more controversial concept of the equant point. The point on the deferent moves with a constant angular velocity with respect to the equant and not the center of the deferent, which was in contradiction with Aristotle's ideal uniform celestial rotation.

The dichotomy between the circular "celestial" motion and straightforward "terrestrial" motion existed until the Persian astronomer Nasir al-Din al-Tusi (1201–1274) demonstrated that straightforward motion can be presented as a superposition of two rotational motions: a small circle rotating inside a larger circle of twice the same radius (Figure 1.2). This so-called "Tusi couple" could substitute for the concept of equant and was eventually adopted by Nicolaus Copernicus (1473–1543) in his revolutionary heliocentric system (Veselovsky 1973). Historians of astronomy hypothesized that Copernicus could have been aware of Tusi's result (Saliba 1996). Several possible effects on European science have been considered, including the works of Byzantine Greek scientists Proclus Lycaeus (412–487), Nicole Oresme (1320–1382), and Abner of Burgos (1270–1347). It is noted also that a crankshaft or crank-slider mechanism converting reciprocating linear motion into rotation was known to the Romans at least since the third century. This mechanism essentially provides the same function as the Tusi couple. However, a satisfactory engineering solution for the conversion of the rotational motion into a straight line was not proposed until James Watt invented the straight-line linkage mechanism for his steam engine (Norton 2011).

The Copernican system and new observations made the main impact on the development of mechanics in the seventeenth century. Johannes Kepler (1571–1630) used the astronomical data obtained by Tycho Brahe (1546–1601) to formulate his laws of planetary motion and found that the planetary orbits are elliptical. Galileo Galilei (1564–1642) conducted both astronomical and mechanical studies. He formulated what is now called Galileo's principle of inertia (GPI): "A body moving on a level surface will continue in the same direction at constant speed unless disturbed." Galileo also recognized the importance of friction force and was the first to find that the distance traveled by a body under the action of force is proportional to the square of time, although apparently some earlier authors such as Nicole Oresme suggested this as well.

Galileo also introduced the idea of inertial "force" as well as the idea of frictional force, which in many practical circumstances prevents bodies from their inertial motion with a constant velocity. This was a radical enhancement of the idea of force (the cause of action) as it was understood in Aristotle's physics. Although the philosophical concepts of "inertia" and "impetus" were discussed by John Philoponus (490–570), Avicenna (980–1037), Jean Buridan (1300–1358), and possibly even Chinese philosopher Mo Tzi (470–391 BC), only the GPI contained a clear formulation of this idea (Figure 1.3). Thus, the problem of inertia as a fundamental force of nature found a satisfactory resolution.

The situation was different for friction, another fundamental force from Aristotle's physics. After the discovery of inertia, it was realized that the velocity of motion is not proportional to the applied force, and therefore friction is not proportional to velocity. Instead, the new concept of friction force and its proportionality to the applied normal load was suggested by Leonardo da Vinci (1452–1519), who believed that the friction force is equal to a quarter of the normal load. In a clearer way it was proposed by Guillaume

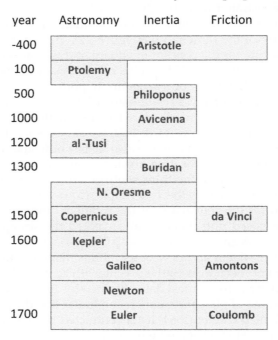

year	Astronomy	Inertia	Friction
-400		Aristotle	
100	Ptolemy		
500		Philoponus	
1000		Avicenna	
1200	al-Tusi		
1300		Buridan	
		N. Oresme	
1500	Copernicus		da Vinci
1600	Kepler		
		Galileo	Amontons
		Newton	
1700		Euler	Coulomb

FIGURE 1.3
Scientists who contributed to the understanding of celestial motion, inertia, and friction force.

Amontons (1663–1705) and later by Charles-Augustin de Coulomb (1736–1806), who formulated the laws of friction, stating that the friction force is proportional to the normal load, does not depend on the area of contact, and is almost independent of the sliding velocity (Truesdell 1968, 1987).

The main breakthrough in mechanics was made by Isaac Newton (1643–1727) who, for the first time, was able to unite the terrestrial and celestial motion by showing that this is the same phenomenon, together with the clear formulation of the second law of Newton, which states that force F is proportional to the acceleration a and not the velocity:

$$F = ma \qquad (1.2)$$

where m is the mass.

This was the abandonment of Aristotle's physics. Furthermore, Newton introduced infinitesimal calculus, which essentially resolved the ancient opposition of motion and rest, showing the state of rest is a special case of the state of motion. Despite that, it should be understood that Newton's mechanics is formulated in such a way that it can best deal with the material points, rather than rotating rigid bodies. Therefore, the opposition between the translational and rotational motion remained to be resolved; this was achieved by L. Euler. Newton's famous saying states that he "has seen further only by standing on the shoulders of giants." While Galileo has always been considered one of these giants, pre-Renaissance developments that resulted in a separation from Aristotelian physics were also significant. In addition to the previously mentioned John Philoponus and Jean Buridan, the group called the Oxford Calculators of the fourteenth century should be mentioned. They discovered properties of uniformly accelerated motion and influenced Galileo, apparently through Giovanni di Casali (d. 1375).

At about the same time, the concept of *vis viva* ("living force") was suggested by Gottfried Leibnitz (1643–1716). He was apparently motivated by the same idea as the "impetus" force of Buridan—namely, the "force of inertia" that acts upon the body during its free motion. While impetus led to the law of conservation of momentum, vis viva yielded the law of conservation of energy, although the latter term was coined in 1802 by Thomas Young (1773–1829), who realized that mechanical energy and internal (heat) energy are quantities of the same nature.

The seventeenth century also was marked by the birth of modern empirical or experiment-driven science, resulting in the separation of science and magic. Up until the Renaissance, rational thinking following the traditions of Plato and Aristotle was thought to be opposed to mystical ways of obtaining "concealed" knowledge using personal experience or a "secret" tradition. Such theories and practices as alchemy, astrology, Kabbalah, and so on belonged to that second type, since they implied the ability to obtain knowledge by revelation rather than from a wise authority. Empirical knowledge was originally perceived as a part of the second paradigm since it implied personal experience as a source of knowledge, rather than studying the legacy of the scholars of the past.

Many natural scientists were also influenced by mysticism. Even Isaac Newton himself put a lot of effort into the research of what would be called today occult studies, which constitutes a significant part of his written legacy. However, in the seventeenth century it was realized that empirical science has a method of its own, which is quite different from that of magic or mystics with their individual visionary "experiences." In mechanics, the works by Galileo constituted this turn to empirical science, as well as those by Isaac Newton (1643–1727), Gottfried Leibnitz (1643–1716), Robert Hooke (1635–1703), and Christiaan Huygens (1629–1695). Remarkably, the first scientific journals, *Philosophical Transactions of the Royal Society* and *Journal des Scavans,* both emerged in 1665 as a result of these developments and the emergence of what was called natural philosophy.

While Newton is considered the founder of classical mechanics, his formulation of the laws of mechanics did not lack shortcomings. First of all, his famous three laws of motion did not constitute a closed set of axioms and dealt with quite different situations. The first law, as it is understood today, postulates the existence of inertial reference frames, relative to which the motion of a particle not subject to forces is a straight line at a constant speed (which is a formulation of the Galileo's principle of inertia). In its original formulation, the first law can be seen as a special case of the second law when $F = 0$.

Newton's second law (Equation 1.2) quantitatively relates acceleration of a point measured in an inertial reference frame to an externally applied force (although it remains to define the force in a rigorous manner). There is an interesting and important argument on whether the second law constitutes a mere definition of the concept of "force" as a cause of acceleration, or whether force can be defined independently and the second law is an empirically verifiable law of nature.

The third law states that the reaction force is always equal in absolute value and opposite in direction to the action force. In other words, force always acts between a pair of particles rather than at a single particle. This leads to the conservation of momentum, which is often treated today as a logical consequence of the translational invariance of space, in accordance with the Noether theorem. While the second law is apparently verifiable and falsifiable by experiments (at least in the opinions of most scientists and philosophers), the third law deals with the logical meaning of the concept of space. It is noted that the rotational invariance of space or the invariance of time did not lead to separate laws in Newton's system, although they lead to the conservation of the angular momentum and energy in the Noether theory. Furthermore, Newton's laws do not say anything about rotational motion.

Leonhard Euler (1707–1783), perhaps the greatest mathematician and mechanician of the eighteenth century, was the first to incorporate the rotational motion into classical mechanics (Truesdell 1968). His two laws of motion were the law of conservation of the momentum $p = mV$, stating that

$$\dot{p} = F \tag{1.3}$$

and the law of conservation of the angular momentum (or the moment of momentum) $L = I\omega$ (in the general case, I is the tensor of the second rank), stating that

$$\dot{L} = M \tag{1.4}$$

where M is the applied torque, ω is the angular velocity of rotation, and I is the moment of inertia. This suggested the parallelism between the translational and rotational motion.

Among the developments that followed, one should mention the ideas of Jean-Baptiste d'Alembert (1717–1783), who tried to formulate the principles of mechanics by excluding the concept of force, which he considered a metaphysical and obscure idea. Despite that, he still used the concept of force quite intensively, but considered it an auxiliary concept without a clear physical meaning. The laws of motion were formulated by d'Alembert in the form of the principle that bears his name and states that the sum of the differences between the forces acting on a system F_i and the time derivatives of the momentum of the system along any virtual displacement δr_i consistent with the constraints of the system is zero:

$$\sum (\dot{p}_i - F_i)\delta r_i = 0 \tag{1.5}$$

Joseph Louis Lagrange (1736–1813) suggested the "analytical mechanics" (*mecanique analytique*) approach to the fundamental laws of mechanics and proposed a new differential equation that dealt with the difference of the kinetic and potential energies $L = T - \Pi$:

$$\frac{d}{dt}\frac{\partial L}{\partial \dot{x}} - \frac{\partial L}{\partial x} = 0 \tag{1.6}$$

This further led to Hamiltonian mechanics and to the variational principle of the least action stating that the time integral of L among the actual trajectory is at minimum:

$$S = \int_{t_1}^{t_2} L\,dt \tag{1.7}$$

Returning to Aristotle, the teleological ideas according to which there is a predetermined objective of motion clearly lie behind the variational principles.

Heinrich Hertz (1857–1894) is famous for his solution of the contact problem between two elastic bodies. The solution states that an elastic sphere (with the elastic modulus E_1 and Poisson's ratio v_1) of radius R pressed by the normal force W into a flat elastic surface (with the elastic modulus E_2 and Poisson's ratio v_2) will have the contact area of

$$A = \pi \left(\frac{3RW}{4E^*} \right)^{2/3}$$

(1.8)

where $E^* = (1 - v_1^2)/E_1 + (1 - v_2^2)/E_2$ is the effective elastic modulus.

Hertz also attempted to formulate classical mechanics in a geometric way, excluding the concept of force. He considered a trajectory of a system of points having the total mass M in a 3N-dimensional space and defined the curvature as

$$ds^2 = \frac{1}{3M} \sum m_j dx_j^2$$

(1.9)

According to Hertz's principle of minimum curvature, the system attains the trajectory of least curvature. This principle is related to the principle of the least action (Equation 1.7) and it influenced general relativity (Merkin 1994).

Among other developments in understanding the fundamentals of classical mechanics it is worth mentioning the theorem by Emmy Noether (1882–1935) that related the laws of conservation of energy, momentum, and angular momentum to the translational symmetry (invariance) of time, translational, and rotational symmetries of space. In this brief overview we are not discussing developments in the mechanics of continuous media, hydrodynamics, and thermodynamics, which also played a role in understanding the logical foundations of mechanics.

The introduction of irreversibility (the second law of thermodynamics) in classical mechanics is of importance for the study of friction, since the latter is an irreversible dissipative process. Lord Rayleigh (1842–1919) introduced the concept of the dissipative function as the quadratic form of velocities; however, it was defined only for linear viscous friction. A. Lurie (1901–1980) generalized the dissipative function for an arbitrary dependence of friction forces on velocity, including the Amontons–Coulomb friction law (Zhilin 2006). However, such an introduction did not occur through any fundamental law of mechanics.

1.2 Certain Philosophical Concepts of Mechanics

It is useful to look at the evolution of classical mechanics from the ancient time until modern times through the development of the concepts of opposing pairs associated with mechanics and resolution of the contradictions related to such pairs. Aristotle's *Physics* brought a number of such oppositions.

1.2.1 Discrete versus Continuum

Although Aristotle himself supported the continuum description, the atomistic idea of matter existed from the earliest times. This contradiction was resolved, to a large extent, by the invention of the infinitesimal calculus by Newton and Leibnitz that led to the modern understanding of continuity. This understanding, however, backfired in the twentieth century when it became clear that many processes do not fit the continuum description. Thus, the concept of fractals was introduced by Benoit Mandelbrot (1924–2010).

Fractals are self-similar or self-affined geometrical objects that retain the same shape when zoomed. In particular, although fractal functions are continuous, they do not have a derivative at every point. It was realized that although many processes are very well described by continuous and differentiable functions implied by infinitesimal calculus, certain processes and objects, usually iterative, do not fit that description well. For example, a rough surface profile reveals more and more details with a smaller roughness scale when zoomed.

In continuum mechanics there are many phenomena where the discrete structure of the material is inherent and neighboring particles of the material do not remain neighboring during the entire process. These include turbulence, certain phase transitions, formation of cracks, and large non-elastic deformations.

1.2.2 Rest versus Motion

Rest and motion in Aristotelian physics were opposite states. Understanding the relationship between rest and motion was a difficult task for the ancient philosophers. One of Zeno's paradoxes cited by Aristotle, the arrow paradox, stated, "If everything when it occupies an equal space is at rest, and if that which is in locomotion is always occupying such a space at any moment, the flying arrow is therefore motionless." In other words, if at any moment of time the arrow is at a particular point of space ("resting"), how can motion occur? Aristotle's solution was that "time is not composed of indivisible nows."

However, rest was not conceived as a special case of motion (with zero velocity). In the Middle Ages, it was significant to point out that the oscillating motion passes through the state of instant zero velocity, although the oscillating body does not spend any finite time at that point. The true solution was developed with the emergence of calculus in the time of Newton and Leibnitz, when it was understood that velocity is a derivative of the coordinate and thus, although an arrow is at particular point at any particular moment of time, what makes it moving is the finite velocity rather than being at a particular spot.

1.2.3 Terrestrial versus Celestial Motion

Terrestrial versus celestial motion was another important distinction in Aristotle's physics. While terrestrial motion is straightforward and has a natural objective or destination corresponding to the nature of an element, celestial motion is circular and eternal. The distinction between terrestrial and celestial motion was criticized by some philosophers, such as John Philoponus, on various grounds. However, the distinction was abandoned only after Newton showed that the motion of the planets and terrestrial bodies (such as an apple falling from a tree) is governed by the same laws. It is noted that the idea that the straightforward motion is the natural one has influenced the GPI and the first law of Newton, as well as some variational principles of mechanics, such as the Hertz's principle of minimum curvature. It is noted also that the distinction may be viewed as an earliest formulation of the distinction of reversible and irreversible processes, which entered modern physics only when the second law of thermodynamics was introduced by Rudolf Clausius (1822–1888).

1.2.4 Translational versus Rotational Motion

Translational versus rotational motion is another important opposition in mechanics. As discussed earlier, Aristotle considered straightforward motion natural for terrestrial objects, whereas circular motion was natural for celestial bodies. The concept of "the Tusi couple" was significant in understanding that straight motion can play a role in celestial mechanics since, fundamentally, it is not different from rotational motion. Newton's mechanics was formulated for material points that were characterized by their position and not by their orientation in space.

It was Euler who realized that the law of momentum and the law of angular momentum (Equations 1.3 and 1.4) are completely independent from each other, and thus he established the parallelism between translational and rotational motion. He developed a way to deal with motion of a rotating rigid body characterized by the so-called Euler's angles, rather than by material points. In the nineteenth century many attempts were made to describe electrical phenomena by translational and rotational motion (dilatational and shear oscillations and waves) of the presumed substance called *ether*, although these attempts were abandoned with the emergence of relativity.

1.2.5 Empty versus Occupied Space

The concept of space is another difficult concept in Aristotle's physics. The preconditions of motion are "place, void, and time." Unlike 3D space occupied by a body, "place" was considered a 2D boundary surface. Aristotle did not believe in empty space and suggested that, in a vacuum, if it could exist, terrestrial bodies would move with infinite velocity to the natural spaces of the element of which they are composed. The concept of space was controversial already in ancient times. Thus, one of Zeno's paradoxes was that "if everything that exists has a place, place too will have a place, and so on ad infinitum." Newton believed in absolute space (and the existence of inertial frames of references was the argument), whereas Leibnitz considered space an abstraction characterizing the relationship between objects.

1.2.6 What Is Force?

The concept of force was conceived in Aristotle's physics as an acting cause of every motion, which was thought to be proportional to velocity. Later, the concept of force evolved into three completely different entities. The *impetus* (momentum) as well as the *vis viva* ("living force" or energy) is responsible for the projectile motion of a body "by inertia." On the other hand, force per se is proportional to acceleration in accordance with the second law of Newton. This made the ontological status of force questionable. Is the only way to determine the value of force from the measurement of acceleration? If so, does the second law constitute only the definition of force, and the concept itself can be eliminated from classical mechanics as redundant? Or is force a separate entity that can be defined independently of the second law of Newton?

If force is defined from external laws (e.g., the law of gravity, Coulomb's law of electricity, Hooke's law of elasticity, or the Archimedes law), then Newton's second law becomes hard to falsify (in the sense of Karl Popper), since any experimentally detected inconsistency could be treated as falsification of such external law, rather than the second law itself. On the other hand, the more extreme position of d'Alembert considered the second law a tautological definition of force that could be excluded from mechanics. This led to the definition of laws of motion using extremal and variational principles, rather than force. Besides inertia, another important finding was the existence of friction force, which is proportional to the normal load, rather than the sliding velocity, as Aristotle's physics stated.

1.2.7 The Law of Identity

There are other important philosophical concepts that can be related to the logical foundations of mechanics. Among them are space, time, and identity, which, according to some ancient philosophical theories (e.g., the kabbalistic *Sefer Yetzirah*), define the texture of the events in our world. While the "space–time" continuum (x, y, z, t) constituting an $\mathbf{R}^4$ manifold is a very common concept for a modern scientist, the presence of the third element— "identity" (or "individuality")—requires some explanation. Indeed, we do not normally use a separate variable to identify objects. Instead, the objects are identified by names or labels, which, at first glance, have little to do with the space–time continuum. However, the concept of motion implies space, time, and identity. The trajectory of the point (a particle) called R is character- ized by the vector function

$$\vec{R}(t) = \vec{\mathbf{i}}x(t) + \vec{\mathbf{j}}y(t) + \vec{\mathbf{k}}z(t) \tag{1.10}$$

where $\vec{\mathbf{i}}$, $\vec{\mathbf{j}}$, and $\vec{\mathbf{k}}$ are the unit vectors in the x-, y-, and z-directions). Here the variables x, y, and z characterize space; t is time; and $\vec{R}$ is, in a sense, the identity of the point.

In order to speak about a trajectory of a particle $\vec{R}(t)$, we must be confident that at any arbitrary moment of time, t_1, we deal with the same particle as at the moment t_0. In other words, the procedure of identifying an object is not less important than the procedure of measuring time intervals and lengths. Apparently these three categories reflect the ability of the human mind to identify objects and events intuitively and to ascribe time and space to them. The procedures of measurement of time intervals, spatial lengths, and establishing identities of material points are needed to define the concept of motion. "The fact that a thing is itself" constitutes "the law of identity," as formulated by Plato and Aristotle, which is usually considered a part of metaphysics, rather than physics. Note also that in Equation (1.10) $\vec{R}(t)$ is a "left value" or "signifier," whereas $x(t)$, $y(t)$, and $z(t)$ are the "right value" or "signified" of the motion in the sense in which these terms are used to define sign as a combination of the signifier and signified.

1.3 Hilbert's Sixth Problem

The problem of adequately formulating the logical foundations of classi- cal mechanics is related to the so-called Hilbert's sixth problem. In 1900, the famous mathematician David Hilbert (1862–1943) delivered a lecture in front of the International Congress of Mathematics in Paris. In that lecture he formulated 23 problems that, in his opinion, would constitute the major challenge for mathematicians of the twentieth century. Among other prob- lems, which were purely mathematical, he listed the problem of axiomatiza- tion of physics:

The supply of problems in mathematics is inexhaustible, and as soon as one problem is solved numerous others come forth in its place. Permit me in the following, tentatively as it were, to mention particular definite problems, drawn from various branches of mathematics, from the discussion of which an advancement of science may be expected.

The investigations on the foundations of geometry suggest the problem: To treat in the same manner, by means of axioms, those physical sciences in which mathematics plays an important part; in the first rank are the theory of probabilities and mechanics.

As to the axioms of the theory of probabilities, it seems to me desirable that their logical investigation should be accompanied by a rigorous and satisfactory development of the method of mean values in mathematical physics, and in particular in the kinetic theory of gases.

Important investigations by physicists on the foundations of mechanics are at hand; I refer to the writings of Mach, Hertz, Boltzmann and Volkmann. It is therefore very desirable that the discussion of the foundations of mechanics be taken up by mathematicians also. Thus Boltzmann's work on the principles of mechanics suggests the problem of developing mathematically the limiting processes, there merely indicated, which lead from the atomistic view to the laws of motion of continua. Conversely one might try to derive the laws of the motion of rigid bodies by a limiting process from a system of axioms depending upon the idea of continuously varying conditions of a material filling all space continuously, these conditions being defined by parameters. For the question as to the equivalence of different systems of axioms is always of great theoretical interest.

If geometry is to serve as a model for the treatment of physical axioms, we shall try first by a small number of axioms to include as large a class as possible of physical phenomena, and then by adjoining new axioms to arrive gradually at the more special theories. At the same time Lie's principle of subdivision can perhaps be derived from profound theory of infinite transformation groups. The mathematician will have also to take account not only of those theories coming near to reality, but also, as in geometry, of all logically possible theories. He must be always alert to obtain a complete survey of all conclusions derivable from the system of axioms assumed.

Further, the mathematician has the duty to test exactly in each instance whether the new axioms are compatible with the previous ones. The physicist, as his theories develop, often finds himself forced by the results of his experiments to make new hypotheses, while he depends, with respect to the compatibility of the new hypotheses with the old axioms, solely upon these experiments or upon a certain physical intuition, a practice which in the rigorously logical building up of a theory is not admissible. The desired proof of the compatibility of all assumptions seems to me also of importance, because the effort to obtain such proof always forces us most effectually to an exact formulation of the axioms. (Hilbert 1902)

Clearly, for Hilbert, this was a part of the effort for geometrization of physics (i.e., presenting physics as an extension of geometry). Numerous achievements in that direction have been made since 1900, including special and general relativity, quantum mechanics (using as the axiomatic basis the Hilbert space), and axiomatic presentation of the probability theory on the basis of measure theory, as well as various "theories of everything" that emerged in physics.

As we discussed earlier, in a broader sense, time, space, and identity (or "the subject") define three categories that are intuitively perceived by all humans. Since ancient times, science has tried to eliminate the subject from consideration and present everything occurring in the world as objective processes. The laws of nature are also supposed to be independent of the signifier. For example, it does not matter in Equation (1.10) which letter is used to letter the point; is it the letter "R" or, for example, the letter "P"? The process of identification of the point R in classical mechanics is completely independent of its motion, in the same manner in which space and time are thought to be independent of the physical objects and their motion.

Another task of objective science that was implemented very successfully was geometrization. This is an attempt to eliminate time and to present physics in a geometric way, with axioms and postulates, as it was first done by Euclid in ancient times and presented in a rigorous way by the Bourbaki school in the twentieth century.

1.4 Noll's Axiomatic Mechanics

The most consistent attempt to formulate classical mechanics in an axiomatic way was made by Clifford Truesdell (1919–2000) and his student, Walter Noll (b. 1925). Truesdell worked on what he called "rational mechanics," or mechanics as a deductive, mathematical science. He also called for a revival of Newton's term "natural philosophy," noting,

> In all of natural philosophy, the most deeply and repeatedly studied part, next to pure geometry, is mechanics. The resurgence of rational mechanics, after half a century of drowsing, has signalled and led the rediscovery of natural philosophy as a whole, just beginning in our time.... Rational mechanics was the first domain of natural philosophy on which modern mathematics was brought to bear so as to form a real theory, comparable in generality and precision to classical geometry. (Truesdell 1966)

Noll (1973) suggested a mathematical formulation of mechanics, including continuum thermodynamics into it (Noll and Seguin 2010). Certain central ideas of Noll's approach are outlined next.

1.4.1 The Laws of Mechanics

Noll explained why Newton's laws are not adequate for rational mechanics:

> When most physicists today hear the term "classical mechanics," they tend to think of Newtonian particle mechanics. In this context, Wilczek's statement, "Newton's second law of motion, F = ma, is the soul of classical mechanics," may have some merit. However, Newtonian particle mechanics is only a very small and relatively trivial part of classical mechanics. When applied to the rest of classical mechanics, Wilczek's statement is absurd. For example, engineering students often take a course called "Statics," which deals with forces in systems having no moving parts at all, and hence accelerations are completely absent. *The beginning of a textbook on statics often contains a statement of Newton's laws, but this functions like a prayer before a business meeting;* it is almost totally irrelevant to the substance of the subject. The substance of statics consists in singling out parts of the system under consideration by drawing "free-body diagrams." Engineering students often also take a course called "Dynamics." Its basic structure differs from the course in statics only by including the inertial forces among the forces considered. (I have taught courses on Statics and Dynamics in the late 1950s, and this experience has influenced my analysis of the foundations of mechanics.)
>
> The two basic principles of classical mechanics are these:
>
> 1) Balance of forces: The total force acting on a physical system and each of its parts is zero.
> 2) Balance of torques: The total torque acting on a physical system and each of its parts is zero.
>
> ...In order to give a precise formulation of these principles, an axiomatic mathematical description of the concepts of *physical systems, force-systems,* and *systems of torques* must be supplied.... Newton's third law, the law of action and reaction, is a logical consequence of the law of balance of forces stated above. (Noll 2007, pp. 1–2)

1.4.2 Space and Frame of Reference

Noll (2004) pointed out:

> The concept of a *frame of reference* is not the same as the concept of a *coordinate system,* as some people seem to believe. For a precise definition and discussion...I believe that the use of coordinate systems is an impediment to insight in conceptual consideration in all of physics, classical or modern.... It contains the following quote by Einstein: "Why were another seven years required for the construction of the general theory of relativity? The main reason lies in the fact that it is not so easy to free oneself from the idea that coordinates must have an immediate metrical meaning." Unfortunately, most physicists (and many mathematicians) are still stuck with outdated mathematical infrastructures, which make it difficult to get away from coordinate systems. (2007, p. 2)

To state his principle of material frame-indifference, Noll develops a "frame-free formulation of elasticity." Furthermore, he considers physical space "an illusion" and provides an interesting parallelism with the Chomskyan psycholinguistic theories of language acquisition:

> It should be clear from the above that the existence of a physical space, apart from any frame of reference, is an illusion. The mystery is that so many intelligent people, Newton included, fell victim to this illusion.... I submit that a solution of the mystery of the illusion of physical space does not come from physics but from psychology and neural science, and in particular from the way our brain processes visual information....
>
> Thus, it seems that the predisposition to fixate on a particular frame of reference at any given situation is hardwired into our brain at birth, just as is the ability to acquire language. Which particular frame we fixate on (or which particular language we learn) depends on the environment....
>
> The fixation on a frame is an involuntary unconscious process, and hence we may fall victim to the illusion that the frame is independent of the presence of any objects around us and hence becomes "physical space." In a way, we have been brain-washed by our own brains into the belief in a physical space. (Noll 2004, pp. 7–9)

1.4.3 Forces

By claiming that space is "an illusion," Noll has to deal with the same problem as Newton some 300 years before him—namely, to explain the existence of inertial frames of reference that seem to be chosen of all other frames of reference and thus strongly suggest absolute space. Noll (2007) revives an old argument by Mach (1838–1916):

> Ernst Mach has suggested that inertia may be considered to be an action of the far away parts of the universe on the bodies in our nearby environment, including the solar system. I believe that, in classical mechanics, it is always useful to consider inertial forces to be exerted by the far outside world on whatever system is considered....
>
> In classical particle mechanics, inertial forces are paramount and the subject would collapse if they were neglected. In continuum mechanics, inertial forces are sometimes unimportant and can be neglected, for example when considering the motion of toothpaste as it is extruded from a tube. The soul of continuum mechanics is the analysis of contact forces. (p. 3)

Noll (2005) further introduced the general concept of surface interaction that can be characterized by a so-called contactor. In the case of force interactions, torque interactions, and heat transfers, the contactor is the field of stress tensors, couple stress tensors, and heat flux vectors, respectively.

1.4.4 Logic and Mathematical Notation

In his endeavor to present mechanics in a rigorous mathematical way, Noll (1995) uses complicated mathematical formalism ("Bourbaki style") hardly understandable to an engineer or applied mechanician. He calls it "stage 3 infrastructure of mathematics" (with axioms, theorems, and proofs being stage 1 tracing to antiquity, and variables and efficient notation being stage 2 and tracing to the sixteenth century): "Unfortunately, many mathematicians, physicists, and engineering scientists are still stuck in stage 2. In a nutshell, stage 2 is based on the concepts of variables, coordinates, parameters, and constants. Stage 3 is based on the concepts of sets, of mappings with domain and co-domain, and of mathematical structures." The concept of a "materially ordered set" was introduced by Noll (2005) to describe mechanics.

1.5 P. Zhilin's Approach

An interesting and original approach to the foundations of mechanics was developed by Russian scientist Prof. Pavel Zhilin (1942–2005) in a sequence of papers published in the 1990s. Zhilin pointed out that material points of Newton's mechanics are characterized only by their spatial position and are not capable of rotation. As opposed to Newton, the central object of Eulerian mechanics is a rigid body characterized by position vector and rotation tensor. Zhilin's approach is different in a number of instances from that of Noll. In the following, we outline several important points in mechanics by Zhilin (2006).

1.5.1 Time, Space, and Coordinates

Time is one of the most difficult concepts in natural philosophy. Isaac Newton did not explain how to define uniform time flow, but he assumed that there are processes that can provide such uniformity. Newton wrote: "And for the better understanding of what follows, you may conceive, that correlate quantity of Time, or rather any other quantity that flows equably, by which time is expounded and measured (the method of fluxions and infinite series: with its application to the geometry of curve-lines)."

The difficulties related to the concept of time were analyzed by H. Poincare in depth. According to Poincare's opinion, the main problem is that it is not possible to guarantee that two time intervals that, according to a chronometer, are equal are indeed equal (Zhilin 1996). This is in contrast with Noll's opinion; he did not see any problem with the concept of time in classical mechanics and noted: "Newton also discussed the idea of absolute time. It is important to understand that there is no parallel between absolute time and absolute space. In classical physics, absolute time is not problematical, and neither Berkeley nor Mach had any quarrel with it" (Noll 2004, p. 9).

The "objectivity principle" suggests that "all physical values and laws are objective and do not depend on the choice of the coordinate system." However, many physical parameters such as velocities and accelerations depend on the frame of reference. Therefore, one should not confuse the frame of reference and the system of coordinates. According to Zhilin, the objectivity principle is satisfied automatically and no new meaningful information about any system can be extracted from it in classical mechanics. The picture is independent of the canvas on which it is drawn (Zhilin 1996). As explained in detail previously, Noll formulated the "principle of material frame indifference," which apparently has the same meaning. Zhilin, however, goes much further, stating that this principle implies that only invariant differential operators can be used in mechanics, such as

$$\nabla, \quad \frac{d}{dt}, \quad \nabla\frac{d}{dt} = \frac{d}{dt}\nabla \tag{1.11}$$

This is in contrast with relativistic physics, which uses different operators:

$$\nabla, \quad \frac{\partial}{\partial t}, \quad \nabla\frac{\partial}{\partial t} \neq \frac{\partial}{\partial t}\nabla \tag{1.12}$$

where the partial time derivatives depend on the choice of the coordinate system.

These operators do not commute, which makes the principle of relativity much more important in relativity, so that "even new laws can be deduced from it." According to Zhilin, this marks a gap between classical mechanics and modern physics. The fundamental laws of mechanics are logical laws and definitions, which, like any logical laws, cannot be tested or disproved experimentally. Modern physics (since the emergence of relativity) views the same laws as experimental observations of nature rather than logical constructions. Similarly to Noll, Zhilin (2003) developed a reference-free description of mechanics, appropriate for non-elastic processes in which neighboring points of materials do not necessarily remain neighboring. However, unlike Noll, Zhilin does not consider space non-existent and views the fundamental equations of mechanics as purely logical laws.

1.5.2 Fundamental Laws as Definitions

There are four fundamental laws of mechanics: two Euler's laws of dynamics (Equations 1.3 and 1.4) and two laws of thermodynamics (the first and second law). These laws are logical statements that cannot be falsified experimentally. They do not constitute laws of nature, but rather methods to investigate nature. Why are such non-falsifiable statements possible at all? The reason is that each truly fundamental law can be viewed as a definition of some new quantity. The first law of dynamics brings a new concept of force. The second law brings a concept of angular momentum (or torque). The energy balance equation (the first law of thermodynamics) brings a new concept of internal energy. The combination of the three laws allows us to consider temperature, entropy, and chemical potential. The status of the fourth fundamental law of mechanics—that is, the second law of thermodynamics—is more questionable. This law introduces the concept of irreversibility into mechanics. Despite numerous attempts to view it as a fundamental (i.e., purely logical) law, it apparently remains an empirical observation of nature.

The concept of the fundamental laws as definitions can be illustrated by a simple example. Let us consider two laws: the second law of Newton, establishing the relationship between force, mass, and acceleration, $F = ma$, and the law of gravity, establishing the magnitude of the gravity force between masses m and M equal to $F = GmM/R^2$, where G is a gravity constant. One who uses these laws would calculate the F from the latter equation and substitute the result into the former equation to calculate the acceleration. In principle, force can be excluded as a redundant concept by postulating $a = GM/R^2$.

However, since the concept of force is useful and convenient, it is usually not excluded. Then the question can be asked: Which one of the two equations provides the definition of force, and which one reflects the experimental observation? According to Zhilin, the fundamental law, $F = ma$, provides the definition, whereas the gravity law is empirical. This marks the distinction between classical mechanics, which uses fundamental laws as non-falsifiable logical statements, and modern physics, which views such laws as laws of nature and often substitutes the aesthetic beauty of a physical law for logic.

1.5.3 Force versus Torque; Linear versus Rotational Motion

Force and torque represent the action of one body upon another. Thus, inertia forces are not proper forces, since they are not caused by other bodies. Force is an additive quantity; that is, the total force acting upon a system of two bodies, A and B, is equal to the sum of forces acting on A and B.

As discussed earlier, the law of angular momentum is totally independent of the law of linear momentum. This fact was realized by Euler in 1771, whereas Newtonian mechanics did not consider the angular momentum as a fundamental quantity because it concentrated on the material points that cannot rotate. (They do not possess any moment of inertia and rotational degrees of freedom.) Newtonian material points introduced asymmetry between translational and rotational motion. Zhilin suggested investigating a material (which he called Kelvin's medium) consisting of particles—each of which has not only distributed mass but also the moment of inertia—and suggested that such a medium could be used for modeling ferromagnetism (Grekova and Zhilin 2001). Furthermore, he suggested mechanical media that can be described mathematically by electrodynamics and quantum mechanics equations, somewhat reviving the idea of Ether.

1.5.4 Thermodynamic Quantities: Temperature and Entropy

In addition to mechanical degrees of freedom, there are "hidden" or "internal" degrees of freedom of a mechanical system, corresponding to the vibration of atoms and molecules and other microscopic effects. One cannot observe motion corresponding to these degrees of freedom; however, their energy constitutes the internal energy of the body.

1.6 The Study of Friction

In the preceding sections, we reviewed the development of fundamental concepts of mechanics and showed that the friction force was originally viewed as fundamental as the inertia force. Furthermore, these two forces could not be considered without each other, since friction is what prevents the inertial motion in Galileo's mechanics. The lack of attention to friction force in Aristotle's physics was due to his misunderstanding of inertia. However, after the problem of inertia was solved, friction did not receive the status of a fundamental force of nature because the available experimental data and conceptual understanding of friction mechanisms did not allow for such understanding.

Friction and lubrication technology was known from ancient times. Some scholars attribute the earliest water lubrication technology to ancient Egyptians, as evidenced by the paintings in Saccara, ca. 2400 BC, and in El Bersheh, ca. 1880 BC, showing transportation of huge stone statues (Dowson 1998). Others historians pointed out that pouring of water during the transportation of a colossus could be a ceremonial rather than a practical act, accompanied also by fanning burning incense in honor of the statue (Figure 1.4). Oil lubrication of leather shields by King Saul (ca. 1020 BC) is mentioned also in the Bible (Nosonovsky 2007). There is also possible evidence of animal fat axle lubrication of chariots' wheels in ancient Egypt and China; a list of lubricants was presented by the Roman author Pliny the Elder (ca. 23–79; Dowson 1998).

The emergence of modern study of friction and lubrication is related to the studies by Leonardo da Vinci (1452–1519), Guillaume Amontons (1663–1705), and Charles August Coulomb (1736–1806), who formulated empirical

(a)   (b)   (c)

FIGURE 1.4
Lubrication in ancient Egypt: (a) a man pouring water (presumably as a lubricant) in front of a sledge with a statue of the god Ti in Saqqara, ca. 2400 BC; (b) painting from El Bersheh, ca. 1800 BC, showing transportation of a giant statue; a man in front of a statue is pouring liquid from a jar (either for ceremonial purposes or for water lubrication); (c) a tomb painting from Theba (second millennium BC) shows manufacturing of leather-covered shields; oil lubrication of such shields is mentioned in the Old Testament. (M. Nosonovsky. 2007. *JAST Tribology Online* 2:44–49.)

observations about friction force and equations to calculate such force. Despite that, for centuries friction remained an ad hoc force that was introduced to solve various problems. Unlike the rest of the entire building of the science of mechanics, it was not embedded into the texture of mechanical concepts or deduced from any fundamental principle. Furthermore, the very compatibility of friction with the rest of mechanics was questionable due to the existence of so-called paradoxes (logical contradictions) discovered in the 1890s by French mathematician (and prime minister of the Third Republic!) Paul Painlevé (1863–1933).

In the mid-twentieth century, the new field of tribology was suggested by D. Tabor and H. P. Jost. Tribology combined friction, wear, lubrication, and other phenomena related to the contact of solid surfaces in relative motion. Many experimental data have been obtained and theoretical models of friction have been developed since then. However, the study of friction remained a highly empirical area with phenomenological laws of friction introduced on the basis of experimental observations rather than as a consequence of fundamental laws of nature.

Among the conceptual developments in the late twentieth and early twenty-first centuries, we should mention the discovery of frictional instabilities and of the fundamental role of friction-induced vibrations, and, in particular, the stick–slip motion (especially in atomic friction); friction-induced self-organization; understanding of the universal critical phenomena related to friction (in particular, the relationship between the instabilities and stick–slip "phase transition"); and the development of thermodynamic methods in friction (Nosonovsky and Bhushan 2009; Nosonovsky 2010a, 2010b). All these developments brought the task of developing a fundamental theory of friction much closer to its realization.

1.7 Summary

As we have seen in this chapter, attempts to formalize mechanics have been made in the literature and these efforts continue a long tradition of presenting mechanics in a rigorous manner, with axioms and theorems making it an extension of geometry. The most interesting results in the past 50 years were obtained by Truesdell and Noll. The latter developed rigorous mathematical (axiomatic) definitions of main mechanical concepts. According to his view, the laws of motion are falsifiable physical laws. Truesdell and Noll further combined mechanics with thermodynamics. Noll also emphasized the importance of contact interactions in classical mechanics. A more extreme view of classical mechanics was developed by Zhilin, who considered the laws of motion logical constructs that cannot be tested experimentally and therefore constitute a method of investigating nature, rather than the laws of nature.

Given this state of research on the foundation of mechanics, the situation with the study of friction is quite grim. Friction force is a fundamental mechanical force that plays a role comparable with that of inertia forces. It also played an important conceptual role for the formulation of the GPI and Newton's laws of motion. However, the study of friction did not go much further beyond the observation by da Vinci, Amontons, and Coulomb that the friction force is linearly proportional to the normal load and almost independent of the sliding velocity and the apparent contact area.

In the literature, friction is usually not considered a general and universal phenomenon. Instead, the theories of friction are viewed as empirical and ad hoc models, despite the prominent role that friction plays in the mechanical processes. Although friction is often introduced into differential equations of mechanics, it is not clear under which circumstances it is compatible—for example, the equations of linear elasticity. Friction and wear reflect the tendency of matter for irreversible deterioration and energy for dissipation, so it is natural to relate the origin of friction and wear to thermodynamics and, in particular, to the thermodynamics of irreversible processes. Given that there are four fundamental laws of mechanics—(1) the law of momentum, (2) the law of angular momentum, (3) the first law of thermodynamics, and (4) the second law of thermodynamics—it is necessary to seek justification of the friction law from these laws and, in particular, the second law of thermodynamics.

References

Dowson, D. 1998. *History of tribology,* 2nd ed. New York: Wiley.
Grekova, E. F., and Zhilin, P. A. 2001. Basic equations of Kelvin's medium and analogy with ferromagnets. *Journal of Elasticity* 64:29–70.

Hilbert, D. 1902. Mathematical problems. *Bulletin of American Mathematics Society* 8:437–479.

Merkin, D. R. 1994. *A short history of Galileo-Newton's classical mechanics.* Moscow: Fizmatlit (in Russian).

Noll, N., and Seguin, B. 2010. Basic concepts of thermomechanics. *Journal of Elasticity* 101:121–151.

Noll, W. 1973. Lectures on the foundations of continuum mechanics and thermodynamics. *Archives of Rational Mechanics and Analysis* 52:62-92.

———. 1995. The conceptual infrastructure of mathematics (http://www.math.cmu. edu/~wn0g/noll/CIM.pdf).

———. 2004. Five contributions to natural philosophy (http://www.math.cmu. edu/~wn0g/noll/FC.pdf).

———. 2005. On the theory of surface interactions (http://www.math.cmu. edu/~wn0g/noll/Surface%20Interactions.pdf).

———. 2007. On the concept of force (http://www.math.cmu.edu/~wn0g/noll/ Force.pdf).

Norton, N. L. 2011. *Design of machinery.* New York: McGraw–Hill.

Nosonovsky, M. 2007. Oil as a lubricant in ancient Middle East. *JAST Tribology Online* 2:44–49.

———. 2010a. Entropy in tribology: In the search for applications. *Entropy* 12:1345–1390.

———. 2010b. Self-organization at the frictional interface for green tribology. *Philosophical Transactions of Royal Society A* 368:4755–4774.

———. 2012. Non-equilibrium thermodynamic approach to Coulomb's friction. Twelfth Pan American Congress of Applied Mechanics, Port of Spain, Trinidad, January 2–6.

Nosonovsky, M., and Bhushan, B. 2009. Thermodynamics of surface degradation, self-organization and self-healing for biomimetic surfaces. *Philosophical Transactions of the Royal Society A* 367:1609–1627.

Saliba, G. 1996. Writing the history of Arabic astronomy: Problems and differing perspectives. *Journal of American Oriental Society* 116:709–718.

Truesdell, C. 1966. *Six lectures on modern natural philosophy.* Berlin: Springer–Verlag.

———. 1968. *Essays in the history of mechanics.* Berlin: Springer–Verlag.

———. 1987. *Great scientists of old as heretics in the scientific method.* Charlottesville, VA: University of Virginia Press.

Veselovsky, I. N. 1973. Copernicus and Nasir al-Din al-Tusi. *Journal for the History of Astronomy* 4:128–130.

Zhilin, P. A. 1996. Reality and mechanics. In *Proceedings of XXIII Summer School—Seminar Nonlinear Oscillations in Mechanical Systems,* St. Petersburg, 6–49 (in Russian) (http://teormeh.spbstu.ru/Zhilin_New/pdf/Zhilin_Reality_rus. pdf).

———. 2003. Mathematical theory of non-elastic media. *Uspehi mechaniki (Advances in Mechanics)* 2:3–36 (in Russian).

———. 2006. *Advanced problems in mechanics* (selection of articles presented at the Annual Summer School Conference, Advanced Problems in Mechanics, vol. 2, St. Petersburg: Edition of the Institute for Problems in Mechanical Engineering of Russian Academy of Sciences (http://teormeh.spbstu.ru/Zhilin_New/pdf/ Zhilin_2006_APM_eng.pdf).

2

Vibrations and Stability at the Bulk and at the Interface

The purpose of this chapter is to introduce the mathematical background and physical concepts for the study of friction-induced vibrations, instabilities, and self-organization. These three phenomena are inter-related. Linear vibrations and waves caused by friction can become unstable under certain circumstances. In that case, the amplitude of vibration may grow unlimitedly within the linear vibration and stability analysis. However, the linear models are based on the assumption that the amplitudes of the vibrations are small. With the growing amplitude, this assumption will not be valid after a certain time, and the vibrations will become non-linear. Non-linear vibrations are complex motions and can often lead to limit cycles such as, in the case of friction, the stick–slip motion. The stick–slip is an organized pattern in the temporal and/or spatial domain.

2.1 Linear Vibrations in Systems with Single, Multiple, and Infinite Number of Degrees of Freedom

A mechanical system with one degree of freedom is described by its Lagrangian $L(x, \dot{x})$, which involves the coordinate x and velocity $\dot{x}$. If displacements of the system and velocities are small, the Lagrangian can be approximated by a series:

$$L(x, \dot{x}) = a_0 + a_1 x + b_1 \dot{x} + a_2 x^2 + b_2 \dot{x}^2 + \cdots \tag{2.1}$$

If the motion near the equilibrium is studied, the constant and linear terms can be assumed equal to zero, so that the Lagrangian

$$L(x, \dot{x}) = \frac{m\dot{x}^2}{2} - \frac{kx^2}{2} \tag{2.2}$$

is substituted into the Lagrange–Euler equation

$$\frac{d}{dt}\frac{\partial L}{\partial \dot{x}} - \frac{\partial L}{\partial x} = \frac{\partial D}{\partial \dot{x}} \tag{2.3}$$

where $D = b\dot{x}/2$ is the dissipative function. Equation (2.3) yields the equation of motion:

$$m\ddot{x} + 2b\dot{x} + kx = 0 \tag{2.4}$$

where m is interpreted as the mass of the system, k is the stiffness or "spring constant," and b is the coefficient of viscous damping (Meirovitch 1986). Note that x is a generalized coordinate (a parameter of the system) and it does not necessarily represent the translational degree of freedom. Thus, if x is the angle of rotation, mass m should be substituted with the moment of inertia I.

The solution of the homogeneous Equation (2.4) can be easily found as

$$x = Ae^{-bt}\cos(\omega t + \varphi) \tag{2.5}$$

where A is the amplitude $\omega = \sqrt{\omega_0^2 - b^2}$ and φ is the phase angle.

When an external force $f = f_0\cos\Omega t$ is applied, the equation of motion is given by

$$m\ddot{x} + 2b\dot{x} + kx = f_0\cos\Omega t \tag{2.6}$$

and the solution of the linear non-homogeneous equation should be sought. If the frequency Ω is close to the natural frequency of the system $\omega_0 = k/m$, resonance (a spontaneous growth of the amplitude) can occur.

For systems with N degrees of freedom x_n, the matrix generalization of Equation (2.4) is available:

$$[M]\ddot{\mathbf{x}} + [B]\dot{\mathbf{x}} + [K]\mathbf{x} = 0 \tag{2.7}$$

where $x = x_n$ is the vector of coordinates and M, B, and K are square mass, damping, and stiffness matrices with the dimension of N × N (we use here bold fonts for vectors). The solution is sought in the form $x_n = \text{Re}[A_n\exp(zt)]$ and on the substitution into Equation (2.7) yields

$$[Mz^2 + Bz + K]\mathbf{A} = 0 \tag{2.8}$$

which can be solved by setting the characteristic equation to zero

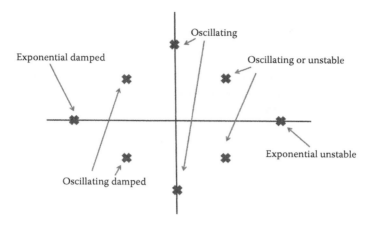

FIGURE 2.1
Stable, unstable, oscillatory, and damped roots of the characteristic equation for linear vibrations.

$$\det[Mz^2 + Bz + K] = 0 \tag{2.9}$$

For real coefficients of the matrices M, B, and K, the solutions for z are either real or complex conjugates. There are N roots of the characteristic equation (some of which may be equal), and real roots normally have a non-positive real part corresponding to decaying motion $\exp(zt)$; pure imaginary conjugate roots correspond to non-decaying oscillating motion and conjugate roots with negative real parts correspond to decaying oscillating motion. The roots with positive real parts are unstable (Figure 2.1).

A mode of vibration A_{mn} corresponds to each root z_n, and an arbitrary motion can be presented as a superposition of modal motion:

$$x_n = \sum_{m=1}^{n} A_{mn} e^{-b_m t} \cos(\omega_m t + \varphi_m) \tag{2.10}$$

Note that, in the general case, modes are not necessarily real, although in many practical situations, when damping is proportional to the mass and/or stiffness of the system, the modes are real. For a complex mode, all points do not pass simultaneously through their maxima or zeros (Meirovitch 1986; Rao 2010).

When the number of degrees of freedom grows and approaches infinity ($N \rightarrow \infty$), the system becomes a continuum or distributed system and it is described by partial derivatives equations. For example, an equation for the displacement of a vibrating string $w(x,t)$ is given by

$$\frac{\partial^2 w}{\partial t^2} + b\frac{\partial w}{\partial t} + \frac{P}{\rho}\frac{\partial^2 w}{\partial x^2} = 0 \tag{2.11}$$

where P is the tension force in the string, ρ is the density per unit length and b is the damping coefficient. For the frictionless motion ($b = 0$), the solution is usually sought by separating variables in the form

$$w(x,y) = X(x)T(t) \tag{2.12}$$

which yields

$$\frac{X}{\dfrac{dX}{dx}} + \frac{P}{\rho}\frac{T}{\dfrac{dT}{dt}} = 0 \tag{2.13}$$

and an eigenvalue problem

$$w = (A\cos\lambda x + B\sin\lambda x)\cos(\omega t + \varphi) = 0 \tag{2.14}$$

to which the boundary conditions should be applied.

The details of linear vibration analysis can be found in numerous books on the topic (Meirovitch 1986; Rao 2010). Although the analysis can be tedious, the equations of motion can be reduced to systems of linear ordinary differential equations (for a finite number of degrees of freedom) or partial differential equations (for continuous systems) that have a general solution.

2.2 Stability Analysis

There are several approaches to the stability analysis of a mechanical system. The simplest is the analysis of the stability of the static equilibrium. The dynamic analysis of the stability is more complicated. For the static analysis, three major methods have been suggested (Panovko and Gubanova 1987; Merkin 1994). *First* is Euler's method of the analysis of whether the equilibrium forms can coexist with the given solution. Usually, the state of the system is characterized by a control parameter, such as the compressive load P in the case of buckling of a beam. When the load exceeds the critical value $P > P_{cr}$, a bent shape of the beam is possible, and buckling occurs (Figure 2.2a). The critical value is determined by writing the differential equation for bending of a beam:

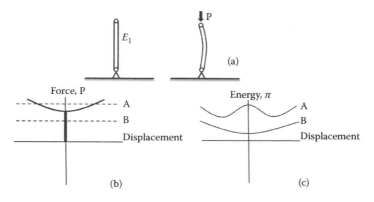

FIGURE 2.2
(a) Buckling of a beam under axial load $P > P_{cr}$; (b) force-displacement diagram showing bifurcation of the equilibrium at $P = P_{cr}$; (c) potential energy profile showing one stable equilibrium for $P < P_{cr}$ (case A) and two stable and one unstable equilibria for $P > P_{cr}$ (case B).

$$EI\frac{d^2w}{dx^2} + Pw = 0 \tag{2.15}$$

where EI is the stiffness of the beam, and solving the eigenvalue problem

$$w(x) = A\cos\sqrt{\frac{P}{EI}}x + B\sin\sqrt{\frac{P}{EI}}x \tag{2.16}$$

for the boundary conditions. Thus, the pinned beam ($w(0) = 0$, $w(L) = 0$) yields $A = 0$.

$$P_{cr} = \frac{\pi^2 EI}{L^2} \tag{2.17}$$

For the loads $P > P_{cr}$, solutions with a bended profile can exist with an arbitrary amplitude of the deflection A, and therefore the initial configuration is unstable.

The *second* static method involves obtaining an explicit relationship between the control parameter and the state of the system, such as the diagram in Figure 2.2(b). The *third* static method is the energetic method. It is based on the idea that the stable equilibrium corresponds to the minimum of the potential energy of the system. For the buckling beam, the elastic energy that corresponds to the deflected state of the beam should be calculated as well as the work of the externally applied force P. The beam remains stable while the energy of beam deformation is greater than the work of the force P.

The force-displacement diagram (or, in the general case, the control parameter versus state of the system diagram) is an important tool to study the stability of the system. One can observe that there are several types of the transition to instability (the loss of stability of the equilibrium state). These may involve the formation of new equilibria, the jump to another state of the equilibrium, or just the absence of the state of equilibrium for $P > P_{cr}$. The energy analysis is an important complement to the force-displacement diagram, as shown in Figure 2.2(b).

All static methods of stability analysis are approximate and a dynamic approach provides a more accurate analysis. For dynamical systems, the so-called Lyapunov stability implies that a small perturbation in the initial conditions near a solution will cease (in the case of asymptotic stability) or at least remain small with time. The so-called structural stability implies that when the system itself is changed and its parameters (such as the stiffness, mass) are modified slightly, the behavior of the system does not change qualitatively. Both approaches to stability deal with the reaction of the system for a small perturbation. In the case of the Lyapunov stability, this is the perturbation of the initial conditions, while in the case of the structural stability this is the perturbation of the parameters of the system itself.

Note that the stability analysis is related to the vibration analysis. When motion of the system is studied near the state of equilibrium, it may be stable or unstable. The stable motion is usually the vibrating motion. The analysis is significantly simplified for the mechanical systems that are conservative (the energy is conserved, no dissipation), stationary (i.e., their parameters do not change with time), and holonomic (no constraints on the velocities or direction of the velocities). For such systems, the energy method can be used to determine whether the equilibrium corresponds to the minimum of the potential energy of the system.

In accordance with Lagrange's theorem, the state of equilibrium of a conservative, stationary, and holonomic system is stable if (and only if) the potential energy $\Pi(x_n)$ is at minimum. In general, the potential energy is given by the series

$$\Pi(x_k) = \Pi(0) + \sum_{i=1}^{n} \frac{\partial \Pi}{\partial x_i} x_i + \frac{1}{2} \sum_{i=1}^{n} \sum_{j=1}^{n} \frac{\partial \Pi}{\partial x_i} \frac{\partial \Pi}{\partial x_i} x_i x_j + \cdots \qquad (2.18)$$

The first and second terms are zero in the state of equilibrium, so the stability can usually be studied by analyzing the second degree term, although in some cases higher order terms are relevant. The stability of a quadratic form

$$\Pi(x_k) = \sum_{i=1}^{n} \sum_{j=1}^{n} C_{ij} x_i x_j, \quad \text{where } C_{ij} = \frac{\partial^2 \Pi}{\partial x_j \, \partial x_j} \qquad (2.19)$$

is determined from Sylvester's criterion stating that all principal minors of the matrix formed by C_{ij} should be positive in order for the matrix to be positive definite and for the potential energy to be positive $\Pi(x_k)$ for $x_k \neq 0$. Namely,

$$C_{11} > 0, \quad \begin{vmatrix} C_{11} & C_{12} \\ C_{21} & C_{22} \end{vmatrix} > 0, \quad \begin{vmatrix} C_{11} & C_{12} & C_{13} \\ C_{21} & C_{22} & C_{23} \\ C_{31} & C_{32} & C_{33} \end{vmatrix} > 0, \quad \dots \qquad (2.20)$$

An important extension is a system that has some coordinates cyclic and some coordinates linear. In such systems the motion in the coordinates of different types is usually studied separately and a so-called Routh function is constructed.

In many cases the characteristic equation of the system can be obtained in the same manner as in the linear stability analysis. The roots of the characteristic equation with a positive real part correspond to exponentially growing modes, and therefore they are unstable. Thus, the characteristic equation of the fourth order can be reduced by changing the scale of x and by division by the coefficient with the fourth power, to the form of

$$x^4 + ax^3 + bx^2 + 1 = 0 \qquad (2.21)$$

The stability condition is $a > 0$, $ab > 1$. For higher order polynomials, the Routh–Hurwitz criterion should be used (Meirovitch 1986). The absolute value of the real part of the root is sometimes called the degree of stability (for a stable root) or degree of instability (for an unstable root).

The true dynamic method of stability analysis is Lyapunov's method. It is a generalization of the idea that energy has its minimum at the stable equilibrium. Lyapunov's function is any function with behavior similar to the energy function; that is, it is zero at the equilibrium point and positively determined in all other states. One should prove by combining the expression for Lyapunov's function and the equations of motion of the system that the Lyapunov's function decreases with time. In that case the motion is stable.

In addition to the Lyapunov method, dynamic stability criteria were developed for the theory of control. Such methods, including the Nyquist frequency criterion, are based on the idea of a transfer function transforming input into the output.

2.3 Non-Linear Vibration and Stability Analysis

Non-linear vibrations are much more complex and diverse than linear vibrations. There is no universal method to solve non-linear equations of motion for the general case. In the theory of non-linear vibrations, certain general qualitative conclusions about non-linear vibrations are formulated, solutions of many special equations and equation types of significant importance are studied, and certain approximated methods are formulated. Unlike for linear vibrations, the frequency of vibrations in the general case depends on their amplitudes; furthermore, they do not have distinct modes.

The qualitative tool used to analyze the stability of a non-linear system is the phase plane plot (Rand 2005; Malkin 1952). Since the dynamics of a system is governed by the differential equation of the second order, the state of the system at any instant of time is defined by the coordinate and the velocity, which corresponds to a point in the phase diagram and determines the future behavior of the system. The equilibrium points in a phase plane plot can be classified as saddles (example is the equation $\ddot{x} - x = 0$), stable nodes ($\ddot{x} + 3\dot{x} + x = 0$), stable spirals ($\ddot{x} + \dot{x} + x = 0$), and centers ($\ddot{x} + x = 0$) (Figure 2.3a).

Note that since the linear equation of the second order $\ddot{x} + b\dot{x} + kx = 0$ can be presented as a system of two equations of the first order

$$\dot{x} = y$$
$$\dot{y} = -by - kx$$

(2.22)

it may be convenient to consider the system

$$\dot{x} = ax + by$$
$$\dot{y} = cx + dy$$

(2.23)

The trace of the matrix is tr = $a + d$, while the determinant is det = $ad - bc$. The type of the equilibrium point as a function of the trace and the determinant is presented in Figure 2.3(b) (Rand 2005). For non-linear systems, the equilibrium points have a similar classification.

The non-linear systems can be studied from the point of view of their "motion in the small" near the points of equilibrium or so-called "motion in the large" away from the equilibrium. In the first case, the behavior of a non-linear system can often be approximated by a linear system with similar properties. Such an approach is referred to as the linearization. In the second case, the trajectories of the system in the phase space are studied.

Many non-linear systems have so-called limit cycle attractors. This means that a system starting at an arbitrary point of the phase space (arbitrary

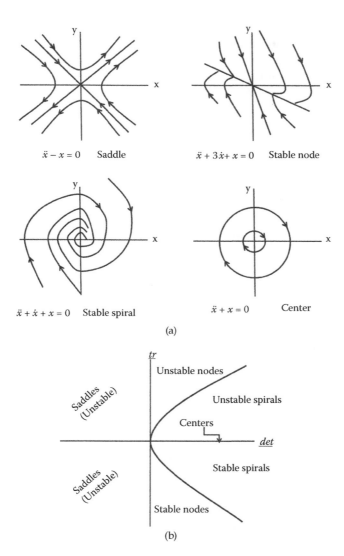

FIGURE 2.3
Classification of equilibrium states: (a) different types; (b) their locations.

initial conditions) tends to approach asymptotically the limit cycle, which is observed as periodic self-excited vibrating motion. It is difficult to establish whether a system has limiting cycle attractors. Although mathematical criteria by Poincare and Bendixson have been proposed, their usefulness is limited (Meirovitch 1986).

Two broad classes of non-linear systems that should be mentioned are the conservative systems (which do not have equilibrium point attractors due to the conservation of energy) and the quasi-harmonic systems.

Among the special types of equations that are extensively studied in the course of non-linear vibrations are

the Duffing oscillator

$$\ddot{x} + x + \varepsilon x^3 = 0 \tag{2.24}$$

the van der Pol oscillator

$$\ddot{x} - \varepsilon(1 - x^2)\dot{x} + x = 0 \tag{2.25}$$

and the Mathieu equation

$$\ddot{x} + (1 - \varepsilon \cos \Omega t)x = 0 \tag{2.26}$$

where $\varepsilon \ll 1$ is the small parameter characterizing the extent of non-linearity (the equations become linear for $\varepsilon = 0$). The Duffing oscillator serves as an approximation for the vibration of a non-linear pendulum $\ddot{x} + \sin x = 0$, while the van der Pol oscillator has a stable attractor (Rand 2005).

When the non-linear term of the system is small, the perturbation method can be used. The idea of the perturbation method is that the solution of the equations of motion,

$$\ddot{x} + \omega_0^2 x = \varepsilon f(x, \dot{x}) \tag{2.27}$$

where ε is the small parameter, is sought in the form of

$$x(t, \varepsilon) = x_0(t) + \varepsilon x_1(t) + \varepsilon^2 x_2(t) + \varepsilon^3 x_3(t) + \dots$$

where $x_0(t)$ is the solution at $\varepsilon = 0$. The solution can be found iteratively by grouping the terms with the same power of ε:

$$\ddot{x}_0 + \omega_0^2 x_0 = 0 \tag{2.28}$$

$$\ddot{x}_1 + \omega_0^2 x_1 = f(x_0, \dot{x}_0) \tag{2.29}$$

$$\ddot{x}_2 + \omega_0^2 x_2 = x_1 \frac{\partial f(x_0, \dot{x}_0)}{\partial x} + \dot{x}_1 \frac{\partial f(x_0, \dot{x}_0)}{\partial \dot{x}} \tag{2.30}$$

and so on.

The idea of Lindstedt's method is in presenting the frequency in the form of

$$\omega = \omega_0 + \varepsilon\omega_1 + \varepsilon^2\omega_2 + \varepsilon^3\omega_3 + \ldots \tag{2.31}$$

and obtaining the iterative equations for the frequency in addition to those for x. Many other methods for non-linear vibration analysis are discussed in the literature (Rand 2005; Meirovitch 1986).

2.4 Bifurcations, Catastrophes, and Chaos

As we discussed in the preceding section, there are several qualitative ways in which a system can become unstable. There is a "soft" loss of stability when the state of equilibrium bifurcates and adjacent states become new stable states of equilibrium. A different scenario is a "hard" loss of stability, when the stable state disappears and the system suddenly jumps to the new stable state (Figure 2.4). The theory that studies qualitative topological behavior of stable and unstable systems and their classification and typology is called the theory of catastrophes.

It should be mentioned that in addition to stable and unstable states of equilibrium, physicists often study the so-called metastable equilibrium. From the mathematical point of view, the metastable equilibrium is stable; however, the energy barrier separating it from other states of equilibrium is so small that the system can drift from one metastable state of equilibrium to another even under the influence of small perturbations (e.g., thermal fluctuations).

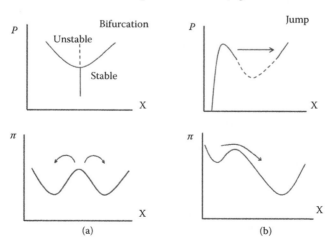

FIGURE 2.4
(a) Bifurcation and (b) jump instabilities.

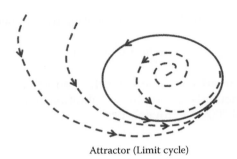

Attractor (Limit cycle)

FIGURE 2.5
An attractor (limit cycle).

Many dynamical systems have attractors. An attractor can be either an equilibrium point or a limit cycle (Figure 2.5). The region in the phase space of the system from which the trajectories are attracted to a given attractor is called the basin of attraction of the attractor. Most attractors are either equilibrium points or one-dimensional (1D) limit cycles to which the trajectories tend to approach in the asymptotic limit of infinite time.

However, an unusual type of attractor, called a "strange attractor," was discovered in the 1970s for some systems of low dimension. The strange attractor is a fractal set with self-similar properties and it demonstrates a chaotic (i.e., unstable) behavior when initially close trajectories tend to diverge from each other with time. Attracting trajectories and repelling them at the same time seems contradictory; however, the strange attractor is anisotropic in the sense that it attracts neighboring trajectories in one direction and repels them in the other direction. The strange attractors are thought to play a role in the onset of turbulence. Many systems also have no attractor at all, demonstrating either true turbulent (unstable) or chaotic behavior. Statistical methods of analysis are used to study the chaotic systems.

2.5 Multi-Scale Systems

Many mechanical systems have only a few degrees of freedom, although they describe complex systems consisting of a large number of parts, such as atoms or molecules. This is possible due to the reduction of the number of degrees of freedom. In a mechanical system, many degrees of freedom are not important and can be ignored—for example, degrees of freedom corresponding to the motion particular molecules. Only some degrees of freedom describing the collective motion of the system, such as its center of mass, are essential.

This reduction in the number of degrees of freedom is possibly due to the hierarchical structure of many objects found in nature. When we look at a

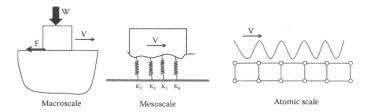

FIGURE 2.6
Multi-scale nature of friction and reduction of degrees of freedom: (a) only one degree of freedom observed at the macroscale, while (b) each elastically deformed asperity constitutes a
degree of freedom at the meso-scale; (c) molecular motion at the atomic scale.

certain object, such as a mechanical body, we see it with a certain resolution and do not observe small details. As a result, the motion that is below
the scope of resolution can be ignored, as well as myriad degrees of freedom associated with such motion. This is possible due to the hierarchical
or multi-scale organization of many systems. As will be discussed, the self-
organization often occurs with the reduction of the number of degrees of
freedom, and new order parameters or degrees of freedom corresponding to
higher level of hierarchy emerge (Figure 2.6).

In many physical systems various length scales can be defined. A length
scale less than 1 mm but larger than 100 nm is considered micro-scale,
while a length scale less than 100 nm but larger than the atomic scale is
considered nano-scale (Anisimov 2004). According to another definition, a
scale length larger than an atomic scale (i.e., 1 nm or less) but smaller than
a macro-scale (that is, between 1 nm and 1 mm) may be called a meso-scale.
While much attention has been paid to the macro- and atomic/molecu-
lar scale, there is growing interest in meso-scale studies. This is because
in many physical systems macro-scale properties, such as the modulus of
elasticity, yield strength, fracture toughness, and coefficient of friction,
cannot be deduced directly from the molecular scale properties. Such
properties are often defined by meso-scale objects, such as defects, grains,
and asperities.

There are several types of objects that are usually considered by physicists
as meso-scale objects. These include systems that are of sub-micron size in at
least one dimension, such as nano-particles; soft condensed matter materials
(foams, gels, polymer melts), characterized by a mesoscopic length scale; and
systems in a near-critical state (near a phase transition point), which have
mesoscopic "correlation length" characterizing spontaneous fluctuations
(Anisimov 2004).

A nano-scale system with a size of several nanometers still involves hun-
dreds and thousands of molecules and can often be considered as a continuum system. However, many physical properties in such a system are
different from macro-scale bulk properties. For example, the yield strength
and hardness are known to be higher when measured at the nano-scale

compared with macro-scale values (Nosonovsky and Bhushan 2008a; Nix and Gao 1998; Hutchinson 2000; Bhushan and Nosonovsky 2003). The reason for that is believed to be because a solid material consists of a large number of sub-micron-sized grains, domains, and defects. It is much easier to deform material when the size of the deformation is greater than the size of a grain. At the nano-scale, there are no such defects and material is much stronger (Hutchinson 2000).

Another case when nano-scale properties are different from those at the micro-scale is during the phase transition, such as boiling/condensation and melting/freezing. While at the macro-scale water is known to boil at 100° C and to freeze at 0° C, the nano-scale water phase diagram may be quite different (Yang et al. 2007; Nosonovsky and Bhushan 2008b). This is because in order to transform into a different phase, an interface should be created (e.g., a vapor bubble inside the bulk of liquid), which requires additional activation energy. However, in order to grow, the size of a bubble must be greater than a certain critical size, so at the nano-scale the bubble would not be formed. Furthermore, due to the capillary effects and small radii of the curvature of nano-droplets and nano-scale water columns, the pressure inside water volumes (the so-called Laplace pressure) may be significantly different from the ambient (e.g., may be negative [tensile strength]; Yang et al. 2007; Nosonovsky and Bhushan 2008a, 2008b).

Although many concepts of meso-scale physics and thermodynamics were developed a long time ago, only recently has it been recognized that diverse meso-scale phenomena possess a high degree of generality. There are several universal methods that allow physicists to deal with the meso-scale. These methods include the renormalization group theory, scaling theory, Landau–Ginzburg mesoscopic functional, Ising and Potts lattice approach, percolation, etc. (Anisimov 2004). A characteristic feature of meso-scale systems is that fluctuations, or spontaneous deviations from equilibrium, can play a significant role in them (Anisimov 2004). For example, consider a small micro-scale pendulum of length l with a mass $m = V\rho$ with its position characterized by the angle θ (Figure 2.7). Collision with molecules will result in the mean square position of the micro-pendulum being equal to

$$\theta^2 = kT/(V\rho g l) \tag{2.32}$$

where k is the Boltzmann constant and T is temperature. Thus, unlike the macro-scale pendulum, the typical position of the micro-pendulum will be different from $\theta = 0$ due to the thermal random fluctuation (Nosonovsky and Bhushan 2008a).

The meso-scale phenomena are closely related to the critical phenomena (i.e., phase transition). Close to a critical point of any kind, the fluctuations become so large that they exhibit macro-scale behavior (Anisimov 2004). The critical point is a point at the phase diagram where the distinction between two phases vanishes. For example, the critical point of water is at around

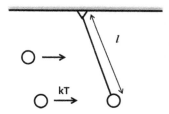

FIGURE 2.7
Meso-scale pendulum; due to thermal fluctuation its average absolute value of deflection is not zero.

374°C and 218 atm, and the distinction between liquid and the difference between gas and water at these conditions disappear. Asymptotically close to the critical point, the physical properties obey simple power laws, called the scaling laws. The physical basis for the scaling theories is in the divergence of a mesoscopic characteristic length scale known as the correlation length of the fluctuation. Powerful physical techniques, such as the renormalization-group theory and the Landau–Ginzburg functional, which were originally formulated for phase transition, have been proposed to calculate the critical exponents of the scaling laws and to study the near-critical behavior.

The phase transition between "disordered" and "ordered" states implies that an order parameter can be identified that is equal to zero for the disordered phase and different from zero for the ordered phase. Then, power exponents for the scaling laws can be determined. The phase transition approach has also been applied to the molecular-scale friction (Einax, Schulz, and Trimper 2004). In solid mechanics, the common "phase transition" is that between the elastic and plastic phases. Modern theories of plasticity intended for the micron and sub-micron scale (the strain-gradient plasticity) postulate meso-scale length parameters, which result in the scale dependence of the yield strength and hardness at the meso-scale (Nix and Gao 1998; Hutchinson 2000). The strain-gradient plasticity approach can be applied for the study of the scale effect on friction (Bhushan and Nosonovsky 2003; Nosonovsky and Bushan 2008a).

The concept of hierarchy is different from the concept of scale in a sense that hierarchy implies a complicated structure and organization. Studying hierarchical systems often requires breaking the boundaries between particular disciplines and applying a multi-disciplinary approach. Investigation of hierarchical surfaces involves mechanics, physics, chemistry, biochemistry, and biology. Hierarchical surfaces are built of elements of different characteristic length, organized in a certain manner. This organization leads to certain functionality. Many examples of these surfaces are found in biology and will be considered in this book (Nosonovsky and Bhushan 2008a).

An important class of hierarchical systems is fractal objects. The so-called self-similar or fractal structures can be divided by parts, each of which is a reduced-size copy of the whole. Unusual properties of the self-similar curves and surfaces, including their non-integer dimensions, were studied by mathematicians in the 1930s. The word *fractal* was coined in 1975 by Benoit Mandelbrot, who popularized the concept of self-similarity and showed that fractal geometry is universal in nature and in engineering applications (Mandelbrot 1982). The fractal concepts were applied to rough surfaces. In the late 1980s through the early 1990s, the fractal geometry approach was introduced into the study of engineering rough surfaces by Gagnepain and Roques-Carnes (1986), Ling (1989), Majumdar and Bhushan (1990), and others (Nosonovsky and Bhushan 2008a).

2.6 Self-Organization: Different Types

The modern concepts of self-organizing systems were developed in the second half of the twentieth century, at first with the field of the cybernetics. Although today cybernetics is commonly understood in the popular culture as a field related to computers and the Internet as well as science fiction (compare with such pop culture terms and brands as "cybercafe," "cyborg," Cyberchase"), the original meaning of the term when it was suggested by Wiener (1948) was the "science of control and communication in animal and machine." Self-organization played a significant role in the ideas of Wiener, who used the term *teleology* for the behavior of a system directed toward a final cause or purpose. Originally, this was a philosophical and theological term. According to Rosenblueth, Wiener, and Bigelow (1943), the feedback (a central concept in cybernetics) can lead to teleology in machinery. Particular theories of self-organization were developed by Ashby (1947) and von Foerster (1992), who studied the dynamical systems that tended to evolve toward attractors as self-organized states.

Later, a thermodynamic approach was suggested by Prigogine and his followers, who studied non-equilibrium thermodynamic systems. H. Haken (1983) suggested the term *synergetics* for the study of self-organizing systems in chemistry and biology. The very word *synergetics* was apparently coined by the architect and philosopher R. B. Fuller (the same person after whom the C_{60} molecule was later called *fullerene*). This term suggests that parts of a self-organizing system work together in synergy in order to achieve a self-organized state. The principle of "order from noise" was suggested by cyberneticians and later formulated as the "order through fluctuations" principle by Nicolis and Prigogine (1977). According to this idea, random fluctuations can lead a physical system to the state where it will be captured by an attractor. Therefore, attractors are associated with self-organized states. Systems with spontaneous symmetry breaking and with structural phase transitions can serve as examples of self-organization. Prigogine (1976) called the self-organized structures "dissipative structures."

Self-organization is a non-linear phenomenon and, as it is with many non-linear effects, it is difficult to provide a comprehensive classification of self-organizing systems. Several principles of self-organization have been suggested and extensively discussed in the literature. Sometimes, self-organization involves phase transitions, both of the first order and of the second order. The first-order phase transitions involve latent heat, and an example of a self-organizing process is the crystallization of an ordered solid crystal during freezing. Thermodynamically, the entropy of a crystal is lower than that of liquid, as will be discussed in the next chapter.

The second-order phase transitions do not involve latent heat and do not require energy input. Similarly, first-order transitions may have a critical point at which the transition becomes a second-order transition. Thus, for water at the temperature of $T_{cr} = 373.9°C$ and pressure $P_{cr} = 217.7$ atm, the liquid–vapor transition occurs smoothly without any latent heat involved. An important feature (not at all obvious intuitively) of the near critical point behavior is the presence of critical exponents and the growth of correlation lengths. The correlation length ξ is a measure of the order in a system, which describes how microscopic variables at different positions are correlated. The temperature can be viewed as a control parameter $\tau = (T - T_{cr})/T_{cr}$, non-dimensional for convenience, and power exponent relationships, such as $\xi/\xi_0 = \tau^\gamma$; similarly, other parameters (e.g., surface energy, heat capacity, etc.) can be related through the power exponents. Interestingly and not intuitively obvious, the values of the exponents often do not depend on the chemical composition of the material, demonstrating the property referred to as the universality of the critical behavior. This universality is explained by a mathematical approach called the renormalization group theory.

A way to visualize the critical behavior is to consider the so-called Ising model (or a similar Potts model) that represents a two-dimensional (2D) surface filled by two phases (say, "A" and "B"). The surface consists of cells (or pixels), and each cell can be in either the A or the B state (Figure 2.8). There are certain laws of evolution of the systems, so that at every time step the

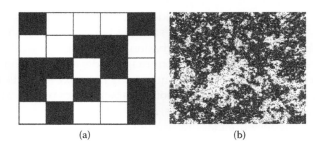

(a)   (b)

FIGURE 2.8
Ising (cellular automata) model showing (a) cells that can be in the state A or B. (b) At the critical state when neither A (white) nor B (black) dominates, the distribution is fractal, and the average size of the spots is infinite. In that state, universal critical behavior is observed.

state of a pixel changes with a certain probability depending on the states of its closest neighbors. Furthermore, there is a control parameter τ (the "temperature") such that for small τ, the state A dominates (there are some islands or domains of B on the A background), whereas for large τ, the state B dominates (islands of A on the B background). The correlation length ξ in this case is the average size of the islands of a non-dominating phase. There is also a critical value $\tau = \tau_{cr}$ at which both states are equally probable. From simple symmetry and scaling arguments, one concludes that

(a) At $\tau = \tau_{cr}$, $\xi \to \infty$ (no dominating phase).

(b) The spots of the two phases have fractal (i.e., scale-invariant) shape.

(c) Various properties depend on the difference in the frequency of the occurrence of the two phases or in the relative size of the domain's scale as a power law of $(\tau - \tau_{cr})^{\alpha}$.

(d) The critical properties dependent on the fractal structure of the near-critical state are independent of the particular laws of evolution (in particular, these laws do not have to be symmetric with respect to A and B).

Note that the second-order phase transitions are quite common in surface science and materials science, including, for example, the wetting transitions between the homogeneously wetted (Wenzel) and heterogeneously wetted (Cassie) states or the elastic–plastic transition. The stick and slip states of friction can also be viewed as phase transitions, and thus the study of the universality of critical behavior can bring insights about the universality of friction.

The study of phase transition is related to the study of instability. The critical point corresponds to the point where the stability of system is lost. There is a significant class of dynamical systems that tend to tune themselves to the critical state, yielding so-called self-organized criticality. Usually these are systems with an input of energy, where energy is accumulated until the critical stated is reached and then suddenly released. Examples are various avalanche systems, including those describing landslides, earthquakes, and frictional stick–slip. A random perturbation can trigger an avalanche (or a slip event) in such a system of a random magnitude, after which the system returns to the stable state for some time until the next event is triggered. The amplitudes of the events have various characteristics of critical behavior, such as universality, critical exponents, and fractality.

Haken (1983) suggested a classification of self-organizing processes from the viewpoint of the number of degrees of freedom involved in the non-organized and in the self-organized states. According to this classification, self-organization of the first kind involves the reduction of the number of degrees of freedom, whereas self-organization of the second kind does not involve such a reduction.

Haken (1983) suggested the following general scheme for the analysis of self-organizing systems. For given value of the control parameter or a set of control parameters, one assumes that a stationary solution exists. Then the stability of the solution is checked by linear stability theory when one or several control parameters are changed. The solutions of the linear stability problem are typically exponential. The exponentially increasing or neutral solutions characterize the "unstable modes." Their amplitudes or phases become the order parameters in the fully non-linear treatment that also considers fluctuations. The equations of motions are then transformed to these new variables, amplitudes, and phases defining order parameters, and the still stable modes. Then, taking into account the fluctuations, the damped (stable) modes are eliminated (slaving principle). The resulting equations for the order parameters are, in general, low dimensional.

A number of self-organizing processes have been also studied in chemistry, including periodical (or autocatalytic) chemical reactions, self-assembly of mono-layers, molecular self-assembly, self-assembled colloidal crystal and quasi-crystal structures with long order, micelles, diffusion limited aggregation, etc. Many of these processes are governed by non-linear equations (such as the reaction-diffusion equations first suggested by Alan Turing in 1952) and can explain patter formation and other related effects.

2.7 Example: Benard Cells

A classical example of a self-organized system in physics that has been studied very extensively is the Benard cells or convection cells occurring in a plane horizontal layer of fluid heated from below (Figure 2.9). The elevated temperature of the bottom plane yields a flow of thermal energy conducted through the liquid. The temperature, density, and pressure with it will vary linearly with a uniform gradient between the bottom and top

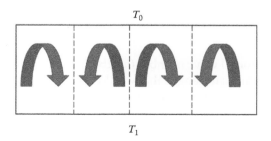

FIGURE 2.9
Water circulation in Benard cells.

planes. However, microscopic random movements spontaneously become ordered on a macroscopic level, forming hexagonal prismatic convection cells with a characteristic correlation length through a spontaneous symmetry-breaking. A further increase of the temperature gradient yields turbulent chaotic motion.

The convection is characterized by the non-dimensional Rayleigh number, meaning the balance between the gravity and viscous damping:

$$R_a = \frac{g\beta\Delta T}{\alpha\nu}L^3 \qquad (2.33)$$

where
L is the height of the container
ΔT is the temperature difference
g is the gravity acceleration
α is the thermal diffusivity
β is the coefficient of thermal expansion
ν is the kinematic viscosity

With the increase of the Rayleigh number, the gravitational force dominates over viscosity.

The stability problem for convection cells was solved by Rayleigh, who studied small perturbations of the Navier–Stokes and heat conduction equations near the steady-state solution (the linear solution with a constant gradient) of the velocity (z-component) and temperature field in the form of

$$w(x,y,z,t) = f(x,y)W(z)e^{\sigma t}$$
$$T(x,y,z,t) = f(x,y)\Theta(z)e^{\sigma t} \qquad (2.34)$$

When subjected to the free boundary conditions, the system yields a characteristic equation with the eigenvalue problem

$$(D^2 - a^2)^3 W = -a^2 R_a W \qquad (2.35)$$

where a is the horizontal wave number (for the free–free boundary, the problem has no horizontal scale length that would define a), and D is the operator of the material derivative. Substituting

$$W = \sin(n\pi z) \qquad (2.36)$$

yields for the lowest mode ($n = 1$) the Rayleigh number as a function of the unknown horizontal wave number

$$R_a = \frac{(\pi^2 + a^2)^3}{a^2} \tag{2.37}$$

that has its minimum at $a = \pi/\sqrt{2}$ equal to

$$R_a = \frac{27}{4}\pi^4 \approx 657.51 \tag{2.38}$$

The result is interpreted in such a way that when the Rayleigh number reaches the critical value, the steady-state solution becomes unstable and the convection cells emerge. Note that the linear analysis of the small perturbation does not provide the solution for the cells; however, it indicates the condition under which the perturbations grow. These growing perturbations eventually emerge into the cells, which constitute a non-linear solution.

The Benard cells problem possesses features typical for many self-organization problems: The stationary (steady-state) solution destabilizes at a certain critical value of the controlled parameter (R_a). The symmetry breaking occurs, since the symmetric equation has now a non-symmetric solution. The stability can be investigated within the linear theory of small perturbations using the eigenvalue analysis. The perturbations are characterized by a new parameter, a—the size of the non-linear structure that serves as a limit cycle attractor of the system.

2.8 Self-Organization in Tribology

Friction and wear are non-equilibrium processes that normally result in energy dissipation and material deterioration despite the fact that friction and wear can lead to self-organization of new structures at the frictional interface. Several attempts to study friction-induced self-organization were made in the former Soviet Union in the 1970s.

First, B. Kostetsky and L. Bershadsky (Bershadsky 1992; Bershadsky and Kostetsky 1993) in Kiev investigated the formation of the so-called self-organized "secondary structures" during friction and the regime of "structural dissipative adjustment." According to Bershadsky, friction and wear are two sides of the same phenomenon and they represent the tendency of energy and matter to achieve the most disordered state. However, the synergy of various mechanisms can lead to self-organization of the secondary structures, which are "nonstoichiometric and metastable phases," whereas "the friction force is also a reaction for the informational (entropic) excitations, analogous to the elastic properties of a polymer, which are related mostly to the change of entropy and have the magnitude on the order of the elasticity of a gas" (Bershadsky and Kostetsky 1993). Bershadsky also formulated entropic variational principles that governed friction and wear and a number of important ideas on the structural dissipative nature of friction.

The books by Prigogine (1983) and Haken (1983) were published in the USSR in Russian translations and became very popular in the 1980s, since the "synergetic" studies claimed to suggest a general methodology to investigate physical, biological, information, and social phenomena, which, in a sense, was (or at least was perfected in this way by some scientists) an alternative to the official Soviet Marxist methodology. This interest affected tribologists. According to Bershadsky (1992),

> Tribosystems apparently constitute the most diverse objects capable for self-organization, which possess all major features of a synergetic system (strong non-linearity, parameter distribution and delay, features of an auto-catalyst and natural regulator, feedback and target function, etc.), as well as a number of specific features, such as the energetic and materials heterogeneity, memory, learning capability, etc. This is why many unsolved tribological problems are in fact general problems of synergetics.

The students of N. Bushe and I. Gershman at the Research Institute for the Railway Transport in Moscow constitute another group that worked on friction-induced self-organization. To a large extent, they continued the research of the Kiev group—in particular, in the theory of "tribological compatibility" of materials and the synergetic action of the electric current during current collection, as well as wear-resistant composite metallic coatings for heavily loaded cutting tools (Fox-Rabinovich and Totten 2004).

A tribosystem is an open, non-equilibrium system, and its entropy is lower than the equilibrium state and entropy production that asserts an increase in orderliness and self-organization. External elements such as load, velocity, temperature, and humidity push the system to operate far from equilibrium. Therefore, to prevent entropy growth, a highly ordered intermediate body forms on the interface, referred to as tribofilm. Formation of tribofilms is the response of the system to external stimulus to reduce wear rate. If tribofilms are not formed during friction, a tribosystem stops the friction by seizure or jamming (Amiri and Khonsari 2010).

A different approach, which also employed the ideas of synergetics, was suggested by Prof. Garkunov (2000) of the Air Force Academy in Russia, who claimed the discovery of the synergetic "non-deterioration effect," also called the "selective transfer." Together with A. Polyakov, he suggested a concept of dynamically formed protective tribofilms (which they called "servovite films" or "serfing films"). These films are formed due to a chemical reaction induced by friction, and they protect against wear, leading to a dynamic equilibrium between the wear and the formation of the protective film. Garkunov's heavily experimental research also received international recognition when he was awarded the 2005 Tribology Gold Medal "for his achievements in tribology, especially in the fields of selective transfer."

Among other related groups engaged in entropic studies in tribology, the scientists of Rostov and Tomsk University, Prof. Krawczyk of Poland and Dr. Sosnovskiy (2009) from Gomel, Byelorussia, should be mentioned. The latter suggested the concept of "tribofatika" (coupling of wear and fatigue).

In the English-speaking world, an important entropic study of the thermodynamics of wear was conducted by Bryant, Khonsari, and Ling (2008), who introduced a degradation function and formulated the degradation entropy generation theorem in their attempt to investigate friction and wear in complex. They note that friction and wear are often treated as unrelated processes, but that they are in fact manifestations of the same dissipative physical processes occurring at sliding interfaces (Bryant 2009). Their approach is based on classical Clausius's concept of entropy.

A completely different approach is related to the theory of dynamical systems. The results of friction tests can be viewed as a data series characterized by certain statistical parameters, including the entropy of distributions. Since the 1980s, it has been suggested that a very specific type of self-organization, called self-organized criticality, plays a role in diverse "avalanche-like" processes, such as the stick–slip phenomenon during dry friction. The research of Zypman et al. (2003) and Ferrante (Adler et al. 2003; Buldyrev, Ferrante, and Zypman 2006; Fleurquin et al. 2010) and others dealt with this topic. A different approach on the basis of the theory of dynamical systems was suggested by Kagan (2010), based on the so-called Turing systems (the diffusion-reaction systems) to describe formation of spatial and time patterns induced by friction.

Nosonovsky and Bhushan (2009) and Nosonovsky et al. (2009) suggested using entropic methods to describe self-lubrication and surface-healing (self-healing surfaces). The possibility of friction-induced reaction-diffusion patterns was studied (Mortazavi and Nosonovsky 2011) as well as the transient "running-in" process as self-organization (Mortazavi and Nosonovsky 2011) or adjustment of the surfaces to each other. It was also suggested that self-organized spatial patterns (such as interface slip waves) can be studied by the methods of the theory of self-organization.

2.9 Asymptotic Transition from a 3D (Bulk) to a 2D (Interface) System

The asymptotic methods of small parameter are not only used to study the non-linear systems that are close to linear systems. They can also be employed for other purposes, such as the transition from three-dimensional

(3D) equations of elasticity to the 2D equations of mechanics of plates and shells and 1D equations of the elastic deformation of beams and rods.

Consider the following equation:

$$\mathbf{C}\mathbf{x} = \mathbf{f} \tag{2.39}$$

where $\mathbf{x}$ and $\mathbf{f}$ are columns and $\mathbf{C}$ is a matrix presented as a series for the small parameter ε:

$$\mathbf{C} = \mathbf{C}_0 + \varepsilon\mathbf{C}_1 + \varepsilon^2\mathbf{C}_2 + \varepsilon^3\mathbf{C}_3 \tag{2.40}$$

If $\det|\mathbf{C}_0| \neq 0$, the solution is sought as a series

$$\mathbf{x} = \mathbf{x}_0 + \varepsilon\mathbf{x}_1 + \varepsilon^2\mathbf{x}_2 + \varepsilon^3\mathbf{x}_3 + \tag{2.41}$$

and grouping the terms with the same order of ε yields

$$\mathbf{C}_0\mathbf{x}_0 = \mathbf{f}$$
$$\mathbf{C}_0\mathbf{x}_1 = -\mathbf{C}_1\mathbf{x}_0 \tag{2.42}$$
$$\mathbf{C}_0\mathbf{x}_2 = -\mathbf{C}_1\mathbf{x}_1 - \mathbf{C}_2\mathbf{x}_0$$

and so on. However, if the determinant is $\det|\mathbf{C}_0| = 0$, the solution is sought in the form

$$\mathbf{x} = \varepsilon^{-1}\mathbf{x}_0 + \mathbf{x}_1 + \varepsilon\mathbf{x}_2 + \varepsilon^2\mathbf{x}_3 +$$
$$\mathbf{C}_0\mathbf{x}_0 = 0$$
$$\mathbf{C}_0\mathbf{x}_1 = \mathbf{f} - \mathbf{C}_1\mathbf{x}_0 \tag{2.43}$$
$$\mathbf{C}_0\mathbf{x}_2 = -\mathbf{C}_1\mathbf{x}_1 - \mathbf{C}_2\mathbf{x}_0$$

Using these arguments, Eliseev and Zinoveva (2008) showed that the equations for the Euler beam can be derived from the equations of the elasticity for bending of the elastic strip (Figure 2.10), if the change of the coordinates $x \to \varepsilon x$ is performed, or

$$\vec{R} = \varepsilon^{-1}x\vec{i} + x\vec{j} \tag{2.44}$$

$$\nabla = \varepsilon\vec{i}\frac{\partial}{\partial x} + \vec{j}\frac{\partial}{\partial y} \tag{2.45}$$

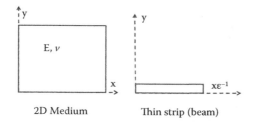

FIGURE 2.10
Asymptotic transition from isotropic 2D elastic medium to 1D elastic beam.

$$\Delta = \varepsilon^2 \frac{\partial^2}{\partial x^2} + \frac{\partial^2}{\partial y^2} \tag{2.46}$$

A similar approach can be used for the transition from the 3D equations of the interactions in the bulk to the interactions at the 2D interface.

2.10 Summary

In this chapter we introduced and discussed the basic concepts related to linear and non-linear vibration and stability analysis for mechanical systems with one, many, and an infinite number of degrees of freedom. The linear problem has a general solution. Linear vibration analysis is based on solving a system of differential equations and finding the roots of the characteristic equation to solve the eigenvalue problem to obtain natural frequencies and mode shapes. Roots with negative real parts correspond to damped (decaying) modes; pure imaginary roots correspond to non-damped oscillating motion, and roots with positive real parts correspond to unstable modes. The stability analysis can also be performed in a static manner (the stability of the equilibrium) and dynamic manner (stability of motion). Energy methods can be used and stable equilibrium corresponds to a local energy minimum of the system.

Non-linear vibrations do not have a general solution and various qualitative and approximate methods can be applied. In addition to simple equilibrium points, non-linear systems have limit cycle attractors. When control parameters of a system change beyond a critical value, the system can become unstable and evolve from the equilibrium point to the limit cycle attractor. Such an attractor can be observed as a self-organized pattern.

The loss of equilibrium and self-organization is related to the phase transitions, and various far-reaching aspects of this relationship can be studied, including the universality of the critical behavior, the reduction in the number of degrees of freedom, and the hierarchical structure of the system.

All the previously mentioned effects have relevance to frictional sliding, which involves instabilities, vibrations, limiting cycles, self-organization, reduction in the number of degrees of freedom, and even phase transitions (between the stick and slip states). Thermodynamic quantities should be introduced in order to account for the effect of ignored degrees of freedom. Non-equilibrium thermodynamic analysis of instabilities, self-organization, and friction will be discussed in the following chapter.

References

Adler, M., Ferrante, J., Schilowitz, A., Yablon, D., and Zypman, F. (2003, January). Self-organized criticality in nanotribology. In *MRS Proceedings* 782 (1). Cambridge University Press.

Amiri, M., and Khonsari, M. M. 2010. On the thermodynamics of friction and wear—A review. *Entropy* 12 (5): 1021–1049.

Anisimov, M. A. 2004. Thermodynamics at the meso- and nanoscale. In *Dekker encyclopedia of nanoscience and technology*, 3893–3904. New York: Dekker.

Ashby, W. R. 1947. Principles of the self-organizing dynamic system. *Journal of General Psychology* 37 (2): 125–128.

Bershadsky, L. I. 1992. On the self-organization and concepts of wear resistance in tribosystems. *Trenie I Iznos (Friction and Wear)* 13:1077–1094 (in Russian).

Bershadsky, L. I., and Kostetsky, B. I. 1993. The general concept in tribology. *Trenie I Iznos (Friction and Wear)* 14:6–18 (in Russian).

Bhushan, B., and Nosonovsky, M. 2003. Scale effects in friction using strain gradient plasticity and dislocation-assisted sliding (microslip). *Acta Materialia* 51 (14): 433–4345.

Bryant, M. D., Khonsari, M. M., and Ling, F. F. 2008. On the thermodynamics of degradation. *Proceedings of the Royal Society A: Mathematical, Physical and Engineering Science* 464 (2096): 2001–2014.

Buldyrev, S. V., Ferrante, J., and Zypman, F. R. 2006. Dry friction avalanches: Experiment and theory. *Physical Review E* 74 (6): 066110.

Einax, M., Schulz, M., and Trimper, S. 2004. Friction and second-order phase transition. *Physical Review B* 70:046113.

Eliseev, V. V., and Zinoveva, T. V. 2008. *Mechanics of thin structural elements: The theory of deformable rods*. St. Petersburg: Polytechnic University (in Russian).

Fleurquin, P., Fort, H., Kornbluth, M., Sandler, R., Segall, M., and Zypman, F. 2010. Negentropy generation and fractality in the dry friction of polished surfaces. *Entropy* 12 (3): 480–489.

Fox-Rabinovich, G. S., and Totten G. 2006. *Self-organization during friction: Advance surface engineered materials and systems design*. Boca Raton, FL: CRC Taylor & Francis Group.

Gagnepain, J. J., and Roques-Carmes, C. 1986. Fractal approach to two-dimensional and three-dimensional surface roughness. *Wear* 109 (1): 119–126.

Garkunov, D. N. 2000. *Triboengineering (wear and non-deterioration)*. Moscow: Agricultural Academy Press (in Russian).

Haken, H. 1983. *Synergetics. An introduction. Nonequilibrium phase transitions in physics, chemistry and biology*, 3rd ed. New York: Springer–Verlag.

Hutchinson, J. W. 2000. Plasticity at the micron scale. *International Journal of Solids and Structures* 37 (1): 225–238.

Kagan, E. 2010. Turing systems, entropy, and kinetic models for self-healing surfaces. *Entropy* 12:554–569.

Ling, F. F. 1989. The possible role of fractal geometry in tribology. *Tribology Transactions* 32 (4): 497–505.

Majumdar, A., and Bhushan, B. 1990. Role of fractal geometry in roughness characterization and contact mechanics of surfaces. *Journal of Tribology* 112 (2): 205–216.

Malkin, I. G. 1952. *Theory of stability of motion*, vol. 3352. United States Atomic Energy Commission.

Mandelbrot, B. B. 1982. *The fractal geometry of nature*. Times Books.

Meirovitch, L. 1986. *Elements of vibration analysis*. New York: McGraw–Hill.

Merkin, D. R. 1994. *Introduction to the theory of stability*, vol. 24. Berlin: Springer.

———. 2011a. Friction-induced pattern formation and Turing systems. *Langmuir* 27 (8): 4772–4779.

Mortazavi, V., and Nosonovsky, M. 2011b. Wear-induced microtopography evolution and wetting properties of self-cleaning, lubricating and healing surfaces. *Journal of Adhesion Science and Technology* 25 (12): 1337–1359.

Nicolis, G., and Prigogine, I. 1977. *Self-organization in nonequilibrium systems: From dissipative structures to order through fluctuations*. New York: Wiley.

Nix, W. D., and Gao, H. 1998. Indentation size effects in crystalline materials: A law for strain gradient plasticity. *Journal of Mechanics and Physics of Solids* 46:411–425.

Nosonovsky, M., Amano, R., Lucci, J. M., and Rohatgi, P. K. 2009. Physical chemistry of self-organization and self-healing in metals. *Physical Chemistry Chemical Physics* 11 (41): 9530–9536.

Nosonovsky, M., and Bhushan, B. 2008a. *Multiscale dissipative mechanisms and hierarchical surfaces: Friction, superhydrophobicity, and biomimetics*. Berlin: Springer.

———. 2008b. *Physical Chemistry Chemical Physics* 10:2137–2144.

———. 2009. Thermodynamics of surface degradation, self-organization and self-healing for biomimetic surfaces. *Philosophical Transactions of the Royal Society A: Mathematical, Physical and Engineering Sciences* 367 (1893): 1607–1627.

Panovko, I. G., and Gubanova, I. I. 1987. *Stability and oscillations of elastic systems: Paradoxes, fallacies, and new concepts*. Moscow: Nauka (in Russian).

Prigogine, I. 1976. Order through fluctuation: Self-organization and social systems. In: *Evolution and consciousness: Human systems in transition*. eds. E. Jantsch and C. H. Waddington, 93–130. Reading, MA: Addison Wesley.

Rand, R. H. 2005. *Lecture notes on nonlinear vibrations*. Ithaca, NY: Cornell University (http://www.tam.cornell.edu/randdocs/).

Rao, S. 2010. *Mechanical vibrations*, 5th edition. Englewood Cliffs, NJ: Prentice Hall.

Rosenblueth, A., Wiener, N., and Bigelow, J. 1943. Behavior, purpose and teleology. *Philosophy of Science* 10:18–24.

Sosnovskiy, L. A., and Sherbakov, S. S. 2009. Surprises of tribo-fatigue; magic book: Minsk, Belorussia.

Turing M. 1952. The chemical basis of morphogenesis. *Philosophical Transactions of Royal Society*, B 237: 37.

von Foerster, H. 1992. Cybernetics. In *The encyclopedia of artificial intelligence,* 2nd ed., ed. S. C. Skapiro, 309–312. New York: John Wiley & Sons.

Wiener, N. 1948. *Cybernetics, or control and communication in the animal and the machine.* New York: John Wiley & Sons, Inc.

Yang, S. H., Nosonovsky, M., Zhang, H., and Chung, K. H. 2007. Nanoscale water capillary bridges under deeply negative pressure. *Chemical Physics Letters* 451:88–92.

Zypman, F. R., Ferrante, J., Jansen, M., Scanlon, K., and Abel, P. 2003. Evidence of self-organized criticality in dry sliding friction. *Journal of Physics: Condensed Matter* 15 (12): L191.

3

Principles of Non-Equilibrium Thermodynamics and Friction

In the preceding chapter we discussed mechanical foundations of vibrations, stability, and self-organization. Here we will introduce the principles of reversible and irreversible thermodynamics that are of relevance to friction-induced self-organization. In particular, we will discuss the entropic stability criterion. In addition to the degrees of freedom that participate in the mechanical motion and are studied by mechanics, physical objects also have internal degrees of freedom. These degrees of freedom contain thermal energy that, under certain circumstances, can be exchanged with the mechanical degrees of freedom. This is why thermodynamic analysis should eventually be integrated with mechanics.

3.1 Thermodynamic Potentials and Equations

In this section we will introduce the basic principles of conventional or equilibrium thermodynamics, including thermodynamic potentials and equations. The starting point to introduce thermodynamic potentials is the concept of internal energy, U. Internal energy is the kinetic energy of molecules constituting the physical object or the energy of "ignored degrees of freedom" corresponding to particular molecules. As discussed in the preceding section, when self-organization occurs, the number of degrees of freedom in the system can reduce significantly because only a small number of degrees of freedom are observed directly at a higher level of hierarchy.

For example, when a rigid body slides on a flat surface, usually it is sufficient for the mechanical analysis to consider only the macro-scale variables, the coordinates of the center of the body mass. However, if we look at the sliding interface with larger resolution, we observe that a nominally smooth surface consists of a large number of deformable asperities, so that the contact occurs on the tips of the asperities. These asperities can be assumed to be elastically deformed and can be modeled by elastic springs. Therefore, with meso-scale resolution, the body will have additional degrees of freedom corresponding to the asperities. The vibrational motion of the asperities is dissipated due to damping. Furthermore, at the molecular scale resolution, single molecules can be investigated, where the dissipation results in the thermal motion of the molecules.

For macro-scale thermodynamic analysis, the atomic degrees of freedom can usually be ignored; however, it is assumed that they contain internal energy U. The state of a thermodynamic system is characterized by so-called "state variables," such as pressure P, volume V, temperature T, and the number of molecules in the system N. Usually, one has to deal with the change of internal energy between states (the finite change ΔU or the infinitesimal change dU), rather than with the absolute value of U.

Typically, $U(V,S,N)$ depends on the "coordinates" volume V, the "coordinate" called entropy S, and on the number of particles N (or the amount of substance) in the system. If we know the values of these coordinates, this uniquely characterizes the state of the thermodynamic system. Furthermore, if we know how U depends on V, S, and N, then we know everything about the system's macro-scale thermodynamic properties and behavior. We can write that for a small change of the state variables of the system, its internal energy would change as (Sandler 1999)

$$dU = -PdV + TdS + \mu dN \qquad (3.1)$$

where P, T, and μ are coefficients called the "pressure," "temperature," and "chemical potential," correspondingly. Note that the value of N is very large and for practical purposes it can be treated as a real (and not integer) number. If species of more than one type are in the system, the term μdN should be replaced with the sum of chemical potentials and change of particles of each substance $\Sigma \mu_k dN_k$.

The coefficients in Equation (3.1) are defined at this point as partial derivatives (Sandler 1999):

$$P = -\frac{\partial U}{\partial V} \qquad (3.2)$$

$$T = \frac{\partial U}{\partial S} \qquad (3.3)$$

$$\mu = \frac{\partial U}{\partial U} \qquad (3.4)$$

However, it is easy to show that P is the force per unit area (i.e., the pressure). Indeed, the change of the volume is equal to the area times the normal distance, $dV = Adx$, so that $dU = -PdV = -APdx$, and thus, since the change of internal energy is equal to the work of force $dU = -Fdx$, we conclude that $AP = F$.

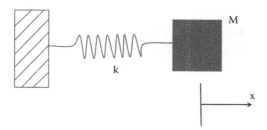

FIGURE 3.1
A simple mechanical system (spring-mass) with one degree of freedom.

Beginning students of thermodynamics sometimes find it difficult to understand the concept of entropy. Here, we just assume that there is a parameter S that we call "entropy" that characterizes the effect of hidden degrees of freedom on internal energy. Then we call the partial derivative $T = \partial U/\partial S$ "temperature." After that, one can show that T is indeed the same temperature that we measure with a thermometer. If two thermodynamic systems are at equilibrium, there is no heat flow and they have equal temperature.

It may be useful to consider a simple mechanical analogy. In a system with one degree of freedom involving a spring with the spring constant of k, the force is $F = -kx$, while the differential of the internal energy is (Figure 3.1)

$$dU = -Fdx \tag{3.5}$$

because $U = kx^2/2$. Note that one can also express energy as a function of force. This is the so-called complementary energy (strain energy) $\tilde{U} = F^2/(2k)$ with the differential $d\tilde{U} = -xdF$. For a linear system (when $k = const$), the complementary energy is equal to the energy $U(F)$ at the same value of $x = -F/k$. However, for a general case of non-linear dependency of the force on the displacement $F(x)$, they are not equal and the expression becomes

$$\tilde{U} = Fx - U \tag{3.6}$$

The parameters x and F are the so-called conjugate variables. Each of them can be used to uniquely characterize the state of the system. The transition from U to $\tilde{U}$ is called the Legendre transformation, and it is useful for many purposes (Tschoegl 2000).

In the thermodynamic equation of state (Equation 3.1), V, S, and N are called *extensive* properties, since they are proportional to volume or mass of the system (Balmer 2010). The variables P, T, and μ are called *intensive* properties, because they are not proportional to the volume or mass of the system. The pairs P and V, T and S, and μ and N are called "conjugate variables." The internal energy $U(V,S,N)$ is a function of volume, entropy, and the number of particles.

Often it is more convenient to present the system as a function of P and T. For that purpose, the Legendre transformation can be applied and the following potentials introduced:

Gibbs free energy

$$G(P,T,N) = U + PV - TS \tag{3.7}$$

$$dG = -SdT + VdP + \mu dN \tag{3.8}$$

Helmholtz free energy

$$A(V,T,N) = U - TS \tag{3.9}$$

$$dA = -SdT - PdV + \mu dN \tag{3.10}$$

Enthalpy

$$H(P,T,N) = U + PV \tag{3.11}$$

$$dH = -TdS + VdP + \mu dN \tag{3.12}$$

The following differential relationships are valid:

$$V = \frac{\partial G}{\partial P} = \frac{\partial H}{\partial P} \tag{3.13}$$

$$S = -\frac{\partial G}{\partial T} = -\frac{\partial A}{\partial T} \tag{3.14}$$

$$\mu = \frac{\partial G}{\partial N} = \frac{\partial A}{\partial N} = \frac{\partial H}{\partial N} \tag{3.15}$$

The following laws of thermodynamics can be formulated. Suppose that a thermodynamic system consists of two sub-systems. The total number of particles in two systems remains constant, $dN_1 = -dN_2$. The law of mass conservation in a closed system can therefore be formulated as (Figure 3.2)

$$dN = 0 \tag{3.16}$$

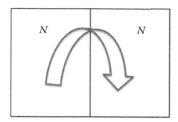

FIGURE 3.2
The law of mass conservation in a closed system: the total mass of the system is conserved.

The law of energy conservation, also referred to as the first law of thermodynamics, states that in an isolated system

$$dU = -PdV + TdS + \mu dN = 0 \tag{3.17}$$

The second law of thermodynamics states that entropy of an isolated system remains the same $dS = 0$ (reversible process) or grows $dS > 0$ (irreversible process):

$$dS = \frac{P}{T}dV - \frac{\mu}{T}dN + \frac{1}{T}dU \geq 0 \tag{3.18}$$

It is easy to demonstrate that the consequence of the second law is that heat flows from a hot body to a cold body. Let us again consider a system consisting of two parts (Figure 3.3). Suppose the volume and the number of particles do not change $dV = 0$, $dN = 0$, but heat dU flows from the sub-system 1 to the sub-system 2:

$$dS = \left(\frac{1}{T_2} - \frac{1}{T_1} \right) dU > 0 \tag{3.19}$$

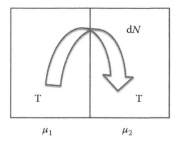

FIGURE 3.3
The consequence of the second law: heat flows from the hotter body to the colder body.

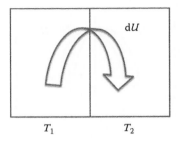

T_1 T_2

FIGURE 3.4
The consequence of the second law: mass flow from a system with higher chemical potential to that with lower chemical potential.

Therefore, $T_1 > T_2$—in other words, heat flows from the hotter body to the colder body.

Suppose now that there is no heat flow ($dU = 0$); however, the mass flow dN exists between two parts of the system (Figure 3.4). The temperatures of these two sub-systems are equal $T_1 = T_2 = T$. Then, from the second law,

$$dS = \left(\frac{\mu_1}{T} - \frac{\mu_2}{T} \right) dN > 0 \qquad (3.20)$$

Therefore, $\mu_1 > \mu_2$ and the mass flows from a system with higher chemical potential to that with a lower chemical potential. Chemical potentials are widely used to study whether a chemical reaction or a phase transition is energetically profitable.

The rate of entropy production per unit time can be calculated from $dU = TdS + \mu dN$ with the assumption that the volume of the system does not change as

$$\frac{dS}{dt} = \frac{1}{T}\frac{dU}{dt} - \frac{\mu}{T}\frac{dN}{dt} \qquad (3.21)$$

The first term in Equation (3.21) is proportional to dU and corresponds to heat flow from the system, while the second term is proportional to dN and corresponds to the mass flow.

The following definitions are used in thermodynamics:

- A thermodynamic system is called *isolated* if it does not exchange mass, heat, or mechanical energy (no work performed) with the surroundings (Figure 3.5). If no mass exchange occurs, but there is heat flow from or to the system and/or mechanical work is done by the system, such a thermodynamic system is called *closed*. An

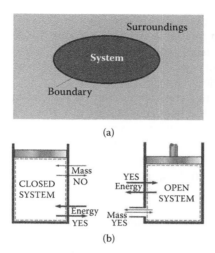

FIGURE 3.5
(a) A thermodynamic system as a quantity of matter or a region, separated from its surroundings; (b) closed and open thermodynamic systems.

open thermodynamic system can exchange mass and heat with the surrounding environment and it can produce/consume mechanical work (Cengel and Boles 2006).

- A *homogeneous* system is a system with no spatial dependence on thermodynamic parameters (pressure, temperature, and concentrations of species remain the same everywhere; Sandler 1999). A *stationary* system is a system with no temporal dependence on thermodynamic parameters (pressure, temperature, and concentrations do not change with time). The system is in *equilibrium* if it is both homogeneous and stationary, and the entropy production is zero. A system close to the equilibrium can often be treated as a *linear* system (Sandler 1999). Note that the system can be in a stationary state with no time dependence of the variables, but entropy can still be produced and dissipation can occur.

3.2 Irreversible Processes and Non-Equilibrium

Now we will consider the situation when thermodynamic parameters, such as the temperature, pressure, or concentrations, change with time or are not constant in space. The processes that occur in such situations are studied by non-equilibrium or irreversible thermodynamics (also called the thermodynamics of irreversible processes; Lavenda 1978). For all intensive variables, we now should consider their dependency on time and coordinates $T(x, y, z, t)$, $P(x, y, z, t)$, and $\mu_k(x, y, z, t)$. For all extensive variables, we consider their special densities per unit volume and use small letters to denote these: the internal energy density $u(x, t)$, the local entropy density $s(x, y, z, t)$, and concentrations or mass densities $c_k(x, y, z, t)$.

In order to investigate time dependency of the variables, their flows should be introduced. Thus, the heat flow vector $\mathbf{q}$ is the amount of energy transferred per unit time per unit area in each direction (x, y, and z). In a similar manner, the vectors of the mass flow, J^c, and the entropy flow, J^s, can be introduced. Now the local law of mass conservation states that the rate of density concentration is equal to the negative of the gradient of mass flow (De Groot and Mazur 2011):

$$\frac{\partial c}{\partial t} + \nabla J^c = 0 \tag{3.22}$$

In a similar manner, the first law of thermodynamics states that the local rate of energy density is equal to the negative of the gradient of heat flow (De Groot and Mazur 2011):

$$\frac{\partial u}{\partial t} + \nabla \mathbf{q} = 0 \tag{3.23}$$

Finally, the second law is formulated as an inequality stating that the rate of entropy density plus the gradient of entropy flow should be equal to zero (for the reversible process) or positive (De Groot and Mazur 2011):

$$\frac{\partial s}{\partial t} + \nabla J^s \geq 0 \tag{3.24}$$

Another important concept is "thermodynamic forces and flows (fluxes)." Let us assume that the state of the system is defined by a number of variables (generalized coordinates) $x_1, x_2, \ldots, x_n$. A non-equilibrium process will continue while dissipation is possible—in other words, until entropy reaches its maximum. For a system in equilibrium, entropy is at the maximum (De Groot and Mazur 2011).

$$dS = \sum \frac{\partial s}{\partial x_k} dx_k = 0 \tag{3.25}$$

The thermodynamic forces X_k can be defined as partial derivatives of

$$X_k = \frac{\partial s}{\partial x} = 0 \tag{3.26}$$

For a system close to equilibrium, the rate of change (the generalized velocity) of x_k is proportional to the linear combination of forces:

$$\dot{x}_k = \sum L_{ik} X_i \tag{3.27}$$

where L_{ik} are the Onsager's phenomenological coefficients. The second law of thermodynamics requires that the matrix L be positive definite. An important result is given by Onsager's reciprocity relationships stating that (De Groot and Mazur 2011):

$$L_{ik} = L_{ki} \tag{3.28}$$

Combining Equations (3.25) and (3.26) and taking the time derivative yield the rate of entropy production, given by

$$\dot{S} = \sum X_k \dot{x}_k \tag{3.29}$$

The same concept can be applied to the local density of entropy production, if flows are taken as local generalized velocities. The gradients of intensive variables can be treated as "thermodynamic forces" in the sense that they cause the flows. It can be shown that appropriate forces responsible for the heat and mass transfer are given by

$$X_q = \nabla \left(\frac{1}{T} \right) = -\frac{\nabla T}{T^2} \tag{3.30}$$

$$X_\mu = \frac{\nabla \mu}{T} \tag{3.31}$$

Equation (3.27) gives the phenomenological (Onsager) linear relationships between the flows and the forces (Prigogine 1968):

$$q = -\gamma T^2 \left(\nabla \frac{1}{T} \right) - \beta T^2 \left(\nabla \frac{\mu_i}{T} \right) \tag{3.32}$$

$$J^c = -\beta T^2 \left(\nabla \frac{1}{T} \right) - \alpha T \left(\nabla \frac{\mu_i}{T} \right) \tag{3.33}$$

where α, β, and γ are coefficients. From the reciprocity relationships (Equation 3.28), $L_{12} = L_{21} = -\beta T^2$ and, using $\nabla(1/T) = -T/T^2$, we can find the relationships (Prigogine 1968):

$$q = -\chi \nabla T + \left(\mu_i + \frac{\beta T}{\alpha} \right) J^c \tag{3.34}$$

$$J^C = -\beta \nabla T - \alpha \nabla \mu_i \tag{3.35}$$

where

$$\chi = \gamma - \frac{\beta^2 T}{\alpha} \tag{3.36}$$

is the coefficient of thermal conductivity, while the earlier introduced α is called the coefficient of diffusion.

An important consequence of Equations (3.34) and (3.35) is that heat transfer and diffusion can be coupled to each other. This is the phenomenon of thermodiffusion (or thermophoresis). When a system exchanges mass and energy with its surroundings, various irreversible processes inside the system may interact with each other. These interactions are called thermodynamic couplings and they provide a mechanism for a process without its primary driving force or even may move the process in a direction opposite to the one imposed by its own driving force.

For example, in thermodiffusion (thermophoresis, Soret effect) a substance diffuses because of a temperature gradient rather than a concentration gradient (Figure 3.6). When a thin hot wire is inserted into tobacco smoke, the

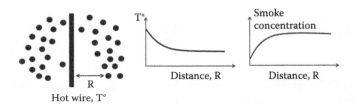

FIGURE 3.6
Thermodiffusion: the concentration of smoke is reduced near the hot wire because hot air molecules hit smoke particles harder than cold air molecules. It is suggested that the heated frictional interface can have a similar effect.

smoke particle concentration will be reduced in the vicinity of the rod (the scale of the effect is less than 1 mm, but it can be seen with the naked eye). This is because hotter air molecules hit smoke particles from the side of the wire, pushing the particles in the opposite direction away from the wire. When a substance flows from a low to a high concentration region, it must be coupled with a compensating process. The principles of thermodynamics allow the progress of a process without or against its primary driving force X_n only if it is coupled with another process.

It is expected that friction may result in an effect similar to thermodiffusion. The sliding interface is a source of heat creating temperature gradients, like the heated wire in the Soret effect. If material is composed of two fractions (such as air and smoke particles), temperature gradients can result in separation of these fractures, leading to in situ tribofilm formation.

Many linear empirical laws of physics, such as Ohm's law of electrical resistance, the Fourier law of heat conduction, Fick's law of diffusion, and linear viscosity, are consequences of the empirical equations of the linear thermodynamics (Nicolis and Prigogine 1977; Gershman and Bushe 2006) given by Equation (3.27).

3.3 Extremal Principles and Stability of Frictional Motion

As discussed in the preceding section, many processes tend to reach a stationary (or steady-state) regime that is not an equilibrium regime because entropy is still produced and dissipation occurs; however, the parameters are stabilized at a certain level. An example can be a steady-state frictional sliding of a slider on a flat deformable surface. Such a sliding, when described in the coordinate system attached to the slider, is stationary since the coordinates of the sliding bodies do not change; however, dissipation occurs.

The question of whether a general principle exists that governs the evolution of non-equilibrium processes and what constitutes such a principle remains a controversial topic in non-equilibrium thermodynamics. Several concepts involving the extremal principle have been suggested. One of them is the minimum entropy production principle, which states that the stationary process tends to be that with a minimum possible entropy production (Maes and Netocny 2006; Jaynes 1980; Prigogine and Stegers 1984; Haken 1983; Nicolis and Prigogine 1977; Prigogine 1968, 1980). The principle applies to systems driven away from the state of equilibrium by an externally applied force. Due to various interactions within the system between its parts, the system can pass through various states. However, due to redistribution of energy between various degrees of freedom, the system will tend to attain the state where damping is at a minimum.

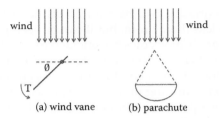

FIGURE 3.7
Schematic for (a) "wind vane" (minimum resistance to fluid flow) and (b) "parachute" (maximum resistance to fluid flow) structures. It is speculated that "wind vane" structures are much more common than the "parachute" structures. As a result, many dynamical systems evolve to the state of minimum resistance/dissipation/entropy production.

Imagine a system with several thermodynamic forces that are all allowed to equilibrate in a time-dependent way except that at least one is maintained so that the system cannot get to equilibrium. The entropy production will start at some rate and because the forces are progressively depleted, the rate of the entropy production will steadily (monotonically) go down until the system gets as close to equilibrium as it can and then stay in that state as long as the one force is maintained. So the entropy production starts at one place and goes down to a minimum, where it stays.

The principle of minimum entropy production was suggested by Prigogine (1968) and its validity and the conditions of its applicability remain the subject of heated discussions. It is perhaps more justified to view the minimum entropy production principle as a phenomenological observation of many processes, rather than as a fundamental law of nature.

A simple and speculative mechanical model that illustrates the concept is the following (Figure 3.7). Consider a rotating flat body that can rotate about the central axis moving in viscous media with constant velocity, V. This rotating system forms a "wind vane." The rotation angle θ is an internal degree of freedom in the system. The viscous friction force, F, and therefore the amount of dissipated energy per unit time, FV, as well as the rate of entropy production, FV/T, are all proportional to the cross-sectional area of the body. For $\theta = 0$, the friction force is at maximum, while for $\theta = 90$, the friction force is at minimum. If the center of rotation of the body (point O) exactly coincides with its geometric center, then the forces acting on the left and on the right side of the structure are equal. However, in a more realistic case, a minor imbalance exists that will result in the rotation of the structure until the position of minimum dissipation is attained.

This example is obviously not universal. One can imagine a structure intentionally designed to maintain maximum dissipation. A common example would be a parachute designed to attain a shape of maximum viscous friction. However, statistically, a "wind vane" type of structure is more common than a "parachute" structure. This is because the resistance of the parachute structure to the fluid flow involves a stable equilibrium and a negative feedback, whereas the corresponding equilibrium state of the wind vane is unstable and involves a positive feedback. Unstable systems are more common in nature because of the "principle of the fragility of good" (Arnold 1992) in the theory of linear programming; this principle states that in order to be stable the system should satisfy *all* stability conditions simultaneously, while in order to be unstable it is sufficient that only *one* condition is violated. Thus, the transition from stability to instability is more likely than the opposite one. It is therefore expected that in a large complex system consisting of many parts, the "wind vane" type of behavior would dominate, and the complex system will drift to the state of minimum, rather than maximum, dissipation (Arnold 1992).

Note also that the principle of minimum entropy production applies to stationary but non-equilibrium systems. For systems that are far from equilibrium, one cannot apply the phenomenological linear Onsager relationships, so it is not possible to minimize the rate of entropy production as given by Equations (3.26–3.29).

To put it simply, the minimum entropy production principle states that a non-equilibrium system is more likely to drift into the direction of "minimal resistance" or lower entropy production than in the opposite direction (Nosonovsky 2010). For a system with friction, one possible example is the transient "running-in" process after friction is initiated (Mortazavi and Nosonovsky 2011a; Mortazavi et al. 2010). Usually, due to the simultaneous action of many causes (for example, fracture of high asperities, tribofilm formations), the initially high friction tends to decrease to the stationary value during the running-in stage. The overall trend for friction to decrease with time, including the difference between static and kinetic friction and the tendency of friction to decrease with the age of contact or with increasing sliding velocity, can perhaps be explained as a part of this phenomenon.

3.4 Stability Criterion for Frictional Sliding

In this section we formulate a thermodynamic stability criterion that will be applied to various frictional situations. This criterion, which is based on the minimum entropy production principle, has been successfully applied to tribosystems by Gershman and Bushe (2006), Nosonovsky and Bhushan (2009), Nosonovsky (2010), and others.

3.4.1 Stability Criterion for the Stationary Regime $\delta^2 \dot{S} > 0$

If entropy production rate is at minimum in the stationary non-equilibrium state, the variation of entropy production should be zero, $\delta \dot{S} = 0$. Furthermore, the stability criterion should apply, stating that the second variation should be positive:

$$\delta^2 \dot{S} > 0 \tag{3.37}$$

The stability criterion is a generalization of the mechanical stability criterion discussed in the preceding chapter, which stated that energy should reach its minimum in the state of equilibrium. Furthermore, similarly to pure mechanical systems described in the preceding sections, the onset of instability can lead to self-organization. The stability criterion given by Equation (3.37) is a powerful tool that can be applied to many tribological processes, so let us discuss it in more depth.

The entropy production rate is a function of many parameters of the system $\dot{S}(q_1, q_2, \ldots, q_N)$. These parameters $q_1, q_2, \ldots, q_N$ are not necessarily independent and can include, for example, velocities of various processes or externally applied force. Let us assume that the system is in its stationary state. Suppose that one parameter, q_m, changes slightly—for example, due to a random perturbation. Then the criterion states that

$$\delta^2 \dot{S} = \frac{\partial^2 S}{\partial q_m^2}(\delta q_m)^2 > 0 \tag{3.38}$$

Since $(\delta q_m)^2 > 0$ always, the stability criterion reduces to

$$\frac{\partial^2 S}{\partial q_m^2}(\delta q_m)^2 > 0 \tag{3.39}$$

Suppose now that two parameters, q_n and q_m, can change; however, these parameters are not independent, $q_n(q_m)$. Then the criterion states that

$$\delta^2 \dot{S} = \frac{\partial^2 S}{\partial q_n^2}(\delta q_n)^2 + 2\frac{\partial^2 S}{\partial q_n \, \partial q_m}\delta q_n \delta q_m + \frac{\partial^2 S}{\partial q_m^2}(\delta q_m)^2 > 0 \tag{3.40}$$

The variations are related as

$$\delta q_n = \frac{dq_n}{dq_m}\delta q_m \tag{3.41}$$

that yields

$$\delta^2 S = \left[\frac{\partial^2 S}{\partial q_n^2} + 2\frac{\partial^2 S}{\partial q_n \, \partial q_m}\frac{dq_n}{dq_m} + \frac{\partial^2 S}{\partial q_m^2}\left(\frac{dq_n}{dq_m}\right)^2\right](\delta q_m)^2 > 0 \tag{3.42}$$

or

$$\frac{\partial^2 S}{\partial q_n^2} + 2\frac{\partial^2 S}{\partial q_n \partial q_m}\frac{dq_n}{dq_m} + \frac{\partial^2 S}{\partial q_m^2}\left(\frac{dq_n}{dq_m}\right)^2 > 0 \qquad (3.43)$$

When the entropy rate depends linearly on q_n and q_m, the first and the third term in this expression are zero, $\partial^2 S/\partial q_n^2 = 0$, while $\partial^2 S/\partial q_n \partial q_m$ is constant, so the stability criterion is reduced to the sign of dq_n/dq_m. For more complex cases, certain conclusions about the sign of Equation (3.43) can be made as well.

3.4.2 Instabilities during Velocity-Dependent Friction

Note that the criterion $\delta^2\dot{S} > 0$ given by Equation (3.37) is universal and the expression for the entropy rate can include mechanical, thermal, chemical, electric, micro-/nano-structural, and other components. First, let us assume that only mechanical interactions occur in a frictional system with one degree of freedom. The rate of entropy is then given by the product of sliding velocity V and the friction force μW:

$$\dot{S} = \frac{\mu W V}{T} \qquad (3.44)$$

If only one of these parameters is varied, this would lead to $\delta^2\dot{S} = 0$. However, if two parameters are varied and inter-related, we obtain

$$\delta^2\dot{S} = \delta^2\left(\frac{2\mu W V}{T}\right) = \frac{2W}{T}\delta V\delta\mu = 2W\frac{d\mu}{dV}(\delta V)^2 > 0 \quad \text{or} \quad \frac{d\mu}{dV} > 0 \quad (3.45)$$

In other words, if the coefficient of friction increases with the sliding velocity, the system is unstable.

3.4.3 Friction-Induced Instabilities in Deformable Bodies

For a system of deformable bodies, a more complex form of the entropy rate should be investigated. The mathematical analysis of the stability can be quite complicated. The procedure is the following. The local energy dissipation involves the products of stresses and strain rates, which can be viewed as generalized forces and flows. The strains are related to the components of the displacement u_k, while stresses are related to strains through the constitutive laws, such as the equations of the elasticity. A small variation of the displacement, δu_k, results in the change of the overall entropy production rate. The displacement should not be studied locally; instead, a mode shape of the displacement $\delta u_k(x,y)$ is investigated, yielding a characteristic equation for frequencies that can have stable or unstable modes. Frictional heat generation and conduction can be incorporated into the model, leading to the thermoelastic instabilities. The stability analysis of the frictional contact of elastic bodies will be discussed in detail in Chapter 6.

3.4.4 Instability during Friction Heat Conduction

Heat generated by friction flows away from the frictional interface into the bulk of the contacting materials and can result in interesting phenomena even in the case of non-deformable bodies in contact. To investigate the stability of this effect, one can apply the criterion Equation (3.37) to the entropy per surface area at the frictional interface. The flow of entropy away from the flat infinite interface in the steady-state situation is equal to entropy generation due to frictional heat $dQ = \mu W dx$ (Nosonovsky and Rohatgi 2012; Nosonovsky 2010; Nosonovsky and Bhushan 2009, 2010a, b; Nosonovsky et al. 2009). Assuming, for simplicity, that all generated heat is dissipated in one of the two contacting bodies and thus ignoring the division of heat between the two bodies, the heat flows away from the interface in accordance with the heat conduction equation:

$$\lambda \frac{\partial T}{\partial z} = \mu W V \tag{3.46}$$

where z is the vertical coordinate (distance from the interface) and λ is the heat conductivity.

Consider a thin layer near the interface with the thickness dz. The temperature drop across the layer is $dT = (\mu W V / \lambda) dz$. The ratio of the heat released at the interface, dQ, to that radiated at the bottom of the layer, dQ', is equal to the ratio of the temperatures at the top and at the bottom of the layers:

$$\frac{dQ'}{dQ} = \frac{T - \mu W V dz / \lambda}{T} \tag{3.47}$$

Therefore, the energy released at the sub-surface layer of depth dz is given by Fox-Rabinovich et al. (2007):

$$dq = \frac{dQ - dQ'}{dz} = dQ \frac{\mu W V}{\lambda T} = \frac{(\mu W V)^2}{\lambda T} \tag{3.48}$$

Thus, the entropy in the sub-surface layer, $dS/dt = dq/T$, is given by

$$\frac{dS}{dt} = \frac{(\mu W V)^2}{\lambda T^2} \qquad (3.49)$$

Note that S in Equation (3.49) is entropy per unit surface area and thus it is measured in $JK^{-1}m^{-2}$, unlike the total entropy, which is measured in JK^{-1}. Equation (3.49) takes into account the thermal conductivity.

Suppose that one contacting material has a micro-structure characterized by a certain parameter ψ, such as, for example, the size of reinforcement particles in a composite material. Such values of ψ in which $\mu'_V(\psi) > 0$ correspond to steady-state sliding. However, $\mu'_V(\psi) = 0$ corresponds to the destabilization of the steady-state solution. As a result, a new equilibrium position will be found with a lower value of μ (Nosonovsky 2010). Suppose now that the coefficient of friction depends also on a micro-structure parameter ϕ, such as the thickness of the interface film (Figure 3.8a). The difference between ψ and ϕ is that the parameter ψ is constant (the composition of the material does not change during friction), whereas the parameter ϕ can change during friction (the film can grow or decrease due to a friction-induced chemical reaction or wear). The stability condition is now given by

$$\frac{1}{2}\delta^2 \dot{S} = \delta\left(\frac{\mu W}{T}\right)\delta(V) = \frac{W}{T}\frac{\partial \mu}{\partial \phi}\frac{\partial V}{\partial \phi}(\delta\phi)^2 \geq 0 \qquad (3.50)$$

If the stability condition is violated for a certain value of ϕ, then further growth of the film will result in decreasing friction and wear, which will facilitate the further growth of the film. The destabilization occurs at $\mu'_\phi(\psi,\phi) = 0$. Note that Equation (3.19) becomes Equation (3.18) if $\phi = V$. At this point, we are not discussing the question of which particular thermodynamic force is responsible for the growth of the film (Nosonovsky 2010).

Since we are interested in the conditions of the formation of such a protective film, consider now the limit of the thin film ($\phi \to 0$). With increasing film thickness, the value of μ changes from that for the bulk composite material to that of the film material. On the other hand, the value for the bulk composite material depends also on its micro-structure ψ (Figure 3.8b). The critical value, ψ_{cr}, corresponds to $\mu'_\phi(\psi,0) = 0$. For the size of reinforcement particles finer than ψ_{cr}, the bulk (no film, $\phi = 0$) values of the coefficient of friction are lower than the values of the film. This can lead to a sudden destabilization (formation of the film with thickness ϕ_0) and reduction of friction to the value of $\mu(\psi, \phi_0)$ as well as wear reduction. Here we do not investigate the question of why the film would form and how its material is related to the material of the contacting bodies. However, it is known that such reaction occurs in a number of situations when a soft phase is present in a hard matrix, including Al–Sn- and Cu–Sn-based alloys (Fox-Rabinovich and Totten 2006; Bushe and Gershman 2006).

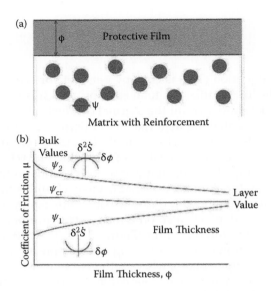

FIGURE 3.8
(a) Self-organized protective film at the interface of a composite material. (b) The coefficient of friction as a function of film thickness for various values of the microstructure parameter, y. If film growth is coupled with friction, spontaneous in situ film formation can result in friction reduction.

3.4.5 Diffusion, Mass Transfer, and Wear during Friction

Wear often accompanies friction, leading to material debris transfer. Furthermore, in materials with a complex composition (alloys, composite materials), even when wear is absent, transfer of some phases of the materials can occur. Therefore, the mass transfer processes governed by the diffusion equations can be incorporated into the stability analysis.

The reaction-diffusion systems were among the first systems with a pattern (Kagan 2010; Mortazavi and Nosonovsky 2011b). Frictional interaction can also lead to pattern formation. Note that, mathematically, there is a significant difference between the systems with friction-induced vibrations in elastic media and diffusion. While elastic instabilities involve wave propagation and hyperbolic partial differential equations (PDEs), which describe the dynamic behavior of elastic media, the reaction-diffusion systems involve parabolic PDEs, typical for diffusion and heat propagation problems.

Wear can also have an effect on surface roughness, which in turn can affect friction. Thus, a small variation of the wear rate, $\delta \dot{V}$, can lead to the variation of the surface roughness, δR, and the coefficient of friction, $\delta \mu$. One can relate the second variation of the entropy rate $\delta^2 \dot{S}$ to the positive $(\delta \dot{V})^2$ in order to apply the stability criterion Equation (3.37) to such systems. The analysis is similar to the velocity-dependent coefficient of friction

(Equation 3.45) and the coefficient of friction dependent on the tribofilm thickness (Equation 3.50).

3.4.6 Tribochemical Reactions

Tribochemical reactions are those caused or catalyzed by friction, including the reactions of oxidation, formation of chemical films, and so on. In order to incorporate tribochemical reactions into the scheme of the stability analysis, entropy production due to a chemical reaction should be included in the total entropy rate. The chemical affinity is related to the chemical potentials of the products, μ_k, and the corresponding stoichiometric coefficients, ν_k, as

$$A = -\sum \nu_k \mu_k \tag{3.51}$$

The rate of entropy production is given by

$$dS = \frac{A d\xi}{T} \tag{3.52}$$

where ξ is the "extent of reaction," the quantity of substance (in moles) participating in the reaction, normalized by the stoichiometric coefficients. The chemical reaction proceeds while the affinity is positive. The chemical affinity can also be defined as a partial derivative of the Gibbs free energy by the extent of reaction

$$A = \frac{\partial G}{\partial \xi} \tag{3.53}$$

The contribution of the chemical reactions to the entropy rate is found by

$$\dot{S} = \frac{1}{T} \sum A_l \dot{\xi}_l \tag{3.54}$$

Therefore, in order to take the chemical reactions into account for the stability criterion, the dependency of $\delta^2 \dot{S}$ in the variations of the extent of the reaction should be calculated.

3.4.7 Electrical Contact and Other Effects

Electric current can often go through the interface between the contacting bodies. Due to the coupling of the electric, mechanical, thermal, and chemical phenomena, the current can lead to the reduction of friction through the formation of self-organized structures, referred to as "electrolubrication" (Gershman 2006). The contribution of the electromagnetic field into the entropy production is

$$\dot{S} = \frac{1}{T}\mathbf{E}\cdot\mathbf{i}$$

(3.55)

where $\mathbf{E} = \mathbf{E_0} + \mathbf{V} \times \mathbf{B_0}$ is the electric field in the coordinate system moving with the velocity $\mathbf{V}$.

Other types of interactions may be included, such as the plasma effect, thermolubricity, atomic friction, and so on. In the general form, the local rate of entropy production should take into account all processes that occur simultaneously:

$$\dot{S} = \mathbf{q}\cdot\nabla\frac{1}{T} - \sum \mathbf{J}^k\cdot\nabla\frac{\mu_k}{T} - \frac{p^v}{V}\nabla\cdot\mathbf{V} - \frac{1}{T}\mathbf{P}^{0v}:\mathbf{V}^0 + \frac{\rho}{T}\sum A_l\dot{\xi}_l + \frac{1}{T}\mathbf{E}\cdot\mathbf{i} \quad (3.56)$$

where
$\mathbf{q}$ is the heat flow vector
$\mathbf{J}^k$ are the diffusion flow vectors of various components of the material
μ_k are corresponding potentials
the viscous stress tensor is presented as the sum of the spherical part p^v
    and the deviator $\mathbf{P}^{0v}$
the strain rate tensor is presented as the sum of the velocity $\mathbf{V}$ and the
    deviator $\mathbf{V}^{0v}$
sign ":" is the dual tensor product
A_l and ξ_l are the chemical affinity and the extent of the reaction l
$\mathbf{E}$ is the vector of electric field
$\mathbf{i}$ is the vector of current density

Note that Equation (3.56) is not the most general and complete formula for the entropy rate. It does not include the effects of friction, other surface effects, and the phase transition effects. The thermodynamic forces and flows for some of these processes are summarized in Table 3.1, which includes, in addition to the flows that were considered previously, the volume rate for the plastic yield, $\dot{V}$; the rate of the growth of the crack, $\dot{a}$; and the volume of the phase transition.

The general scheme of the stability analysis is the following. First, it is necessary to identify which processes participate in the system and the expression for the entropy rate written. Second, for coupled processes, the variations of the corresponding variables should be related (for example, the velocity δV and the coefficient of friction $\delta\mu$). Third, the second variation of the entropy rate is calculated and the stability criterion Equation (3.37) is applied. If the process is unstable (as in the case of a decreasing coefficient

TABLE 3.1

Typical Thermodynamic Forces and Flows during Various Processes

Effect	Sliding	Heat Transfer	Diffusion	Volume Viscosity	Electric Field	Shear Viscosity	Chemical Reaction	Dry Friction	Phase Transition	Fracture	Plastic Yield
Force	$\dfrac{\mathbf{F}}{T}$	$\nabla\dfrac{1}{T}$	$-\nabla\dfrac{\mu_k}{T}$	$-\dfrac{1}{T}\nabla\cdot\mathbf{V}$	$-\dfrac{1}{T}\mathbf{E}$	$-\dfrac{1}{T}\mathbf{V}^0$	$-\dfrac{1}{T}A_l$	$\dfrac{\sigma_{xy}}{T}$	$\dfrac{\Delta H}{T}$	$\dfrac{\frac{\partial U}{\partial a}-2\gamma}{T}$	$\dfrac{U_p}{T}$
Flow	$\mathbf{V}$	$\mathbf{q}$	$\mathbf{J}^k$	p^v	$\mathbf{I}$	$\mathbf{p}^{ov}$	$\rho\dot{\xi}_l$	$\dot{u}_x$	$\dot{V}$	$\dot{a}$	$\dot{V}$
Rank	Vector, volume density			Tensor, volume density			Scalar, volume density		Scalar (volume)	Scalar (length)	Scalar (volume)

of friction with increasing velocity), a self-organized process (e.g., stick–slip motion) can occur.

3.5 Summary

In this chapter we introduced three important concepts from non-equilibrium thermodynamics. The first is the phenomenological linear relationships between the thermodynamic flows and forces (Equation 3.27). These are equations that drive thermodynamic processes that are close to the equilibrium back to the state of equilibrium. The linear viscosity is the consequence of these phenomenological relationships. In subsequent chapters we will try to deduce the linear law of dry friction (the Coulomb–Amontons law) from these relationships, taking into account an asymptotic transition from the three-dimensional bulk material to the two-dimensional frictional interface. The second law of thermodynamics requires that the matrix of the phenomenological coefficients be positive definite or that the directions of flows and the forces should coincide.

The second concept is the principle of minimum entropy production, which states that for a stationary non-equilibrium process, the dissipation tends to reach its minimum. In other words, the system finds the "state of minimum resistance." Unlike the phenomenological relationships, this principle is supposed to apply to processes far from the equilibrium, and the validity and the limits of applicability of the principle remain controversial. However, it may play a role in many tribological processes, including the transient processes of running-in, and it can justify state-and-rate-dependent models of friction.

The third concept is the entropic stability criterion for the stationary non-equilibrium process (Equation 3.37). It is a generalization of the mechanical stability criterion discussed in the preceding chapter, which stated that energy should reach its minimum in the state of equilibrium. The stability criterion states that the second variation of the entropy production rate should be positive, $\delta^2 \dot{S} > 0$, for the stationary state to be stable. The criterion can be applied in tribology to study the onset of the instability in various systems ranging from purely mechanical to those in which friction is coupled with another process such as wear, a tribochemical reaction, tribofilm formation, electrical current, and plasma. If the onset of the instability is detected, the unstable state can lead to vibrations and the formation of self-organized structures.

References

Arnold, V. I. 1992. *Catastrophe theory.* Berlin: Springer–Verlag.

Balmer, R. T. 2010. *Modern engineering thermodynamics.* San Diego, CA: Academic Press.

Barber, J. R. 1969. Thermoelastic instabilities in the sliding of conforming solids. *Proceedings of Royal Society London A* 312:381–394.

Bershadski, L. I. 1992. On the self-organization and concepts of wear-resistance in tribosystems. *Trenie I Iznos (Friction and Wear)* 13:1077–1094 (in Russian).

———. 1993. B. I. Kostetski and the general concept in tribology. *Trenie I Iznos (Friction and Wear)* 14:6–18 (in Russian).

Bhushan, B., and Nosonovsky, M. 2003. Scale effects in friction using strain gradient plasticity and dislocation-assisted sliding (microslip). *Acta Materialia* 51:4331–4340.

———. 2004. Comprehensive model for scale effects in friction due to adhesion and two- and three-body deformation (plowing). *Acta Materialia* 52:2461–2474.

Bushe, N. A., and Gershman, I. S. 2006. 3 Compatibility of tribosystems. In *Self-organization during friction. Advanced surface-engineered materials and systems design,* ed. G. S. Fox-Rabinovich and G. E. Totten, 31–59. Boca Raton, FL: CRC Taylor & Francis.

Cengel, Y. A., and Boles, M. A. 2006. *Thermodynamics: An engineering approach.* New York: McGraw–Hill Higher Education.

De Groot, S. R., and Mazur, P. 2011. *Non-equilibrium thermodynamics.* Mineola, NY: Dover Publications.

Fox-Rabinovich, G., & Totten, G. E., eds. 2006. *Self-organization during friction. Advanced surface-engineered materials and systems design,* vol. 31. Boca Raton, FL: CRC Press.

Fox-Rabinovich, G. S., Veldhuis, S. C., Kovalev, A. I., Wainstein, D. L., Gershman, I. S., Korshunov, S., Shuster, L. S., and Endrino, J. L. 2007. Features of self-organization in ion modified nanocrystalline plasma vapor deposited AlTiN coatings under severe tribological conditions. *Journal of Applied Physics* 102:074305.

Gershman, I. S. 2006. Formation of secondary structures and self-organization process of tribosystems during friction with the collection of electric current. In *Self-organization during friction. Advanced surface-engineered materials and systems design,* ed. G. S. Fox-Rabinovich and G. E. Totten, 197–230. Boca Raton, FL: CRC Taylor & Francis.

Gershman, I. S., and Bushe, N. 2006. Elements of thermodynamics of self-organization during friction. In *Self-organization during friction. Advanced surface-engineered materials and systems design,* ed. G. S. Fox-Rabinovich and G. E. Totten, 13–58. Boca Raton, FL: CRC Taylor & Francis.

Jaynes, E. T. 1980. The minimum entropy production principle. *Annual Review of Physical Chemistry* 31 (1): 579–601.

Kagan, E. 2010. Turing systems, entropy, and kinetic models for self-healing surfaces. *Entropy* 12 (3): 554–569.

Lavenda, B. H. 1978. *Thermodynamics of irreversible processes.* London: Macmillan.

Maes, C., and Netocny, K. 2006. Minimum entropy production principle from a dynamical fluctuation law (http://arxiv.org/abs/math-ph/0612063).

Menezes, P. L., Nosonovsky, M., Kailas, S. V., and Lovell, M. 2011. Friction and wear. In *Tribology for engineers: A practical guide*, ed. J. P. Davim. Cambridge, England: Woodhead Publishing.

Mortazavi, V., and Nosonovsky, M. 2011a. Friction-induced pattern formation and Turing systems. *Langmuir* 27 (8): 4772–4779.

———. 2011b. Wear-induced microtopography evolution and wetting properties of self-cleaning, lubricating and healing surfaces. *Journal of Adhesion Science and Technology* 25 (12): 1337–1359.

Nicolis, G., and Prigogine, I. 1977. *Self-organization in nonequilibrium systems.* New York: John Wiley & Sons.

Nosonovsky, M. (2010). Entropy in tribology: In the search for applications. *Entropy* 12 (6): 1345–1390.

Nosonovsky, M., and Bhushan, B. 2009. Thermodynamics of surface degradation, self-organization and self-healing for biomimetic surfaces. *Philosophical Transactions of the Royal Society A: Mathematical, Physical and Engineering Sciences* 367 (1893): 1607–1627.

Prigogine, I. 1968. *Thermodynamics of irreversible processes.* New York: John Wiley & Sons.

Sandler, S. I. 1999. *Chemical and engineering thermodynamics,* 3rd ed. New York: Wiley.

Tschoegl, N. W. 2000. *Fundamentals of equilibrium and steady-state thermodynamics.* New York: Elsevier Science Limited.

Bibliography

Garkunov, D. N. 2000. *Triboengineering (wear and non-deterioration).* Moscow: Moscow Agricultural Academy Press (in Russian).

———. 2004. *Scientific discoveries in tribotechnology.* Moscow: MSHA (in Russian).

Haken, H. 1983. Synergetics. An introduction. Nonequilibrium phase transitions in physics, chemistry and biology, 3rd ed. New York: Springer–Verlag.

Mortazavi, V., Wang, C., and Nosonovsky, M., 2010. Shannon entropy as a characteristic of a rough surface: Why the running-in transient process leads to friction reduction. *Proceedings of STLE/ASME 2010 International Joint Tribology Conference* IJTC2010-41135. New York: ASME.

———. 2012. Stability of frictional sliding with the coefficient of friction depended on the temperature. *ASME Journal of Tribology* 134:041601.

Nosonovsky, M., Amano, R., Lucci, J. M., and Rohatgi, P. K. 2009. Physical chemistry of self-organization and self-healing in metals. *Physical Chemistry Chemical Physics* 11:9530–9536.

Nosonovsky, M., and Bhushan, B. 2010a. Introduction to green tribology: Principles, research areas, and challenges. *Philosophical Transactions of Royal Society A* 368:4677–4694.

———. 2010b. Surface self-organization: From wear to self-healing in biological and technical surfaces. *Applied Surface Science* 256:3982–3987.

Nosonovsky, M., and Esche, S. K. 2008. A paradox of decreasing entropy in multiscale Monte Carlo grain growth simulation. *Entropy* 10:49–54.

Nosonovsky, M., and Rohatgi, P. K. 2012. *Biomimetics in materials science: Self-healing, self-lubricating, and self-cleaning materials* New York: Springer Series in Materials Science.

Prigogine, I. 1980. *From being to becoming.* San Francisco, CA: WH Freeman and Company.

Prigogine, I., and Stengers, I. 1984. *Order out of chaos.* New York: Bantam.

4

Fundamentals of Friction

In the preceding chapters we reviewed the foundations of mechanics and thermodynamics that are of relevance to friction. We discussed the basics of the theory of mechanical vibrations and stability, non-linear effects and self-organization, thermodynamic kinetic relationships, and stability of the stationary states. In this chapter we will discuss friction in depth: the empirical data about friction, its laws and mechanisms, thermodynamics of friction, and various models of friction. We will identify properties of friction that are *invariant for different friction mechanisms* and thus constitute the essence of friction as a physical phenomenon. The main characteristic of friction is that it is an interface phenomenon. The interactions across the interface are usually weaker than those in the bulk of the material. As a result, there is always a small parameter present in systems with friction. When the friction force is presented as a series by the small parameter, the dominating term is linear. This is why various mechanisms of friction lead to the same linear dependency of the friction force on the normal load. We will discuss also how linear phenomenological coefficients of non-equilibrium thermodynamics are related to viscous friction and how three-dimensional (3D) viscous friction can be reduced to dry Coulomb friction as a limiting case of the asymptotic transition to the two-dimensional (2D) interface. All of these considerations are intended to answer the question of why friction is so universal and what the general features of friction are.

4.1 Empirical Laws of Friction

Empirical observations of both dry and lubricated frictional sliding of solid surfaces have resulted in the formulation of the phenomenological or empirical laws of friction. Although these laws are sometimes partially attributed to Leonardo da Vinci, they became well known in the scientific community due to the work of Guillaume Amontons. Some scholars prefer to call them the "rules" rather than the "laws" of friction, stressing their approximate and empirical character, as opposed to the fundamental laws of nature. The three Amontons (or Amontons–Coulomb) laws are the following:

1. The friction force, F, is directly proportional to the applied normal load, W:

$$F = \mu W \tag{4.1}$$

   where μ is a constant coefficient referred to as the coefficient of friction. In an alternative formulation, the coefficient of friction is independent of the normal load, W. This law effectively states that the dependency of the friction force on the normal load is almost linear. The higher degree terms can usually be ignored.

2. The friction force, F, is independent of the nominal (or apparent) area of contact, or μ is independent of A_a. As will be discussed later, the real (or true) area of contact can be only a small fraction of the nominal area of contact because the surfaces are rough with asperities and the contact occurs only on the tips of the asperities.

3. The friction force, F, is independent of the sliding velocity V, or μ is independent of V.

The three laws state that the coefficient of friction is independent of W, A_a, and V. Sometimes the first two laws are attributed to Amontons, whereas the third one is attributed to Coulomb. In the original formulation, the authors of the laws admitted that the friction force can slightly depend on the sliding velocity; in particular, the force tends to increase for small but increasing velocities, remain constant for higher velocities, and decrease for even higher velocities. Summarizing the three laws, the coefficient of friction is independent of the normal load, the nominal area of contact, and the sliding velocity.

The three Amontons–Coulomb laws of friction should not be viewed as logically independent "axioms." On the contrary, the first and the second laws can be treated as a logical consequence of each other. If the friction force F is independent of the nominal area of contact A_a, dividing the contact region into two equal parts (each having the area of $A_a/2$) will result in the normal load $W/2$ supported by each half. Since the total force is equal to the sum of forces acting at each part of the contact region, one concludes that the friction force $F/2$ acts at each half and thus the friction force is proportional to the normal load. Certainly the same simple scaling argument can be used if the area is changed by an arbitrary factor β as $A_a \rightarrow \beta A_a$.

Speaking more formally, suppose that the friction force depends on the normal load and the nominal area of contact as $F = f(W, A_a)$, so that changing the scale of W and A_a by the factors of α and β results in the change of friction with the power exponents of n and m, respectively:

$$f(\alpha W, \beta A_a) = \alpha^n \beta^m f(W, A_a) \tag{4.2}$$

Since force is an additive function, we have also

$$f(\alpha W, \beta A_a) = \alpha f(W, A_a) \tag{4.3}$$

Setting $\alpha = \beta$ and combining Equations (4.2) and (4.3) yields $n + m = 1$. Therefore, the validity of the first law ($n = 1$) implies also the validity of the second law ($m = 0$) and vice versa (Nosonovsky 2007b).

The third empirical law of friction, which states that the friction force does not depend on the sliding velocity, is logically independent of the first two laws. Despite that, the velocity dependence of dry friction is also related to the dependence on the size of contact. Consider the contact of a deformable rough surface with a rigid flat surface. An important characteristic of such contact is the average size of the individual asperity contacts, a, which depends upon the separation (or penetration) distance between the two bodies, z. The separation distance is the extent of the contact and this parameter controls friction. The simplified models of contact predict that a is independent of the normal load and the size of the contact, and therefore, the average area of the individual asperity contact, τa^2, is also independent of W and z.

The supporting argument is that with an increasing load, the separation between the contacting surfaces grows as does the size of the individual asperity contacts; however, more new small contacts are created, so the average contact size does not change. However, more accurate analysis shows that the average size of the contact depends slightly on the separation between two bodies: For a larger separation, the contact size is smaller.

The separation, z (defined in such a way that $z = 0$ for $W = 0$), can change due to the change of the load, the nominal area of contact, and the sliding velocity (longer existing contacts tend to provide lower separation due to the creep and visco-plastic deformation), which makes the load, contact size, and velocity dependence of friction (Figure 4.1). The following scaling relationships are valid:

$$z(\alpha W, \beta A_a, \gamma V) = \alpha^p \beta^q \gamma^s z(W, A_a, V) \tag{4.4}$$

By setting $\alpha = \beta$ and noting that a simultaneous increase of the apparent area of contact and the normal load should not affect z (since the pressure does not change), we conclude that $p + q = 1$. Furthermore, $p \neq 0$ because the separation distance decreases with the normal load, and $s \neq 0$ because the separation increases with the sliding velocity. By taking $\gamma = \alpha^{-p/s}$, we conclude that the effect of W on the separation distance can be compensated for by the effect of V. Because the separation distance is the parameter controlling friction, the effect of W on the friction force can be compensated for by the effect of V. We conclude that if the coefficient of friction depends on the normal load, it also depends on the sliding velocity. Thus, the third law cannot be valid separately from both the first and the second laws.

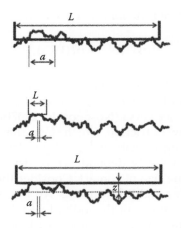

FIGURE 4.1
The average size of asperity contact, a, is dependent on the separation (penetration) z, as well as on the nominal (apparent) size of contact, L. For larger sliding velocity, V, separation increases. Therefore, the effect of the sliding velocity on friction is equivalent to the effect of the nominal area of contact. (Nosonovsky 2007a)

Tribology textbooks usually emphasize that the three Amontons–Coulomb empirical laws of friction are only approximations, and there are many situations when these laws are not valid. For this reason, some tribologists prefer to speak about the "rules of friction" rather than laws, reserving the word *law* for fundamental laws of nature, such as Newton's laws of mechanics. On the other hand, the Amontons–Coulomb laws are valid for an amazingly broad range of material combinations, friction mechanisms, and loads ranging from nanonewtons to thousands of tons.

The preceding applies to the dry sliding friction. For lubricated friction, the friction laws tend to be the same for a thin layer of lubricant. For a thick lubricating film, the regime of hydrodynamic lubrication can apply and the coefficient of friction can deviate from Equation (4.1). The laws of static friction are similar, with the obvious exception of the velocity law, so that Equation (4.1) is normally valid for static friction as well as for kinetic friction. The value of the coefficient of static friction is usually greater than that of kinetic friction. Another type of friction is rolling friction, which essentially results in the same frictional law.

4.2 Mechanisms of Friction

Despite the simplicity of the empirical laws, friction is a very complex phenomenon that involves various mechanisms of a different physical nature, apparently unrelated. These mechanisms act independently and often simultaneously. The question can be asked why these diverse mechanisms lead to the same simple linear Amontons–Coulomb law (Equation 4.1). Are there any common features in all these mechanisms that constitute the essence of friction? Why do they all lead to the simple friction laws? The central common feature of all diverse mechanisms of friction is that they involve interactions through the interface, which are much weaker than the bulk interaction forces. Consequently, a "small parameter"—the ratio of the interface-to-bulk forces—is present in the systems of all mechanisms for friction.

4.2.1 Adhesive Friction and Surface Roughness

4.2.1.1 Adhesive Contact

When two surfaces are brought into contact, adhesion or bonding across the interface can occur, and a normal force, called the adhesion force, is required to pull apart these two solids (Bowden and Tabor 1986). The adhesion force is a general term for an attractive force of any physical nature. The four main types of adhesion forces are (1) chemical covalent forces, (2) electrical van der Waals interactions, (3) electrostatic forces, and (4) capillary forces. Since the typical range of the adhesion force is in nanometers (with the exception of the capillary force having a larger range), the role of adhesion is particularly important at the nano-scale.

Every nominally flat surface in reality is rough. The real area of contact is only a small fraction of the nominal area of contact because the contact takes place only at the summits of the asperities. Bowden and Tabor (1986) suggest that the friction force F is directly proportional to the real area of contact A_r and shear strength due to adhesion at the interface τ_f:

$$F = \tau_f A_r \tag{4.5}$$

Various statistical models of contact of rough surfaces show that, for small and moderate loads, A_r is almost directly proportional to the applied normal load W for elastic and plastic surfaces. This explains the empirically observed linear proportionality of F and W given by Equation (4.1), assuming constant τ_f.

4.2.1.2 Greenwood–Williamson Approach to Surface Roughness

Greenwood and Williamson (1966) modeled a random rough surface as a flat surface with asperities of a certain statistical distribution of heights—for example, an exponential distribution:

$$p(h) \sim e^{-h/h_0} \tag{4.6}$$

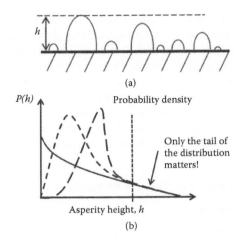

(a)

(b)

FIGURE 4.2
The Greenwood–Williamson model. Asperities with a random height h over the surface plane have certain probability distribution of heights, $p(h)$. Since only the highest asperities participate in contact (at low and moderate loads), only the tail of the distribution matters.

For a given penetration depth d of two surfaces in contact, only asperities with the height $h > d$ participate in contact (Figure 4.2a). Furthermore, for an asperity of height h, the penetration is $z = h - d$, the contact radius is proportional to $(h - d)^{1/2}$, or just $z^{1/2}$, while the contact area is proportional to z. The elastic force, P, is proportional to the power 3/2 of the contact area; therefore, $P \sim z^{3/2}$. The total contact area and the total normal force are found by integration of all asperities participating in contact:

$$A_r \sim \int_0^\infty p(z - d)z\, dz = \int_0^\infty ze^{-\frac{z-d}{h_0}}\, dz = e^{\frac{d}{h_0}} \int_0^\infty ze^{-\frac{z}{h_0}}\, dz \qquad (4.7)$$

$$W \sim \int_0^\infty p(z - d)z^{3/2}\, dz = \int_0^\infty z^{3/2}e^{-\frac{z-d}{h_0}}\, dz = e^{\frac{d}{h_0}} \int_0^\infty z^{3/2}e^{-\frac{z}{h_0}}\, dz \qquad (4.8)$$

It is clear from Equations (4.7) and (4.8) that the integrals do not depend on the value of d. Furthermore, the ratio A_r/W does not depend on d. In other words, the ratio A_r/W is constant for any value of load/penetration. Therefore, for a given surface with an exponential distribution of asperity heights, the real area of contact is linearly proportional to the normal load. This accounts for Equation (4.5), which explains the linear proportionality of the friction force with respect to the normal load.

In their original paper, Greenwood and Williamson (1966) showed that if a Gaussian statistical distribution is considered instead of the exponential distribution (Equation 4.6), the dependency of the friction force on the normal load remains almost linear. This is understandably so, since for small or moderate loads only the highest asperities participate in the contact, so only the "tail" of the statistical distribution, which is similar for the exponential and Gaussian distributions, contributes into W and F (Figure 4.2b). A stronger statement can be made: For any statistical distribution, if the extent of contact is moderate, the ratio of the real area of contact to the normal load is constant:

$$\lim_{A_r/A_n} \to 0 \quad W/A_r = const \tag{4.9}$$

This is because if d is large enough, the probability density in Equations (4.7) and (4.8) can always be presented as $p(z - d) = p(z)p(d)$, since the probability of rare events depends exponentially on the magnitude of these events (frequency and height of asperities in our case).

Note that the linearity of the Amontons–Coulomb law (Equation 4.1) turned out to be the consequence of a small extent of contact (small A_r/A_n).

4.2.1.3 Statistical Models of Contact of Rough Surfaces

A more general approach treats a random surface as a random signal, rather than as a flat with asperities. There are several statistical parameters that can characterize surface roughness. For a 2D roughness profile $z(x)$, the centerline average is defined as the arithmetic mean of the absolute value of the vertical deviation from the mean line of the profile:

$$R_a = \frac{1}{L} \int_0^L |z - m| \, dx \tag{4.10}$$

where L is the length of the interval (the sampling length) and

$$m = \frac{1}{L} \int_0^L z \, dx \tag{4.11}$$

is the mean value. An alternative parameter is the root mean square RMS given by

$$\sigma^2 = \frac{1}{L} \int_0^L (z - m)^2 \, dx \tag{4.12}$$

In many practical cases, the random data tend to have the so-called Gaussian or normal distribution with the probability distribution function given by

$$p(z^*) = \frac{1}{\sqrt{2\pi}} \exp\left(-\frac{z^{*2}}{2}\right) \tag{4.13}$$

where the normalized variable $z^* = (z - m)/\sigma$, m is the mean, and σ is the standard deviation.

The Gaussian distribution is found in nature and technical applications when the random quantity is a sum of many random factors acting independently of each other. When an engineering surface is formed, many random factors simultaneously contribute to the roughness, and thus, in many cases, the roughness height is governed by the Gaussian distribution. Such surfaces are called Gaussian surfaces.

Note that the standard deviation σ characterizes the distribution of roughness in the normal (vertical) direction or, in other words, the height of the asperities. It does not say anything about the typical width of the asperities. In order to represent spatial distribution of random roughness, the autocorrelation function is

$$C(\tau) = \lim_{L\to\infty} \frac{1}{\sigma^2 L} \int_0^L [z(x) - m][z(x+\tau) - m]\,dx \tag{4.14}$$

The autocorrelation function characterizes the correlation between two measurements taken at the distance τ apart, $z(x)$ and $z(x + \tau)$, and it is obtained by comparing the function $z(x)$ with a replica of itself shifted for the distance τ. The function $C(\tau)$ approaches zero if there is no statistical correlation between values of z separated by the distance τ; in the opposite case, $C(\tau)$ is different from zero. Many engineering surfaces are found to have an exponential autocorrelation function

$$C(\tau) = \exp(-\tau/\beta) \tag{4.15}$$

where β is the parameter called the correlation length or the length over which the autocorrelation function drops to a small fraction of its original value. At the distance β, the autocorrelation function falls to the $1/e$. In many cases the value $\beta^* = 2.3\beta$ is used for the correlation length, at which the function falls to 10% of its original value (Bhushan 2002).

For a Gaussian surface with the exponential autocorrelation function, σ and β^* are two parameters of the length dimension that conveniently characterize the roughness. While σ is the height parameter, which characterizes the height of a typical roughness detail (asperity), β^* is the length parameter, which characterizes the length of the detail. The average absolute value of the slope is proportional to the ratio σ/β^*, whereas the average curvature is proportional to β^*/σ^2. For a Gaussian surface, σ is related to the RMS as $\sigma = (\sqrt{\pi}/2)R_a$ (Nosonovsky and Bhushan 2008). These two parameters, σ and β^*, are convenient for the characterization of many random surfaces. Note that a Gaussian surface has only one inherent length scale parameter, β^*, and one vertical length scale parameter, σ, and thus it cannot describe the multiscale roughness.

When two rough surfaces come into a mechanical contact, the real area of contact is small in comparison to the nominal area of contact, because the contact takes place only at the tops of the asperities. For two rough surfaces in contact, an equivalent rough surface can be defined of which the values of the local heights, slopes, and local curvature are added to each other. The composite standard deviation of the profile heights is related to those of the two rough surfaces, σ_1 and σ^2 as

$$\sigma^2 = \sigma_1^2 + \sigma_2^2 \qquad (4.16)$$

The composite correlation length is related to those of the two rough surfaces, β^*_1 and β^*_2 as

$$1/\beta^* = 1/\beta^*_1 + 1/\beta^*_2 \qquad (4.17)$$

Using the composite rough parameters allows effectively reducing the contact problem of two rough surfaces to the contact of a composite rough surface with a flat surface (Bhushan 2002; Nosonovsky and Bhushan 2008).

Two parameters of interest during the elastic and plastic contact of two rough surfaces are the real area of contact, A_r, and the total number of contact spots, N. In most cases, only the highest asperities participate in the contact. This situation allows linearizing the dependence of A_r and N upon the roughness parameters during the elastic contact as

$$A_r \propto \frac{W\beta^*}{\sigma E} \qquad (4.18)$$

$$N \propto \frac{1}{\sigma\beta^*} \qquad (4.19)$$

where W is the normal load force and E is the composite elastic modulus. Qualitatively, the higher the asperities are, the larger is σ and the smaller is A_r; the wider the asperities are, the larger is β^* and smaller is A_r. The larger and wider the asperities are, the smaller is A_r (Bhushan and Nosonovsky 2004b).

For plastic contact, N, which depends upon the contact topography and thus is independent on whether the contact is elastic or plastic, is still given by Equation (4.19) for a given separation between the surfaces (Bhushan and

Nosonovsky 2004a), whereas the real contact area is found by dividing the load by the hardness.

$$A_r \propto W / H \tag{4.20}$$

Again, the linearity of the Amontons–Coulomb law (Equation 4.1) turned out to be the consequence of a small extent of contact (small A_r/A_n).

4.2.1.4 Fractal and Hierarchical Roughness

Models of a fractal rough surface provide an alternative description of roughness. Long before the discovery of fractals by mathematicians, Archard (1957) studied multi-scale roughness with small asperities on top of larger asperities, with even smaller asperities on top of those, and so on (Figure 4.3). According to the Hertzian model, for the contact of an elastic sphere of radius R with an elastic flat with the contact radius a, and the contact area $A = \pi a^2$ are related the normal load as $A_r \sim W^{2/3}$, while the pressure distribution as the function of the distance from the center of the contact spot, r, is given by $p(r) \sim W^{1/3}\sqrt{1-(r/a)^2}$.

For a big spherical asperity covered uniformly by many asperities with a much smaller radius forming individual contacts, the dependency of the total contact area upon W is then given by integration of the individual contact areas by r as

$$A_r \sim \int_0^a \left(W^{1/3}\sqrt{1-(r/a)^2}\right)^{2/3} 2\pi r\, dr \sim W^{2/9}a^2 \sim W^{2/9}W^{2/3} \sim W^{8/9} \tag{4.21}$$

If the small asperities are covered by the "third-order" asperities of an even smaller radius, the total area of contact can be calculated in a similar way as

$$A_r \sim \int_0^a \left(W^{1/3}\sqrt{1-(r/a)^2}\right)^{8/3} 2\pi r\, dr \sim W^{8/27}a^2W^{8/27}W^{2/3} \sim W^{26/27} \tag{4.22}$$

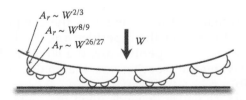

FIGURE 4.3
The Archard (1957) model of a hierarchy of smaller asperities on tops of larger asperities.

while further increasing the number of asperities, one can show that $A_r \sim W$. Later, more sophisticated fractal surface models were introduced, which led to similar results (Nosonovsky and Bhushan 2008).

Both statistical and fractal models of rough surfaces, for elastic and plastic contact, combined with the adhesive friction law (Equation 4.5) result in an almost linear dependency of the friction force upon the normal load in the case of a small extent of contact (Figure 4.4a).

4.2.1.5 Adhesion Hysteresis and Dissipation

The van der Waals adhesion force is conservative and, therefore, by itself cannot provide a mechanism needed for energy dissipation. It was suggested recently (Maeda et al. 2002; Szoszkiewicz et al. 2005; Zeng, Tirrell, and Israelachvili 2006; Ruths, Berman, and Israelachvili 2003; Nosonovsky 2007a) that nano-friction is not related to the adhesion per se, but to the adhesion *hysteresis*. The energy needed to separate two surfaces is always greater than the energy gained by bringing them together. As a result, there is energy loss during the loading–unloading cycle. The adhesion hysteresis or surface energy hysteresis can arise even between perfectly smooth and chemically homogeneous surfaces supported by perfectly elastic materials. The adhesion hysteresis exists due to the surface roughness and the chemical heterogeneity (Maeda et al. 2002). Both sliding and rolling friction involve the creation and consequent fracture of the solid–solid interface. During such a loading–unloading cycle energy is dissipated (Nosonovsky and Bhushan 2008).

Summarizing, the adhesive friction provides the mechanism of energy dissipation due to breaking the strong adhesive bonds between the contacting surfaces and due to the adhesion hysteresis. The linearity of the Amontons–Coulomb law (Equation 4.1) turned out to be the consequence of a small extent of contact (small A_r/A_n).

4.2.2 Deformation of Asperities

Another important mechanism of friction is the deformation of interlocking asperities. As with adhesion, which may be reversible (weak) and irreversible (strong), deformation may be elastic (i.e., reversible) and plastic (irreversible plowing of asperities); see Figure 4.4(b). For elastic deformation, a certain amount of energy is dissipated during the loading–unloading cycle due to the radiation of elastic waves and viscoelasticity, so an elastic deformation hysteresis exists similar to adhesion hysteresis. The value of the deformational friction force is usually higher than that of adhesive friction and depends on the yield strength and hardness, which trigger a transition to plastic deformation and plowing. The transition from an adhesive to a deformational fiction mechanism depends on the load and the yield strength of materials and usually results in a significant increase of the friction force (Nosonovsky and Bhushan 2008).

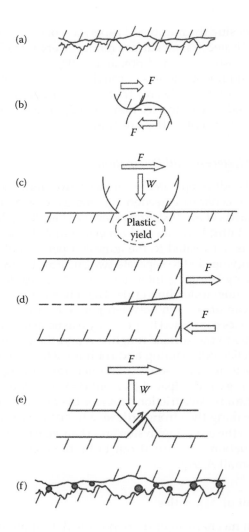

FIGURE 4.4
Schematics of various friction mechanisms: (a) adhesion at asperity contacts; (b) deformation (plowing) of asperities; (c) plastic yield; (d) mode II fracture; (e) ratcheting (climbing of asperities); (f) third-body mechanism (trapped debris).

The contribution of adhesion hysteresis into sliding friction is equal to that of rolling friction, since creation and destruction of the same interface area are involved. However, it is well known from experiments that sliding friction is usually greater than rolling friction because plowing of asperities occurs during sliding. Even smooth surfaces have nano-asperities, and their interlocking can result in plowing and plastic deformation of the material. Usually, asperities of a softer material are deformed by asperities of a harder material. The shear strength during plowing is often assumed to be proportional to the average absolute value of the surface slope (Bhushan 2002). It is therefore assumed that, in addition to the adhesion hysteresis term, there is another component, H_p, which is responsible for friction due to surface roughness and plowing (Nosonovsky 2007a)

$$F = A_r(\Delta W/d + H_p) \tag{4.23}$$

The plowing term may be assumed to be proportional to the average absolute value of the surface slope. Note that the normal load is not included in Equation (4.23) directly; however, A_r depends upon the normal load. The right-hand side of Equation (4.23) involves two terms: a term that is proportional to adhesion hysteresis and a term that is proportional to roughness (Nosonovsky and Bhushan 2008).

Due to the surface roughness, deformation occurs only at small parts of the nominal contact area, and the friction force is proportional to the real area of contact involving plowing. Due to the small size of the real area of contact compared with the nominal area of contact, the plastically deformed regions constitute only a small part of the bulk volume of the contacting bodies.

4.2.3 Plastic Yield

Chang, Etsion, and Bogy (1987) proposed a model of friction based upon plastic yield, which was later modified by Kogut and Etsion (2004). They considered a single-asperity contact of a rigid asperity with an elastic–plastic material. With an increasing normal load, the maximum shear strength grows and the onset of yielding is possible. The maximum shear strength occurs at a certain depth in the bulk of the body (Figure 4.4c). When the load is further increased and a tangential load is applied, the plastic zone grows and reaches the interface. This corresponds to the onset of sliding. Kogut and Etsion (2004) calculated the tangential load at the onset of sliding as a function of the normal load using finite elements analysis and found a non-linear dependence between the shear and tangential forces. This mechanism involves plasticity and implies structural vulnerability of the interface compared with the bulk of the contacting bodies (Nosonovsky and Bhushan 2008).

4.2.4 Fracture

For a brittle material, asperities can break, forming wear debris. Therefore, fracture also can contribute to friction. There is also an analogy between mode II crack propagation and sliding of an asperity (Rice 1992; Gerde and Marder 2001; Kessler 2001; Figure 4.4d). When an asperity slides, the bonds are breaking at the rear, while new bonds are being created at the front end. Thus, the rear edge of asperity can be viewed as a tip of a propagating mode II crack, while the front edge can be viewed as a closing crack. Gliding dislocations, emitted from the crack tip, can also lead to a micro-slip or a local relative motion of the two bodies (Bhushan and Nosonovsky 2003; Nosonovsky and Bhushan 2005). Calculations have been conducted to relate the stress intensity factors to the friction parameters (Rice 1991; Gerde and Marder 2001; Kessler 2001). Crack and dislocation propagation along the interface implies that the interface is weak compared with the bulk of the body (Nosonovsky and Bhushan 2008).

4.2.5 Ratchet and Cobblestone Mechanisms

Interlocking of asperities may result in one asperity climbing upon the other, leading to the so-called ratchet mechanism (Bhushan 2002). In this case, in order to maintain sliding, a horizontal force should be applied that is proportional to the slope of the asperity (Figure 4.4e). At the atomic scale, a similar situation exists when an asperity slides upon a molecularly smooth surface and passes through the tops of molecules and the valleys between them. This sliding mechanism is called a "cobblestone mechanism" (Israelachvili 1992). This mechanism implies that the strong bonds are acting in the bulk of the body, whereas the interface bonds are weak (Nosonovsky and Bhushan 2008).

4.2.6 "Third Body" Mechanism

During the contact of two solid bodies, wear and contamination particles can be trapped at the interface between the bodies (Figure 4.4f). Along with liquid that condenses at the interface, they form the so-called "third body" that plays a significant role in friction. The trapped particles can significantly increase the coefficient of friction due to plowing. Some particles can also roll and thus serve as rolling bearings, leading to a reduced coefficient of friction. However, in most engineering situations, only 10% of the particles roll (Bhushan 2002) and thus the third body mechanism leads to an increase of the coefficient of friction. At the atomic scale, adsorbed mobile molecules can constitute the "third body" and lead to significant friction increase (He, Müser, and Robbins 1999). The third body has much weaker bonds to the surface than those in the bulk of the body (Nosonovsky and Bhushan 2008).

4.2.7 Note on the Linearity of the Amontons–Coulomb Law

As we have discussed, the linearity of the Amontons–Coulomb law (Equation 4.1) is the consequence of the linear (or almost linear) dependency of the real area of contact on the normal load. This linearity, in turn, is the result of small or moderate loads, for which only the tail of the statistical distribution of asperity heights affects the contact area, and, as a result, only the tail of the statistical distribution plays a role.

Another type of model of dry friction is based on the assumption that during sliding, asperities climb upon each other (the ratchet mechanism).

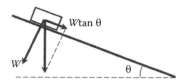

FIGURE 4.5
The slope of a tilted surface is equivalent to the coefficient of friction.

From the balance of forces, the horizontal force, which is required to initiate motion, is given by the normal load multiplied by the slope of the asperities:

$$F = W \tan \theta \qquad (4.24)$$

where θ is the slope angle of the asperities (Figure 4.5). Similar to the ratchet mechanism is the cobblestone mechanism, which is typical for atomic friction (Nosonovsky and Bhushan 2008).

Among other attempts to explain the linearity of the friction force with respect to the load, two modeling approaches are worth mentioning. Sokoloff (2006) suggested that the origin of the friction force is in the hard-core atomic repulsion. The vertical component of the vector of the repulsion force, which contributes to the normal load, is proportional to the horizontal component of the same vector, which contributes to friction, because the vector has a certain average orientation. In a sense, this is still the same slope-controlled mechanism, but considered at the atomic level.

Ying and Hsu (2006) suggest an interesting macro-scale approach. They noticed that for a spherical asperity of radius R, slightly indented into a substrate, the contact radius a is proportional to the second power of the penetration h (Figure 4.6):

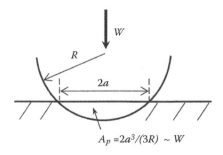

FIGURE 4.6
"Elastic plowing" model by Ying and Hsu (2006). Both the plowed cross-sectional area and the normal load W are proportional to third power of the contact radius a, and thus the load is proportional to the shear force. (Ying, Z. C., and Hsu, S. M. 2006. *First observation of elastic plowing in nanofriction*. ASME.)

$$a \propto W^{1/3} \qquad (4.25)$$

When such an asperity plows the substrate, the cross-sectional plowing area (or projection of the indented part of the sphere upon a vertical plane) A_p is given by a cubic function of a and thus is proportional to the normal load:

$$A_p = \frac{2a^3}{3R} \propto W \qquad (4.26)$$

This is the case of "elastic plowing" the plowing force, which is proportional to A_p, and is linearly proportional to the normal load (Nosonovsky and Bhushan 2008).

4.2.8 Linearity and the "Small Parameter"

We have found that several physical mechanisms result in a linear dependence of the friction force upon the normal load. Mathematically, a linear dependence between two parameters usually exists when the domain of a changing parameter is small, and thus a more complicated dependency can be approximated within this domain as a linear function. For example, if the dependency of the friction force upon the normal load is given by

$$F = f(W) \approx f(0) + f'(0)W + \frac{f''(0)}{2}W^2$$
$$= \mu W + \frac{f''(0)}{2}W^2 \qquad (4.27)$$

the dependency can be linearized as $F = \mu W$ if

$$W \ll \frac{2\mu}{f''(0)} \qquad (4.28)$$

In other words, the ratio of the load W to a corresponding parameter of the system, given by Equation (4.22) (with the dimension of force), is small. That parameter may correspond to the bulk strength of the body.

In summary, all mechanisms of dry friction are associated with a certain type of heterogeneity or non-ideality, including surface roughness, chemical heterogeneity, contamination, and irreversible forces (Table 4.1). All of these mechanisms are also characterized by the interface forces being small compared with the bulk forces.

TABLE 4.1

Entropic Description of Surface Roughness, Friction, and Wear

	Information	Energy	Mass
	Surface roughness	*Friction (dissipation)*	*Wear (mass flow)*
Entropic description	Shannon entropy and entropy rate for a stochastic process $$S = -\sum_{j=1}^{B} p_j \ell n[p_j]$$	Thermodynamic entropy $$dS = \frac{dQ}{T}$$	Entropy of mixing (configurational) $$\Delta S = -R\sum_{i}^{n} \frac{N_i}{N} \ln \frac{N_i}{N}$$

4.3 Non-Linear Character of Friction

In the preceding section we discussed the linearity of friction as a result of the small extent of contact interactions at the interface in comparison with the bulk interactions (and, in particular, the small ratio A_r/A_n). However, friction is non-linear in a number of senses. The Amontons–Coulomb friction law (Equation 4.1) is only apparently linear because the friction force is directed in the same direction as a sliding velocity. Therefore, a more accurate presentation of Equation (4.1) would be $F = \mu W \text{sign}(V)r$, in the vector form,

$$\vec{F} = \mu W \frac{\vec{v}}{|v|} \tag{4.29}$$

Obviously the function $\text{sign}(V)$ is non-linear and it has discontinuity at $V = 0$, making the Amontons–Coulomb law non-linear.

However, friction also leads to a number of non-linear effects:

- Memory of the past states of the system, which is kept in the static "stick" states. The sliding velocity is zero in these states; however, the friction force can have any value in the range $F < \mu |W|$ depending on the past history of the system.
- Hysteresis or dependence of a system with friction not only on its instant state, but also on its past history
- Phase transitions between the stick and slip states as well as the instabilities associated with these states
- Paradoxes leading to non-unique and multiple solutions of problems with friction
- Dynamic instabilities

4.3.1 Friction and Reduction of Degrees of Freedom

As we discussed in the preceding chapters, friction is a multi-scale phenomenon and a result of neglecting many unessential degrees of freedom in the system. Thus, at the molecular scale, atomic and molecular motion is dominant as well as inter-molecular and inter-atomic bonds. However, the effect of the molecular motion is usually observed in the form of temperature and other averaged thermodynamic parameters. At the meso-scale, non-idealities of the material, defects and roughness asperities are observed. Their motion contributes to the macro-scale dissipation of energy. In fact, meso-scale effects are the most important. It is usually impossible to deduce or derive macro-scale properties of a material or a system, such as the yield strength or the coefficient of friction, from the atomic scale properties, because too many interactions occur at the meso-scale.

Different methods of computational modeling are used at different scale lengths. The molecular dynamics (MD) simulation is used at the atomic scale. However, MD can capture only very short time scales (nanoseconds) that may be insufficient to observe processes essential for friction. The meso-scale computational analysis includes Monte Carlo, cellular automata, and other similar methods, whereas standard macro-scale methods involve finite element analysis (FEA) and computational fluid dynamics.

Neglecting of numerous "non-essential" degrees of freedom is an approximation. As with any approximation, it has a price and sometimes can lead to inconsistencies, referred to as frictional paradoxes. More complicated models of dynamic friction should include the internal degrees of freedom or state variables.

4.3.2 Stick–Slip Friction Phases

This stick–slip phenomenon can occur if the coefficient of static friction is greater than the coefficient of kinetic friction. When the stick–slip motion occurs, the frictional force does not remain constant, but rather oscillates significantly as a function of sliding distance or time. During the stick phase, the friction force builds to a critical value. Once the critical force to overcome the static friction has been attained, slip occurs at the interface, and energy is released so that the frictional force decreases (Figure 4.7).

The stick–slip phenomenon is found in many situations, such as car brake vibration and squeal. However, it is particularly common at the atomic scale; for example, during the contact of an atomic force microscope tip with an atomically smooth surface, the energy dissipation occurs through a stick–slip movement of individual atoms at the contact interface (Nosonovsky and Bhushan 2008).

Stick and slip can be treated as two phases at the surface. The definition of a phase in physical chemistry is that properties of a medium do not change within the same phase. The effects related to phase transitions can be studied

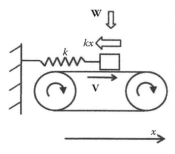

FIGURE 4.7
A typical setup for the study of stick–slip. The conveyer rotates with the velocity V and the mass remains stuck to the conveyer until the elastic force kx exceeds the static friction. (Based on Nosonovsky, M., and Bhushan, B. 2008. *Multiscale Dissipative Mechanisms and Hierarchical Surfaces: Friction, Superhydrophobicity, and Biomimetics.* Berlin: Springer.)

then. Einax, Schulz, and Trimper (2004) used this approach. Their numerical analysis demonstrated that, for a small inter-atomic interaction strength, the system is in a frictionless state, but when the strength is above the critical value, there is static friction. They analyzed scaling behavior of the order parameter near the critical value and calculated corresponding critical exponents. The order parameter and the energy functional can be powerful mathematical tools to study scaling laws of friction and near-critical behavior.

Rubinstein, Cohen, and Fineberg (2004) experimentally studied the onset of frictional slip with a high-speed camera. They observed three fronts propagating with different velocities: sub-sonic, inter-sonic, and an order of magnitude slower. No overall motion (sliding) of the blocks occurs until either of the slower two fronts traverses the entire interface. Their overall conclusion is that the static coefficient of friction is a "myth," while kinetic friction is a result of stick–slip.

4.3.3 Running-in and "Shakedown"

Running-in is the transient process from the moment when sliding starts until the friction force reaches the stationary value. Usually, friction tends to decrease during this transient process (although some opposite examples are known). The exact reason for the running-in remains unclear, and most likely the phenomenon involves several reasons and mechanisms. Running-in can be treated as an adjustment of surfaces and their roughness to each other (Blau 1981).

Another phenomenon related (but not identical) to the running-in transient process is the frictional "shakedown." In a discrete elastic system—a system with a finite number of degrees of freedom with elastic and frictional links between the elements—the steady state (and in particular the steady-state energy dissipation) can depend on the initial condition when there is elastic coupling between the normal and tangential contact problems

(Klarbring, Ciavarella, and Barber 2007; Ahn, Bertocchi, and Barber 2008). In other words, the state of the system depends on its history and on the initial conditions. Because of the history dependence, the steady state of a periodically loaded system may depend on the initial condition or on an initial loading path leading to the periodic state. Shakedown means that in the steady state there is no frictional slip. In other words, slip that has occurred early in the loading process is sufficient to prevent further slip in the steady state.

Jang and Barber (2011) found that this history dependence occurred only for relatively moderate values of the periodic load amplitude. For larger amplitudes, the system converged on a unique steady state. A possible explanation for this behavior was suggested by Barber (2011), who conjectured that the steady state would be unique if the amplitude of the periodic load was sufficient to ensure that all nodes in the discrete system slip at least once during each period. Barber's phenomenological argument was that the history memory of such a system should be at any given time in the tangential displacements of the nodes that are not slipping. However, if all nodes slip at least once during each cycle, the memory of the initial conditions must vanish over time.

Barber compared the frictional system with an elastic–plastic system, for which Melan's theorem states that the system will tend to shake down. According to Barber (2011), the frictional Melan's theorem applies if and only if there is no coupling between normal and tangential loading. According to Barber, there is anecdotal evidence that fretting fatigue tests are very consistent for smooth "Hertzian" contact geometries with no coupling, but indentation tests (where there is significant coupling) give very erratic and irreproducible results. Similarly, tests on the effective damping in bolted joints show that the results are very erratic. Apparently identical systems give different results, and even the same system, if disassembled and then reassembled, can give very different results. Barber also advanced the conjecture that residual stresses would tend to reduce dissipation (with shakedown as a special case) and, hence, that dissipation per cycle might decrease monotonically with successive cycles, although E. Bertocchi demonstrated a counter example where the evolution of dissipation is not monotonic.

4.3.4 Dynamic Friction: State-and-Rate Models

It is known from experiments that the absolute value of the friction force is not completely independent of the sliding velocity. In fact, it was known already to Coulomb, who claimed that for very small velocities, friction force grows with increasing velocity; for moderate velocities, friction force remains constant; and for high velocities, it decreases. Tolstoi (1967) was the first who paid attention to the importance of the normal coordinate during dry sliding, showing that separation distance between the sliding bodies grows with increasing velocity. Since there is less time for the asperity contact and, therefore, less time for asperities to deform, the body elevates. This usually results in a decrease of the real area of contact and a decrease of the friction force.

Various dynamic models have been suggested, based on various physical effects such as, for instance, time-dependent creep-like relaxation and viscosity (Persson 2000). Increasing velocity typically results in a decrease of friction (the so-called "negative viscosity," sometimes also referred to as the Stribeck effect), although for some material combinations and friction regimes the opposite trend is observed. The decrease of friction with increasing velocity may lead to a dynamic instability, since decreased frictional resistance will lead to acceleration, a further increase of velocity, and a decrease in friction.

The state-and-rate models of friction, which introduce internal degrees of freedom, have been successful in modeling this effect (Dieterich 1979; Dieterich and Kilgore 1994; Persson 2000). These models, used at first to study sliding friction for seismic and geophysical applications, showed reasonable agreement with the experimental data. Based on the state-and-rate models, when sliding velocity changes, the friction force always increases at first. After some time, friction decreases due to creep relaxation to a velocity-dependent steady-state value. According to the Dieterich (1979) state-and-rate model, the area of contact is given by

$$A_r = A_{r0}[1 + B\ln(\theta / \theta_0)] \tag{4.30}$$

where θ is a "state" parameter physically equal to the age of contact, B is a constant, and θ_0 is a normalization parameter, such that $\theta = \theta_0$ for $A_r = A_{r0}$. The contact area increases with the age of contact due to the creep. For steady-state sliding with a constant velocity V, θ is inverse proportional to the sliding velocity (Dieterich and Kilgore 1994):

$$A_r = A_{r0}[1 - C\ln(V / V_0)] \tag{4.31}$$

where V_0 is a normalization parameter, such that $V = V_0$ for $A_r = A_{r0}$, and C is a constant. The velocity-dependent friction law has been used to resolve some paradoxes associated with dynamic Coulombian friction, such as the "ill-posedness" (a dynamic instability with an unlimited rate of amplitude increase for small wavelengths) of steady frictional sliding of two smooth elastic half-spaces, or dynamic instabilities studied in the preceding section (Ranjith and Rice 2001).

4.3.5 Modeling Dynamic Friction for Control

In addition to the state-and-rate models of friction, several other models have gained popularity, especially in the context of control engineering (Armstrong). Among them is the Lund–Grenoble (LuGre) model, which is particularly popular in the control engineering community, as it is capable of predicting stick–slip effects (Olsson et al. 1998). The LuGre model's equations are

$$F = \sigma_0 z + \sigma_1 \dot{z} + f(V) \tag{4.32}$$

$$\dot{z} = V - \sigma_0 \frac{|v|}{g(V)} z \tag{4.33}$$

where
   V is the sliding velocity
   z is the internal friction state
   $f(V)$ is the memoryless velocity-dependent term while $g(V)$ is a velocity-dependent function
   σ_0 is the stiffness parameter and σ_1 is micro-damping

The effect of the term $\sigma_0 z + \sigma_1 \dot{z}$ is to incorporate spring-like behavior for small displacements (Åström and Canudas-de-Wit 2008). For constant velocity, the steady-state friction force is given by $F = g(V)\sin(V) + f(V)$. The functions $f(V)$ and $g(V)$ can be determined experimentally by measuring the steady-state friction force for various constant velocities, although a typical choice is

$$g(V) = F_{kinetic} + (F_{static} - F_{kinetic})e^{-[V/V_s]^\alpha} \tag{4.34}$$

$$f(V) = \sigma_2 V \tag{4.35}$$

where σ_2, V_s, and a are parameters of the model.

Various control engineering methods to study frictional systems have been suggested as well (Figure 4.8). One of them is the theory of complex networks. A complex network is a graph with "non-trivial" topological features that do not occur in simple networks such as lattices or random graphs, with patterns of connection between their elements that are neither purely regular nor purely random. A standard example of complex networks is scale-free networks characterized by specific structural features, including power-law degree distributions.

In general, critical phenomena similar to those of the cellular automata (Potts and Ising models) are found in complex networks (Dorogovtsev, Goltsev, and Mendes 2008). A complex network approach was used to study the correlation patterns of void spaces during rough fracture. Using the well-known analogy between sliding and mode II fracture, Ghaffari and Young (2012) developed a complex network model of friction and found interesting

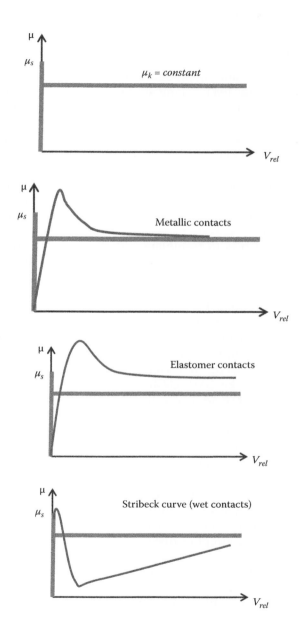

FIGURE 4.8
Typical dependencies of the friction force on the sliding velocities.

universal power law behavior relating the rate of stored energy in asperities and graph node distribution.

4.3.6 Atomic Friction and Kinetics

After the invention of scanning probe microscopy and, in particular, atomic force microscopy in the 1980s, many studies of friction at the atomic and molecular scale (i.e., when the size of a contacting asperity is comparable with the molecular size) have been conducted and much empirical information has been accumulated. Numerous attempts have been made to formulate empirical friction laws at the nano-scale (Wenning and Muser 2001; Stark, Schitter, and Stemmer 2004; Carpinteri and Paggi 2005). Such friction laws are intended to substitute for the classical Amontons–Coulomb empirical laws of friction. This approach, however, does not deal with friction as a universal phenomenon and virtually considers nano-scale and macro-scale friction as unrelated.

A different approach is to formulate scaling laws of friction. It is well known from experiments that the values of the coefficient of friction, when measured at the micro-/nano-scale, are different from those at the macro-scale, and therefore friction is scale dependent (Homola et al. 1990; Bhushan, Israelachvili, and Landman 1995; Schwarz et al. 1997). Various approaches have been proposed to study and explain the scale dependence of friction. Many scholars have considered the so-called *scale effect* on friction or *scaling laws* of friction (Blau 1991; Hurtado and Kim 1999; Niederberger et al. 2000; Zhang, Johnson, and Cheong 2001; Wenning and Muser 2007; Czichos 2001; Bhushan and Nosonovsky 2003, 2004a, 2004b; Nosonovsky and Bhushan 2007; He and Robbins 2003; Adams, Muftu, and Mohd Azhar 2003; Urbakh et al. 2004). While the origin of scaling laws is in geometrical relations, such as surface-to-volume ratios (Carpinteri and Paggi 2005; Pugno 2007), the term "scale effect" implies more general laws dependent upon physical mechanisms rather than on pure geometrical relations.

Johnson (1997) noted the fact that frictional stress is strongly dependent upon the scale of contact and suggested that gliding dislocations at the surface contribute to frictional stress. Hurtado and Kim (HK; 1999) proposed a model of single-asperity contact with a scale-dependent shear stress. Their model is based on the concept of dislocation-assisted sliding with dislocation loop nucleation at the perimeter of a circular contact zone. The model, however, is limited to the case of a commensurate interface between the bodies, which therefore should have the same orientation and spacing of the crystal lattices. This is not a likely situation in most cases. Adams et al. (2003) applied the HK model for multiple-asperity elastic contact with a Gaussian statistical distribution of asperity heights and identified parameters responsible for the scale effect.

Bhushan and Nosonovsky (2003, 2004a, 2004b; Nosonovsky and Bhushan 2007) took a different approach and considered a scale-dependent distribution of surface heights combined with scale-dependent frictional stress due to dislocations' nucleation from Frank–Read sources (rather than at the perimeter of the contact zone), as well as the strain-gradient plasticity. They later included the effect of asperity and particle deformation into their model, with scale-dependent densities of trapped particles at the interface. Zhang et al. (2001) studied scale effects on friction using molecular dynamics simulation. He and Robbins (2003) used MD simulation to study the origin of scale dependence on friction. Deshpande, Needleman, and Van der Giessen (2004) conducted numerical simulation of dislocation motion during frictional plastic deformation and showed that dislocations' nucleation from the sources (rather than at the perimeter of the contact zone) results in scale-dependent frictional stress. Kogut and Etsion (2004) proposed a model of elastic–plastic frictional contact, with scale-independent plasticity, which resulted in the coefficient of friction being strongly dependent upon the apparent area of contact, A_a, and normal load. W. Nosonovsky (2007d) also studied size, load, and velocity dependence of friction in combination. All of these studies investigate some aspects of the scale effect on friction; however, they do not provide a general theory of scale dependence of friction (Nosonovsky and Bhushan 2008).

As far as the origin of atomic friction, several physical models exist for atomic-scale friction, including the Tomlinson (1929) model, the Frenkel and Kontorova (1939) model, and similar models (Muser, Urbakh, and Robbins 2003). These models assume that the substrate atoms form a periodic energy profile with energy barriers that should be overcome during sliding (Figure 4.9). However, for two atomically smooth non-commensurate bodies in contact, positions of the energy barriers and sinks would not coincide, so virtually no friction is expected. This effect is known as "super-lubricity," and it has been argued that super-lubricity is in fact observed for graphite (Dienwiebel et al. 2004).

He et al. (1999) suggested that chemical inhomogeneity, such as hydrocarbon molecules adsorbed at the surface, provides energy barriers that lead to friction. Sokoloff (2002) showed that for static friction to occur, the elastic energy should be higher than the interfacial potential energy barriers, so that the atoms can sink into their interfacial potential minima. Later, the model was extended for kinetic friction, and it was shown that even fluctuations in the concentration of atomic level defects do not account for static and kinetic friction, but that surface roughness (multiple asperity contact) at the

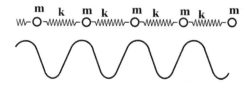

FIGURE 4.9
The idea of the Tomlinson and Frenkel–Kontorova models: the substrate is presented as a periodic energy profile created by atoms; the sliding body is modeled by a spring-mass system.

micron scale can indeed lead to friction (Daly, Zhang, and Sokoloff 2003). Thus, atomic-scale energy barriers combined with micron-scale roughness lead to frictional dissipation. A simple kinetic model may be developed for single asperity contact with a periodic substrate that represents the atomic crystal lattice with the period L and energy amplitude B using the Arrhenius method approach.

4.3.7 Paradoxes: Is Friction Compatible with Elasticity?

Despite the simplicity of the Amontons–Coulomb law, there are several difficulties in integrating friction with the rest of mechanics. Several ways to include dry friction into Lagrangian mechanics have been suggested, but these are quite artificial. These difficulties of integrating friction with the rest of mechanics lead to paradoxes. One such type of paradox is the Painleve paradox that shows that dynamical systems of rigid bodies with friction can have multiple solutions or no solution at all. This is because the process of solution implies that a certain direction of sliding velocities is assumed before the solution is performed and forces are obtained. If the friction force turns out to be perpendicular to the original sliding direction, a paradox occurs.

In continuum mechanics, stress is the measure of force exerted per unit area at a given point. The maximum shear stress τ_{xy} at the interface between two bodies during friction is proportional to the normal stress σ_{yy} at the same point:

$$\tau_{xy} = \mu\sigma_{yy} \tag{4.36}$$

However, the sign of the shear stress depends upon the sign of the local sliding velocity, which leads to friction paradoxes, when formal mathematical solution of the continuum mechanics problem with the boundary condition given by Equation (4.3) leads to a non-unique solution or to the sign of the shear stress not necessarily opposite to the sign of the local velocity of sliding. Adams et al. (2005) demonstrated that dynamic effects lead to new types of frictional paradoxes, in the sense that the assumed direction of sliding used for Coulomb friction is opposite that of the resulting slip velocity. In a strict mathematical sense, Coulomb friction is inconsistent not only with rigid body dynamics (the Painlevé paradoxes), but also with the dynamics of elastically deformable bodies. The paradoxes are related to the frictional dynamic instabilities.

4.3.8 Friction and Fracture: The Role of Heterogeneity

The similarity of frictional sliding with mode II fracture and crack propagation has already been discussed. Fracture involves various types of nonlinearity and singularities at the crack tip. Fracture can also lead to the formation of random and fractal surfaces.

To summarize, in this section we made several important observations about the nature of friction:

- Friction is a result of non-idealities of various kinds. In the world of ideal rigid non-deformable objects with conservative interactions between them, there would be no energy dissipation and no friction. However, surface roughness, heterogeneity, adhesion hysteresis and elastic hysteresis, irreversible bonds, and defects result in energy dissipation and friction (Table 4.2). To relate quantitatively, say, the degree of heterogeneity to the frictional dissipation would be a challenging task. We will attempt to apply non-equilibrium thermodynamics for this purpose.

TABLE 4.2

Dissipation and Friction Mechanisms Corresponding to Different Hierarchy Levels

Ideal Situation (Reversible Process)	Real Situation (Irreversible Process)	Mechanism of Dissipation Leading to Friction	Friction Mechanism	Hierarchy Level
Non-adhesive surfaces	Chemical interaction between surfaces is possible	Breaking chemical adhesive bonds	Adhesion	Molecule
Conservative adhesive forces	Conservative (van der Waals) forces and non-conservative (chemical) bonds	Breaking chemical adhesive bonds	Adhesion	Molecule
Rigid material	Deformable (elastic and plastic) material	Radiation of elastic waves (phonons)	Adhesion	Surface
Smooth surface	Rough surface	Plowing, ratchet mechanism, cobblestone mechanism	Deformation, ratchet, cobblestone mechanisms	Asperity
Homogeneous surface	Inhomogeneous surface	Energy dissipation due to inhomogeneity	Adhesion	Surface

Source: Based on Nosonovsky, M., and Bhushan, B. (2008). *Multiscale Dissipative Mechanisms and Hierarchical Surfaces: Friction, Superhydrophobicity, and Biomimetics.* New York: Springer.

Note: The table shows how non-ideality and heterogeneity at different levels contribute to friction.

TABLE 4.3

Mechanisms of Friction and Linear Dependence of the Friction Force upon the Normal Load

Mechanism		Friction Force and Real Area of Contact as Functions of the Normal Load
Area controlled	Elastic hierarchical "pseudo-fractal" asperities (Archard)	$F = \tau_a A_r \propto W^{\frac{3_n-1}{3^n}}$
	Elastic statistical (Greenwood–Williamson)	$F = \tau_a A_r \propto \dfrac{\beta^*}{E^* \sigma} W$
	Plastic deformation	$F = \tau_a A_r = \dfrac{W}{H_s}$
Slope controlled	Ratchet (climbing of asperities)	$F = W \tan \theta$
Other	Elastic plowing (Ying–Hsu)	$F = \tau_a A_p = \dfrac{2a^3}{3R} \propto W$

Source: Based on Nosonovsky, M., and Bhushan, B. 2008. *Multiscale Dissipative Mechanisms and Hierarchical Surfaces: Friction, Superhydrophobicity, and Biomimetics.* New York: Springer.

- Friction involves weak interactions at the interface. These can be quantified by an appropriate parameter, the extent of contact, and the ratio of real to nominal areas of contact that can serve such a parameter in some cases. Many manifestations of the linearity of friction are the consequence of the extent of contact being a small number (Table 4.3).

- Friction is a result of a reduction of the number of degrees of freedom in a system, when numerous "non-essential" degrees of freedom are disregarded. As with any approximation, this reduction has a price of its own and sometimes it can lead to oversimplifications and even to inconsistencies, referred to as frictional paradoxes. More sophisticated models of dynamic friction include the internal degrees of freedom or state variables.

- Frictional paradoxes are related to frictional dynamic instabilities. An assumption of rigid bodies participating in frictional contact could result in the non-existence or non-uniqueness of a solution constituting a paradox. The same problem with the assumption of deformable bodies may result in unstable solutions.

- Friction involves various kinds of instabilities and phase transitions. The onset of friction and near-critical behavior involves many effects that remain to be understood.

4.4 Thermodynamics of Friction

Traditionally, thermodynamic analysis of friction has concentrated mostly on the frictional rise of temperature due to dissipation, since most frictional work converted into heat raises the interface temperature, as well as on the temperature partition between the contacting bodies. Studies have included temperature distribution analysis within the contacting bodies and particularly the near-surface temperature in sliding systems (Alyabev, Kazimirchik, and Onoprienko 1973; Ling 1973; Kennedy 1980, 1984; Greenwood and Alliston-Greiner 1992; Knothe and Liebelt 1995). A temperature rise at the interface can be high enough to modify the mechanical and metallurgical properties of sliding surfaces, which drastically changes the behavior of tribological systems (Welsh 1957; Blau 1981; Quinn 1992; Amiri and Khonsari 2010). The "flash temperature" in a contact area of a sliding system involving steel can reach 750°C–800°C, which may induce phase transformation, cause oxidation, and reduce wear resistance (Archard 1969). These studies (Blok 1937, 1963; Jaeger 1942; Czichos 1978; Wang et al. 1992) were reviewed by Amiri and Khonsari (2010).

However, we would like to approach friction thermodynamically from a different point of view. Friction is a non-equilibrium dissipative process. First, let us consider a material point moving in a viscous medium. The entropy production rate is equal to the rate of energy dissipation divided by temperature, $\dot{S} = FV/T$, and, in accordance with Equation (3.26), the generalized force is

$$X = \frac{\partial S}{\partial x} = \frac{\partial}{\partial x}\left(\int \frac{F}{T} V \, dt\right) = \frac{\partial}{\partial x}\left(\int \frac{F}{T} dx\right) = \frac{\partial S}{\partial \dot{x}} = \frac{F}{T} \qquad (4.37)$$

Thus, in the proximity of the equilibrium, the linear approximation of Equation (3.27) yields

$$V = \frac{L}{T} F \qquad (4.38)$$

where $\eta = L/T$ is the viscosity. Assuming that the linear approximation remains valid at a certain distance from the equilibrium (which would be the state of rest), we conclude that the phenomenological relationships Equation (3.27) can yield the viscous friction law, but not dry friction.

Let us now consider the motion of a material point in a two-dimensional strongly anisotropic medium with the coordinates

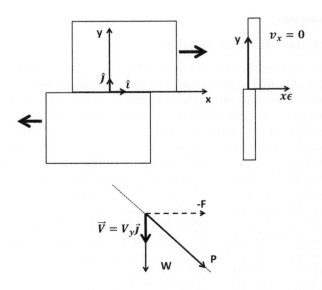

FIGURE 4.10
Introducing a small parameter ε into the system results in a small force in the *x*-direction, causing a large effect. Furthermore, if vectors $\vec{V}$ and $\vec{P}$ are related to each other in a linear way, $\vec{V} = \mathbf{C} \cdot \vec{P}$, the tangential component should be bounded and constrained to $V_x = 0$, thus constraining P to a certain direction, providing $P_x = mP_y$. We interpret $P_y = W$ as the normal load and $P_x = -F$ as the driving force (negative to the friction force).

$$\vec{R} = \varepsilon x \vec{i} + y \vec{j} \tag{4.39}$$

where ε is a small parameter. We will think about the *x*-coordinate as the "main" coordinate of motion of the material point, while *y* is the normal coordinate, representing a hidden degree of freedom. The displacement in the normal direction is measured in small units (e.g., typical value is $y = 1$ μm), while the *x*-displacement is measured in usual units—for example, $x = 1$ mm. By setting, for example, $\varepsilon = 0.001$, we assume that the effect of the *y*-displacement on the overall behavior of our system is on the same order as the effect of the *x*-displacement when the coupling of these effects is investigated (Figure 4.10).

The normal degree of freedom was introduced into the analysis of dynamic friction in the pioneering works of Tolstoi (1967), who discovered the existence of natural normal micro-vibrations coupled with tangential vibrations, which strongly affect the magnitude of the friction force as well as the stability of sliding. The introduction of a small parameter ε and an asymptotic decomposition is the standard way of the transition from a 3D to a 2D problem (for example, from the bulk elastic 3D body to a thin elastic plate). Now, the vector velocity is introduced:

$$\vec{V} = \frac{d\vec{R}}{dt} = \varepsilon V_x \vec{i} + V_y \vec{j} \qquad (4.40)$$

Using the force components,

$$\vec{P} = F\vec{i} + W\vec{j} \qquad (4.41)$$

one can write similarly to the derivation of Equations (4.37) and (4.38):

$$\dot{S} = \vec{P} \cdot \frac{\vec{V}}{T} = (\varepsilon F V_x + W V_y)/T \qquad (4.42)$$

$$\vec{X} = \nabla S = \nabla \left(\int \frac{\vec{P}}{T} \cdot \vec{V} \, dt \right) = \frac{\partial}{\partial x} \left(\frac{\vec{P}}{T} \cdot \vec{dR} \right) = \varepsilon^{-1} \frac{\partial S}{\partial V_x} \vec{i} + \frac{\partial S}{\partial V_y} \vec{j} = \frac{\vec{P}}{T} \qquad (4.43)$$

And, finally, the phenomenological relationships and Equation (3.28):

$$V_x = \varepsilon^{-1} \frac{1}{T} (L_{11} F + L_{12} W) \qquad (4.44)$$

$$V_y = \frac{1}{T} (L_{12} F + L_{22} W) \qquad (4.45)$$

Further resolving Equations (4.44) and (4.45) for the force components:

$$F = \varepsilon \frac{T V_x}{L_{11}} - \frac{L_{12}}{L_{11}} W \qquad (4.46)$$

$$W = \frac{T V_{yy}}{L_{22}} - \frac{L_{12}}{L_{22}} F = V_y \frac{T L_{11}}{L_{11} L_{22} - L_{12}^2} + \varepsilon V_x \frac{T L_{12}}{L_{11} L_{22} - L_{12}^2} \qquad (4.47)$$

The limit of $\varepsilon \to 0$ with *any* finite value of V_x yields

$$F = -\frac{L_{12}}{L_{11}} W \qquad (4.48)$$

$$W = V_y \frac{T L_{11}}{L_{11} L_{22} - L_{12}^2} \qquad (4.49)$$

By defining the ratios

$$\mu = \frac{L_{12}}{L_{11}} \qquad (4.50)$$

we finally obtain that, for any finite velocity V_x,

$$F = -\mu W \qquad (4.51)$$

$$V_y = \frac{1}{T}(L_{12} - \mu^2 L_{11})W \qquad (4.52)$$

The second law of thermodynamics requires that the matrix L in Equation (3.27) and in Equations (4.44) and (4.45) be positive definite or, using Sylvester's criterion,

$$L_{11} > 0 \qquad (4.53)$$

$$\frac{L_{22}}{L_{11}} > \mu^2 \qquad (4.54)$$

One concludes from Equations (4.53) and (4.54) that $V_x F$ and $V_y W$ are always positive and, therefore, the driving force is co-directed with the sliding velocity, while the friction force is opposite to the sliding velocity.

We interpret Equation (4.51) as the Amontons–Coulomb law (Equation 4.1) and Equation (4.45) as a viscosity equation for the internal coordinate of our system y. Thus, we showed that the Amontons–Coulomb friction law (Equation 4.1) can be deduced from the thermodynamic equations of motion (Equation 3.27), provided that two assumptions are made.

1. Motion in the normal degree of freedom (y) is coupled with the tangential degree of freedom (x).
2. The change of coordinates $(x, y) \to (\varepsilon x, y)$ is introduced and the limiting case of $\varepsilon \to 0$ is investigated.

The same procedure can be applied also for continuous surface stress, rather than a material point. In that case, the entropy production density (see Equation 3.56) is given by

$$\dot{S} = -\frac{p^v}{T}\nabla \cdot \mathbf{V} - \frac{1}{T}\mathbf{P}^{0v} : \mathbf{V}^0 \qquad (4.55)$$

where the viscous stress tensor is presented as the sum of the spherical part p^v and the deviator $\mathbf{P}^{0v}$; similarly, the strain rate tensor is presented as sum of the velocity $\mathbf{V}$ and the deviator $\mathbf{V}^{0v}$; sign ":" is the dual tensor product. Thus, if the 2D stress tensor

$$
\tau = \begin{pmatrix} \tau_{xx} & \tau_{xy} \\ \tau_{yx} & \tau_{yy} \end{pmatrix} \tag{4.56}
$$

is introduced and the strain rate tensor involves the small parameter, then the linear equations relating the stress and the strain rate would result in $\tau_{xy} = \mu \tau_{yy}$ (i.e., in the local form of the Amontons–Coulomb law).

4.5 Summary

In this chapter we discussed the empirical laws of friction and the physical mechanisms of energy dissipation, which all remarkably led to the same simple empirical laws. The empirical laws are valid for a remarkable range of material combination loads and frictional mechanisms, covering practically all classes of solid materials, loads from nanonewtons (in nano-tribology) to millions of tons (in seismology), such diverse mechanisms as adhesion of various physical nature, fracture, and elastic wave emission. The question that we asked is what is common and essential among all these mechanisms, leading to these common patterns of behavior?

We found that introducing friction as a separate force is a result of reduction in the number of degrees of freedom in a physical system and ignoring non-essential degrees of freedom (typically at the micro-scale, making friction a multi-scale phenomenon).

In some cases this leads to oversimplification and then internal degrees of freedom should be introduced to obtain models of dynamic friction. In an ideal world of absolutely rigid bodies and conservative interactions, there would be no friction and no energy dissipation. Various friction mechanisms are related to a certain degree of inherent non-ideality.

Another common feature of various friction mechanisms is the presence of an interface characterized by small interaction forces through the interface compared with stronger forces in the bulk. This small ratio of the interface-to-bulk interactions yields the linearity of friction.

After establishing that friction can be studied within the limits of linear approximations, we applied linear phenomenological relationships (which are normally valid only in the proximity of the equilibrium) to friction. This leads to linear viscous friction rather than to dry Coulomb friction. However, if an anisotropic medium $\vec{R} = \varepsilon x \vec{i} + y \vec{j}$ is studied in the limit of $\varepsilon \to 0$ corresponding to the bulk becoming the interface (or, more accurately, to the interface interactions coupled in the same manner as they would be in the bulk), the bulk viscosity, combined with the second law of thermodynamics, yielded Amontons–Coulomb law $F = \mu W$ and arbitrary sliding velocity. Thus, we derived the dry friction law from the thermodynamic relationships.

In the subsequent chapters we will further extend the thermodynamic analysis and consider the stability criterion $\delta^2 \dot{S} > 0$ (Equation 3.37) and various kinds of friction-induced instabilities, some of which can lead to self-organization.

References

Adams, G. G., Barber, J. R., Ciavarella, M., and Rice, J. R. 2005. A paradox in sliding contact problems with friction. *Journal of Applied Mechanics* 72:450.

Adams, G. G., Muftu, S., and Mohd Azhar, N. 2003. A scale-dependent model for multi-asperity contact and friction. *Journal of Tribology* 125 (4): 700–708.

Ahn, Y. J., Bertocchi, E., and Barber, J. R. 2008. Shakedown of coupled two-dimensional discrete frictional systems. *Journal of the Mechanics and Physics of Solids* 56 (12): 3430–3443.

Alyabev, A. Y., Kazimirchik, Y. A., and Onoprienko, V. P. 1973. Determination of temperature in the zone of fretting corrosion. *Materials Science* 6 (3): 284–286.

Amiri, M., and Khonsari, M. M. 2010. On the thermodynamics of friction and wear—A review. *Entropy* 12 (5): 1021–1049.

Archard, J. F. 1957. Elastic deformation and the laws of friction. *Proceedings of the Royal Society of London. Series A. Mathematical and Physical Sciences* 243 (1233): 190–205.

———. 1969. The temperature of rubbing surfaces. *Wear* 2:438–455.

Åström, K., and Canudas-De-Wit, C. 2008. Revisiting the LuGre friction model. *Control Systems, IEEE*, 28 (6): 101–114.

Barber, J. R. 2011. Frictional systems subjected to oscillating loads. *Annals of Solid and Structural Mechanics* 2 (2): 45–55.

Bhushan, B. 2002. *Introduction to tribology.* New York: Wiley.

Bhushan, B., Israelachvili, J. N., and Landman, U. 1995. Nanotribology: Friction, wear and lubrication at the atomic scale. *Nature* 374 (6523): 607–616.

Bhushan, B., and Nosonovsky, M. 2003. Scale effects in friction using strain gradient plasticity and dislocation-assisted sliding (microslip). *Acta Materialia* 51 (14): 4331–4345.

———. 2004a. Comprehensive model for scale effects in friction due to adhesion and two-and three-body deformation (plowing). *Acta Materialia* 52 (8): 2461–2474.

———. 2004b. Scale effects in dry and wet friction, wear, and interface temperature. *Nanotechnology* 15 (7): 749.

Blau, P. J. 1981. Interpretations of the friction and wear break-in behavior of metals in sliding contact. *Wear* 71 (1): 29–43.

————. 1991. Scale effects in steady-state friction. *Tribology Transactions* 34 (3): 335–342.

Blok, H. 1937. Theoretical study of temperature rise at surfaces of actual contact under oiliness lubricating conditions. *Proceedings of the General Discussion on Lubrication and Lubricants,* Institute of Mechanical Engineers, London, 2:222–235.

————. 1963. The flash temperature concept. *Wear* 6:483–494.

Bowden, F. P., and Tabor, D. 1986. *The friction and lubrication of solids,* 374. Oxford, England: Clarendon Press.

Carpinteri, A., and Paggi, M. 2005. Size-scale effects on the friction coefficient. *International Journal of Solids and Structures* 42 (9): 2901–2910.

Chang, W. R., Etsion, I., and Bogy, D. B. 1987. An elastic-plastic model for the contact of rough surfaces. *Journal of Tribology* 109 (2): 257–263.

Czichos, H. 1978. *Tribology—A systems approach to the science and technology of friction, lubrication and wear.* Amsterdam: Elsevier.

————. 2001. Tribology and its many facets: From macroscopic to microscopic and nano-scale phenomena. *Meccanica* 36 (6): 605—615.

Daly, C., Zhang, J., and Sokoloff, J. B. 2003. Friction in the zero sliding velocity limit. *Physical Review E* 68:066118

Deshpande, V. S., Needleman, A., and Van der Giessen, E. 2004. Discrete dislocation plasticity analysis of static friction. *Acta Materialia* 52 (10): 3135–3149.

Dienwiebel, M., Verhoeven, G. S., Pradeep, N., Frenken, J. W., Heimberg, J. A., and Zandbergen, H. W. 2004. Superlubricity of graphite. *Physical Review Letters* 92 (12): 126101.

Dieterich, J. H. 1979. Modeling of rock friction 1. Experimental results and constitutive equations. *Journal of Geophysical Research* 84 (B5): 2161–2168.

Dieterich, J. H., and Kilgore, B. D. 1994. Direct observation of frictional contacts: New insights for state-dependent properties. *Pure and Applied Geophysics* 143 (1): 283–302.

Dorogovtsev, S. N., Goltsev, A. V., and Mendes, J. F. 2008. Critical phenomena in complex networks. *Reviews of Modern Physics* 80 (4): 1275.

Einax, M., Schulz, M., and Trimper, S. 2004. Friction and second-order phase transitions. *Physical Review E* 70 (4): 046113.

Frenkel, Ya., and Kontorova, T. 1939. Fiz. Zh. (Moscow) 1, 137 (in Russian).

Gerde E., and Marder M. 2001. Friction and fracture. *Nature* 413:285–288

Ghaffari, H. O., and Young, R. P. 2012. Topological complexity of frictional interfaces: Friction networks. *Nonlinear Processes in Geophysics* 19:215–225.

Greenwood, J. A., and Alliston-Greiner, A. F. 1992. Surface temperatures in a fretting contact. *Wear* 155 (2): 269–275.

Greenwood, J. A., and Williamson, J. B. P. 1966. Contact of nominally flat surfaces. *Proceedings of the Royal Society of London. Series A. Mathematical and Physical Sciences* 295 (1442): 300–319.

He, G., Müser, M. H., and Robbins, M. O. 1999. Adsorbed layers and the origin of static friction. *Science* 284:1650–1652.

He, G., and Robbins, M. O. 2003. Scale effects and the molecular origins of tribological behavior. In *Nanotribology: Critical assessment and research needs,* ed. S. M. Hsu and Z. C. Ying, 29–44. Dordrecht: Kluwer Academic.

Homola, A. M., Israelachvili, J. N., McGuiggan, P. M., and Gee, M. L. 1990. Fundamental experimental studies in tribology: The transition from "interfacial" friction of undamaged molecularly smooth surfaces to "normal" friction with wear. *Wear* 136 (1): 65–83.

Hurtado, J. A., and Kim, K. S. 1999. Scale effects in friction of single–asperity contacts. I. From concurrent slip to single-dislocation-assisted slip. *Proceedings of the Royal Society of London. Series A: Mathematical, Physical and Engineering Sciences* 455:3363–3384.

Israelachvili, J. N. 1992. *Intermolecular and surface forces*, 2nd ed. London: Academic Press.

Jaeger, J. C. 1942. Moving sources of heat and the temperature at sliding contacts. *Proceedings of Royal Society of New South Wales* 76:203–224.

Jang, Y. H., and Barber, J. R. 2011. Frictional energy dissipation in materials containing cracks. *Journal of the Mechanics and Physics of Solids* 59 (3): 583–594.

Johnson, K. L. 1997. Adhesion and friction between a smooth elastic spherical asperity and a plane surface. *Proceedings of the Royal Society of London. Series A: Mathematical, Physical and Engineering Sciences* 453 (1956): 163–179.

Kennedy, F. E., Jr. 1980. Surface temperatures in sliding systems—A finite element analysis. In *Century 2 International Lubrication Conference* (vol. 1).

———. 1984. Thermal and thermomechanical effects in dry sliding. *Wear* 100 (1): 453–476.

Kessler, D. A. 2001. Surface physics: A new crack at friction. *Nature* 413 (6853): 260–261.

Klarbring, A., Ciavarella, M., and Barber, J. R. 2007. Shakedown in elastic contact problems with Coulomb friction. *International Journal of Solids and Structures* 44 (25): 8355–8365.

Knothe, K., and Liebelt, S. 1995. Determination of temperatures for sliding contact with applications for wheel-rail systems. *Wear* 189 (1): 91–99.

Kogut, L., and Etsion, I. 2004. A static friction model for elastic-plastic contacting rough surfaces. *Transactions of American Society of Mechanical Engineers Journal of Tribology* 126 (1): 34–40.

Ling, F. F. 1973. *Surface mechanics.* New York: Wiley.

Maeda, N., Chen, N., Tirrell, M., and Israelachvili, J. N. 2002. Adhesion and friction mechanisms of polymer-on-polymer surfaces. *Science* 297 (5580): 379–382.

Muser, M. H., Urbakh, M., and Robbins, M. O. 2003. Statistical mechanics of static and low-velocity kinetic friction. *Advances in Chemical Physics* 126:187–272.

Niederberger, S., Gracias, D. H., Komvopoulos, K., and Somorjai, G. A. 2000. Transitions from nanoscale to microscale dynamic friction mechanisms on polyethylene and silicon surfaces. *Journal of Applied Physics* 87 (6): 3143–3150.

Nosonovsky, M. 2007a. Model for solid–liquid and solid–solid friction of rough surfaces with adhesion hysteresis. *Journal of Chemical Physics* 126:224701.

———. 2007b. Modeling size, load and velocity effect on friction at micro/nanoscale. *International Journal of Surface Science and Engineering* 1 (1): 22–37.

Nosonovsky, M., and Bhushan, B. 2005. Scale effect in dry friction during multiple-asperity contact. *Journal of Tribology* 127 (1): 37–46.

———. 2007. Multiscale friction mechanisms and hierarchical surfaces in nano -and bio-tribology. *Materials Science and Engineering: Reports* 58 (3): 162–193.

———. 2008. *Multiscale dissipative mechanisms and hierarchical surfaces: Friction, superhydrophobicity, and biomimetics.* Berlin: Springer.

Olsson, H., Åström, K. J., Canudas de Wit, C., Gäfvert, M., and Lischinsky, P. 1998. Friction models and friction compensation. *European Journal of Control* 4:176–195.

Persson, B. N. 2000. *Sliding friction: Physical principles and applications*, vol. 1.New York: Springer.

Pugno, N. M. 2007. A general shape/size-effect law for nano-indentation. *Acta Materialia* 55 (6): 1947–1953.

Quinn, T. F. J. 1992. Oxidational wear modeling: I. *Wear* 135:179–200.

Ranjith, K., and Rice, J. R. 2001. Slip dynamics at an interface between dissimilar materials. *Journal of the Mechanics and Physics of Solids* 49 (2): 341–361.

Rice, J. R. 1991. Dislocation nucleation from a crack tip: An analysis based on the Peierls concept. *Journal of the Mechanics and Physics of Solids* 40 (2): 239–271.

Rubinstein, S. M., Cohen, G., and Fineberg, J. 2004. Detachment fronts and the onset of dynamic friction. *Nature* 430 (7003): 1005–1009.

Ruths, M., Berman, A. D. and Israelachvili, J. N. 2005. *Nanotribology and nanomechanics: An introduction.* Berlin: Springer–Verlag.

Schwarz, U. D., Zwörner, O., Köster, P., and Wiesendanger, R. 1997. Quantitative analysis of the frictional properties of solid materials at low loads. I. Carbon compounds. *Physical Review B* 56 (11): 6987.

Sokoloff, J. B. 2002. Possible microscopic explanation of the virtually universal occurrence of static friction. *Physical Review B* 65 (11): 115415.

———. 2006. Theory of the effects of multiscale surface roughness and stiffness on static friction. *Physical Review E* 73 (1): 016104.

Stark, R. W., Schitter, G., and Stemmer, A. 2004. Velocity dependent friction laws in contact mode atomic force microscopy. *Ultramicroscopy* 100 (3): 309–318.

Szoszkiewicz, R., Bhushan, B., Huey, B. D., Kulik, A. J., and Gremaud, G. (2005). Correlations between adhesion hysteresis and friction at molecular scales. *Journal of Chemical Physics* 122:144708.

Tolstoi, D. M. 1967. Significance of the normal degree of freedom and natural normal vibrations in contact friction. *Wear* 10 (3): 199–213.

Tomlinson, G. A. 1929. CVI. A molecular theory of friction. *The London, Edinburgh, and Dublin Philosophical Magazine and Journal of Science* 7 (46): 905–939.

Urbakh, M., Klafter, J., Gourdon, D., and Israelachvili, J. 2004. The nonlinear nature of friction. *Nature* 430 (6999): 525–528.

Wang, Y., Lei, T., Yan, M., and Gao, C. 1992. Frictional temperature field and its relationship to the transition of wear mechanisms of steel 52100. *Journal of Physics D: Applied Physics* 25:A165–A169.

Welsh, N. C. 1957. Frictional heating and its influence on the wear of steel. *Journal of Applied Physics* 28:960–968.

Wenning, L., and Muser, M. H. 2007. Friction laws for elastic nanoscale contacts. *Europhysics Letters* 54 (5): 693.

Ying, Z. C., and Hsu, S. M. 2006. *First observation of elastic plowing in nanofriction.* ASME.

Zeng, H., Tirrell, M., and Israelachvili, J. 2006. Limit cycles in dynamic adhesion and friction processes: A discussion. *Journal of Adhesion* 82 (9): 933–943.

Zhang, L. C., Johnson, K. L., and Cheong, W. C. D. 2001. A molecular dynamics study of scale effects on the friction of single-asperity contacts. *Tribology Letters* 10 (1): 23–28.

5

Wear and Lubrication

The studies of wear and lubrication are traditionally included in tribology. According to an accepted point of view, friction and wear constitute two components of the same irreversible process of deterioration during sliding. While wear is the term for the "visible" deterioration of the material resulting in the removal of material from the surface, friction is just the dissipation of energy. Lubrication is used to control both friction and wear.

5.1 Mechanisms of Wear

Wear is the material removal and deterioration during the contact of solid surfaces in relative motion. There are several mechanisms that lead to wear. In general, two thirds of all wear processes encountered in industrial situations occur due to adhesive and abrasive wear mechanisms. In many cases, wear is initiated by one mechanism, and it may proceed by other wear mechanisms. Wear components are generally examined to determine the type of wear mechanism by using microscopy or surface analytical techniques.

Adhesive wear is due to adhesion between the contacting surfaces, which can lead to the removal and transfer of particles of a material and displacement of wear debris from one surface to the other. Adhesive wear occurs when two bodies slide over or are pressed into each other, which promotes material transfer, involving the plastic deformation of very small fragments within the surface layers. When two solid surfaces are brought together, the atoms are in contact at some points. Consequently, van der Waals attractive forces act between the surfaces. At a distance of about or less than 1 nm, strong short-range forces (covalent chemical bonding forces) come into action and strong adhesive junctions may be formed at the spots of the real area of contact. When there is relative motion, the junctions are sheared and fractured. This results in the softer material being transferred to the harder surface.

Soft material can remain adhered to the harder surface or subsequent sliding can produce loose wear debris. The amount of wear depends on the location at the junctions that are sheared. If shear occurs entirely at the original interface, then the wear is zero. However, if shear occurs away from the interface, a fragment of material is transferred from one surface to the other. In practice, the transfer of material is observed from the softer material to the harder material, but occasionally from the harder material to the softer material. Adhesive wear is often called galling, scuffing, cold welding, or smearing.

The second type is *abrasive wear* due to plowing, cutting, and fragmenting of asperities. Abrasive wear is the loss of material due to hard particles or hard protuberances that are forced against and move along a solid surface. The abrasion process includes several deformation modes such as plowing, wedge formation, and cutting. Abrasive wear occurs when a hard, rough surface slides across a softer surface. If a hard material is kept in contact with the soft solid surface, the asperities of the hard material are pressed into the soft surface, causing plastic flow of the soft surface. When shear load is imposed, the hard material will slide and remove the soft material by plowing.

Plowing is the displacement of the material to the side, away from the wear particles, resulting in the formation of grooves that do not involve direct material removal. The displaced material forms ridges adjacent to grooves, which may be removed by subsequent passage of abrasive particles. Cutting is material separation from the surface in the form of debris or micro-chips with little or no material displaced to the sides of the grooves. This mechanism closely resembles conventional machining. During cutting, an abrasive tip cuts a groove and removes the material. Fragmentation is material separation from a surface by a cutting process and the indenting abrasive causes localized fracture of the wear material. During wedge formation, an abrasive tip plows a groove and forms a wedge on the front of the groove.

Abrasion is typically classified as "two-body" or "three-body" abrasion according to the type of contact. During two-body abrasion, hard material slides along a softer surface. During three-body abrasion, wear debris is captured between two surfaces, causing one or both surfaces to be abraded. Wear loss in two-body abrasion is typically 10–1,000 times greater than three-body abrasion for a given load and sliding distance.

In addition to the two main types of wear, other types are delamination, surface fatigue, fretting wear, erosive wear, and cavitation wear:

- Surface fatigue is weakening of the surface due to cyclic loading.
- Fretting or fretting fatigue is a repeated cyclical rubbing of the surface.
- Erosive wear is caused by the impact of solid particles or fluid on the surface.
- Cavitation wear is due to contact with fluid.

Fracture can be another mode of wear. During the contact of two asperities with friction and shear loading, the maximum shear strength is usually achieved beneath the surface, rather than at the surface. As a result, plastic yield and flow starts at the sub-surface zone and, due to repeated loading, a thin "skin-like" layer of material delaminates.

The *delamination* theory of wear was suggested by Suh (1973). His theory describes the production of laminate wear debris due to cracking at a certain depth under the surface. When this crack reaches a critical length, the material between the crack and the surface will shear, yielding long and thin laminated wear debris. This results because the maximum stress during the contact involving the shear force component is not at the surfaces, but rather at a certain sub-surface depth.

Erosion wear is the form of damage experienced by a solid body when liquid or solid particles impinge on a solid surface in the form of solid erosion or fluid erosion. Solid erosion occurs by impingement of solid particles and is a form of abrasive wear, which is, however, treated differently because the contact stress arises from the kinetic energy of particles flowing in air as it encounters the surface. Wear debris is formed in erosion as a result of repeated impacts. As in the case of abrasive wear, erosive wear occurs by plastic deformation and/or brittle fracture, depending on the material being eroded away and on the operating conditions.

There are two types of fluid erosion: liquid impact erosion and *cavitation* erosion. When tiny liquid drops strike the solid surface at high speeds, due to high pressure, the material undergoes plastic deformation or fracture and repeated impacts lead to pitting and erosive wear. In many cases, the combined erosion–corrosion mechanism damages the surface. Cavitation damage occurs when bubbles entrained in a liquid become unstable and implode against the surface of a solid, forming micro-jets of liquid directed toward the solid.

Fatigue wear is important at the macroscopic and microscopic scales. Macroscopic fatigue occurs at non-conforming loaded surfaces—for example, during rolling contact—whereas micro-fatigue occurs at the contacts between sliding asperities. The repeated loading–unloading cycles can induce sub-surface or surface cracks, which after a critical number of cycles will eventually result in the breakup of the surface with the formation of large fragments, leaving large pits in the surface (also called pitting). Before the critical number of cycles, negligible wear occurs, in marked contrast to the wear caused by an adhesive or abrasive mechanism, when wear causes a gradual deterioration from the start of running. Fatigue wear can occur during both sliding and rolling loading.

Fretting wear occurs when a low-amplitude oscillatory motion in the tangential direction occurs between contacting surfaces. This is typical in machinery that is subjected to vibration. Fretting can involve several wear processes. Basically, fretting is a form of adhesive or abrasive wear where the normal load causes adhesion between asperities and oscillatory movement causes rupture, resulting in wear debris. Fretting in a corrosive environment produces wear particles that are harder than their parent metals and thus can lead to abrasion. Since there is no macroscopic sliding at fretting contacts, the fretting wear debris is trapped between the surfaces and cannot escape easily. Consequently, the amount of wear per unit of sliding distance due to fretting may be larger than that of adhesive and abrasive wear. The oscillatory sliding can also cause vibration and thus can lead to fretting fatigue.

Corrosive wear occurs when sliding occurs in a corrosive environment. In air, the dominant corrosive medium is oxygen, and therefore, corrosive wear in air is referred to as oxidative wear. However, the same principle applies to wear in any other type of corrosive medium. In the absence of sliding, oxide films typically less than a micron thick form on the surfaces. Sliding action wears the oxide film away so that corrosive attack can continue. Thus, corrosive wear requires corrosion and rubbing.

However, oxidation on sliding surfaces can be beneficial. Formation of oxide films acts as a solid lubricant and prevents metal–metal contact, thus mitigating against the severe adhesion-enhanced wear that would otherwise occur. Oxidation can reduce the wear rate of metallic pairs by two orders of magnitude, as compared with that of the same pair in an inert atmosphere. When surfaces are oxidized, the wear debris is finely divided oxide, so the rubbing surfaces remain smooth and the rate of loss of material is low. Oxidation is a thermally activated process; the rate of oxidation can increase exponentially with temperature; changing the surface temperature by only few degrees can change the rate of oxidation by an order of magnitude.

Wear is also characterized by the size and intensity of the production of wear debris as mild wear and severe wear. In mild wear, the wear occurs at the outer surface layers and worn debris contains fine oxide particles varying in size from 0.01 to 100 nm. In severe wear, the wear occurs at deep surfaces and the size of the wear debris ranges from 100 nm to 100 μm.

5.2 Empirical Laws of Wear

The quantitative characteristic of wear is the wear volume or volume of worn material, w. The rate of wear is measured in wear volume per unit time. The so-called empirical Archard wear law (or rule) relates the wear rate with the sliding velocity, V; the applied normal load, W; and the hardness, H, of a softer material among the two contacting materials:

$$\dot{w} = k \frac{WV}{H} \tag{5.1}$$

This empirical law of wear, formulated by Archard for abrasive wear, states that the wear rate is linearly proportional to the sliding velocity (or the wear volume is linearly proportional to the sliding distance), the ratio of the normal load to the hardness of the softer material, and the coefficient k referred to as "the wear coefficient," which is a characteristic of tribological systems somewhat similar to the coefficient of friction. A similar approach applies also to adhesive wear. Note that the empirical law of wear, Equation (5.1), is another linear phenomenological law that can be presented as a relationship between thermodynamic forces and flows.

5.3 Thermodynamics of Wear: Entropy Generation and Degradation

Friction and wear are dissipative irreversible processes. Entropy is the measure of irreversibility and dissipation. Therefore, entropy can be used to characterize wear and related degradation processes. Manufacturing, which transforms nature's raw materials into highly organized finished components, reduces entropy. Aging or degradation from friction and wear tends to return these components to natural states. Accordingly, entropy must monotonically increase to be consistent with the second law of thermodynamics.

Many researchers have attempted to use thermodynamic methods for a general theory of wear; however, most of these attempts have had limited success due to the complexity of the equations involved and the difficulty of their solution (Klamecki 1980a, b; Zmitrowicz 1987; Bryant 2009). In Kiev, B. Kostetsky and L. Bershadsky (1992, 1993) investigated the formation of the so-called self-organized "secondary structures." According to Bershadsky (1993), friction and wear are two sides of the same phenomenon and they represent the tendency of energy and matter to achieve the most disordered state. However, the synergy of various mechanisms can lead to the self-organization of the secondary structures, which are "non-stoichiometric and metastable phases," whereas "the friction force is also a reaction on the informational (entropic) excitations, analogous to the elastic properties of a polymer, which are related mostly to the change of entropy and have the magnitude of the order of the elasticity of a gas."

N. Bushe and then I. Gershman, in Moscow, developed the theory of tribological compatibility of materials and wear control using this effect. Their results were summarized in the book edited by Fox-Rabinovich and Totten (2006). In 2002, N. Bushe won the most prestigious tribology award, the tribology gold medal, for his studies of tribological compatibility and other related effects.

D. Garkunov (2004) and co-workers claimed the discovery of the synergetic "non-deterioration effect," also called "selective transfer." They also introduced the concept of protective tribofilm formed in situ, which they called the "servovite film." Garkunov's mostly experimental research also received international recognition when he was awarded the 2005 tribology gold medal "for his achievements in tribology, especially in the fields of selective transfer."

In the English language literature, the works by Klamecki (1980a, 1980b, 1982, 1984) were the first to use the concepts of non-equilibrium thermodynamics, including the minimum entropy production principle by Prigogine, to describe friction and wear. Klamecki's work was extended by Zmitrowicz (1987), Dai et al. (2000), Doelling et al. (2000), and others. Abdel-Aal (2006) developed a model to predict and explain the role of wear as a self-organizing occurrence in tribosystems. Tross and Fleischer in Germany related wear and damage in rail brakes to a so-called critical energy density. Fleischer in Magdeburg, Germany, won the tribology gold medal for developing this concept. His student Sadowski (1990, 1995) from Poland contributed several important ideas to the thermodynamic theory of wear.

Later work by Ling, Doelling, Bryant, and Khonsari relates wear to the transport of entropy (and not the generation of entropy), which is a different approach. An important entropic study of the thermodynamics of wear was conducted by M. Bryant, Khonsari, and Ling (2008), who introduced a degradation function and formulated the degradation-entropy generation theorem in their approach intended to study friction and wear in combination. They note that friction and wear, which are often treated as unrelated processes, are in fact manifestations of the same dissipative physical processes occurring at sliding interfaces. The possibility of reducing friction between two elastic bodies due to a pattern of propagating slip waves was investigated by Adams (1998) and Nosonovsky and Adams (2001), who used the approach of the theory of elasticity.

Many researchers have endeavored to use thermodynamic methods to develop a general theory of wear; however, most of these attempts had limited success due to the complexity of the equations involved and the difficulty of their solution (Klamecki 1980; Zmitrowicz 1987; Bryant 2009). Doelling et al. (2000) experimentally correlated wear with entropy flow, dS/dt, at a wearing surface and found that wear was roughly proportional to the entropy produced by the steady sliding of copper on steel under boundary-lubricated conditions.

Bryant et al. (2006) conducted an interesting entropic study of wear. They started from the assumption that friction and wear are manifestations of the same dissipative physical processes occurring at sliding interfaces. The production of irreversible entropy by interfacial dissipative processes is associated with both friction and wear. Friction force dissipates power and generates entropy, whereas wear irreversibly changes a material's structure, often with a loss of material. Bryant (2009) identified entropy production mechanisms during various dissipative processes relevant to friction and wear, which are summarized in Table 5.1.

It is observed that the change of entropy has the general form of $dS = Yd\xi$—that is, a thermodynamic force Y times the change of the generalized coordinate $d\xi$. Bryant et al. (2006) also formulated the so-called "degradation-entropy generation theorem," which states that for N dissipative processes, characterized by energies $p_i = p_i(\zeta_1{}^i, \zeta_2{}^i, \ldots \zeta_n{}^i)$, where ζ_j are generalized coordinates (or "phenomenological variables") associated with the processes,

TABLE 5.1

Entropy Change during Various Dissipative Processes

Process	Entropy Change
Adhesion	$$dS = \frac{\gamma dA}{T}$$ where γ is surface energy, A area
Plastic deformation	$$dS = \frac{U_p dV}{T}$$ where U_p is the work per volume, V
Fracture	$$dS = \frac{\left(\frac{\partial U}{\partial a} - 2\gamma\right) da}{T}, \text{ where } \frac{\partial U}{\partial a}$$ is the energy release rate, a is crack length
Phase transition	$$dS = \frac{dH}{T}$$ where H is enthalpy
Chemical reaction	$$dS = \frac{\sum_{react} \mu_i dN_i - \sum_{products} \mu_i dN_i}{T}$$ where N_i are numbers of molecules and μ_i are chemical potentials for reactants and products.
Mixing	$$\Delta S = -R \sum_{i}^{n} \frac{N_i}{N} \ln \frac{N_i}{N}$$ where N_i are numbers of molecules and R is the universal gas constant
Heat transfer	$$dS = \left(\frac{1}{T_1} - \frac{1}{T_2}\right) dQ$$ where T_1 and T_2 are temperatures of the two bodies

Source: Based on Bryant, M. D. 2009. *FME Transactions* 37:55–60.

and for a degradation measure $w(p_1, p_2,...p_n)$, which is a non-negative and monotonic function of the process energies p_i, the rate of

$$\dot{w} = \sum_{i}^{n} B_i \dot{S}_i$$

degradation is a linear function of the components of entropy production

$$\dot{S}_i = \sum_{j}^{n} X_i^j J_i^j$$

of the dissipative processes (where X and J are generalized forces and flows of the processes). Degradation components

$$\dot{w}_i = \sum_{j}^{n} Y_i^j J_i^j$$

proceed at rates J_i^j determined by entropy production, whereas the generalized "degradation forces" $Y_i^j = B_i Y_i^j$ are linear functions of $Y_i = B_i X_i$, and degradation coefficients B_i are partial derivatives of w by the entropy.

5.4 Lubricated Contact

Lubrication is interposing a substance called a lubricant between the surfaces to carry or to help carry the load between the sliding surfaces in order to reduce friction and wear. The lubricant can be in various phase states, including a solid (such as graphite or MoS_2), a solid–liquid dispersion, a liquid, a liquid–liquid dispersion, or a gas. In the most typical case, the lubricant is a fluid capable of bearing the pressure between the surfaces. Adequate lubrication allows smooth continuous operation of equipment, with only mild wear and without excessive stresses or seizures at bearings and other contacting surfaces.

There are several lubrication regimes that can be observed, depending on the level of the load:

1. Fluid film lubrication is the regime in which the load is fully supported by the lubricant trapped within the space or gap between the parts in motion relative to one another so that the solid–solid contact is avoided. There are two types of fluid film lubrication: hydrostatic and hydrodynamic. If the fluid film lubrication occurs during the static loading when an external pressure is applied to the lubricant in the bearing, this is referred to as hydrostatic lubrication. Hydrodynamic lubrication occurs when the motion of the contacting surfaces helps to pump the lubricant to maintain the lubricating film.

2. Elastohydrodynamic lubrication is the regime in which the contacting surfaces are separated most of the time; however, interaction between the asperities can occur and an elastic deformation on the contacting surface enlarges the contact area whereby the viscous resistance of the lubricant becomes capable of supporting the load.

3. Boundary lubrication is the regime in which the bodies come into closer contact at their asperities and the load is carried by the surface asperities rather than by the lubricant.

Lubricant also is used to cool the contact areas and to remove wear products. While carrying out this function, the lubricant is constantly replaced from the contact areas either by relative movement (hydrodynamics) or by externally induced forces.

In general, lubrication is used to reduce friction and wear. We are not discussing this topic further in depth, since there is abundant tribological literature devoted to the issue.

5.5 Summary

The energy that is dissipated during frictional sliding can be consumed in different ways. Part of it is spent on thermal heating of the contacting material, which means an increase of the kinetic energy of the molecules. However, another part of energy is spent on breaking the bonds between the molecules, resulting in deterioration of the material. The first part corresponds to friction while the second part corresponds to wear. Wear accompanies frictional sliding and often involves several unrelated mechanisms, such as adhesion, abrasion, and others. The quantitative characteristic of wear is the wear volume (i.e., the volume of worn particles). Typically, the wear rate is linearly proportional to the normal load and sliding velocity. Lubrication is used to decrease friction and wear.

References

Abdel-Aal, H. A. 2000. On the influence of thermal properties on wear resistance of rubbing metals at elevated temperatures. *Wear* 122 (3): 657–660.

Bershadsky, L. I. 1992. On self-organization and wear resistance in tribo-systems. *Trenie i Iznos*, 13:1077–1094 (in Russian).

Bershadsky, L. I. 1993. B. I. Kostetski and the general concept in tribology. *Trenie I Iznos (Friction and Wear)* 14:6–18 (in Russian).

Bryant, M. D. 2009. Entropy and dissipative processes of friction and wear. *FME Transactions* 37:55–60.

Bryant, M. D., Khonsari, M. M. and Ling, F. F. 2008. On the thermodynamics of degradation. *Proceedings of the Royal Society A* 464 (2096): 2001–2014.

Dai, Z., Yang, S. and Q. Xue. 2000. Thermodynamic model of fretting wear. *J. Nanjing Univ. Aeronaut. Astronaut.* 32:125–131.

Doelling, K. L., Ling, F. F., Bryant, M. D., and Heilman, B. P. 2000. An experimental study of the correlation between wear and entropy flow in machinery components. *Journal of Applied Physics* 88 (5): 2999–3003.

Fox-Rabinovich, G. S. and G.E. Totten (Eds.). 2006. *Self-organization during friction: Advance surface engineered materials and systems design;* CRC Taylor and Francis Group: Boca Raton, FL, USA.

Garkunov, D. N. 1985. The selective transfer effect at points of friction (Moscow, 1985, in Russian).

Garkunov, D. N. 2000. *Triboengineering (wear and non-deterioration).* Moscow Agricultural Academy Press: Moscow (in Russian).

Garkunov, D. N. 2004. Scientific Discoveries in Tribotechnology. MSHA: Moscow (in Russian).

Klamecki, B. E. 1980a. Wear—An entropy production model. *Wear* 58 (2): 325–330.

Klamecki, B. E. 1980b. A thermodynamic model of friction. *Wear* 63:113–120.

Klamecki, B. E. 1982. Energy dissipation in sliding. *Wear* 77:115–128.

Klamecki, B. E. 1984. An entropy-based model of plastic deformation energy dissipation in sliding. *Wear* 96:319–329.

Sadowski, J. 1990. Untersuchungen zur maximalen Verschleibetafestigkeit fester korper" (Research into the maximum wear resistance of solids). *Tribologie und Schmierungstechnik,* 37:171–174.

Sadowski, J. 1995. Energie-und Verschleissverteilung zwischen den Festkorpern wahrend der Reibung (Energy and wear division into solids during friction process). *Tribologie und Schmierungstechnik,* 42:131–134.

Suh, N. P. 1973. The delamination theory of wear. *Wear* 25 (1): 111–124.

Zmitrowicz, A. 1987. A thermodynamical model of contact, friction, and wear. *Wear* 114 (2): 135–221.

Bibliography

Archard, J. F. 1953. Contact and rubbing of flat surfaces. *Journal of Applied Physics* 24 (8): 981–988.

Archard, J. F., and Hirst, W. 1956. The wear of metals under unlubricated conditions. *Proceedings of Royal Society London, A* 236:397–410.

Berkovich, I. I., and Gromakovsky, D. G. 2000. *Tribology: Physical fundamentals, mechanical and technical applications.* Samara, Russia: Samara State University (in Russian).

Buckley, D. H. 1978. The use of analytical surface tools in the fundamental study of wear. *Wear* 46 (1): 19–53.

Hornbogen, E. 1975. The role of fracture toughness in the wear of metals. *Wear* 33 (2): 251–259.

Kailas, S. V. 2003. A study of the strain rate microstructural response and wear of metals. *Journal of Materials Engineering and Performance* 12 (6): 629–637.

Kailas, S. V., and Biswas, S. K. 1995. The role of strain rate response in plane strain abrasion of metals. *Wear* 181–183: 648–657.

———. 1997. Strain rate response and wear of metals. *Tribology International* 30:369–375.

Kato, K. 2000. Wear in relation to friction—A review. *Wear* 241 (2): 151–157.

Khruschov, M. M. 1974. Principles of abrasive wear. *Wear* 28 (1): 69–88.

Lim, S. C., and Ashby, M. F. 1987. Overview no. 55 wear-mechanism maps. *Acta Metallurgica* 35 (1): 1–24.

Nosonovsky, M., and Bhushan, B. 2009. Thermodynamics of surface degradation, self-organization, and self-healing for biomimetic surfaces. *Philosophical Transactions Royal Society A* 367:1607–1627.

Prasad, Y. V. R. K., and Sasidhara, S. 1997. *Hot working guide: A compendium of processing maps.* Materials Park, OH: ASM International.

Rabinowicz, E. 1995. *Friction and wear of materials,* 2nd ed. New York:Wiley.

Saka, N., Eleiche, A. M., and Suh, N. P. 1977. Wear of metals at high sliding speeds. *Wear* 44 (1): 109–125.

Rice, S. L., Nowotny, H., and Wayne S. F. 1989. A survey of the development of subsurface zones in the wear of materials. *Key Engineering Materials* 33:77–100.

Suh, N. P. 1986. *Tribophysics.* Englewood Cliffs, NJ: Prentice Hall, Inc.

Suh, N. P., and Sridharan, P. 1975. Relationship between the coefficient of friction and the wear rate of metals. *Wear* 34 (3): 291–299.

6

Friction-Induced Instabilities and Vibrations

In the preceding chapters, we reviewed thermodynamic principles that are relevant to friction and wear, and empirical data on these phenomena. Here we will concentrate on the analysis of frictional sliding as provided by contact mechanics, with the emphasis on friction-induced instabilities, vibrations, and waves that can lead to self-organized structures and friction reduction. We formulate a general stability criterion and apply it to various situations.

6.1 Mechanics of Elastic Contact and Stability of Frictional Sliding

Many stability problems related to friction involve the sliding contact of two linear pure elastic bodies. A linear pure elastic material is an idealized model of real materials, which is very useful for many applications. In this section, we discuss the basics of the contact mechanics of such materials.

6.1.1 Static Problems of Contact Mechanics

Contact mechanics is the branch of continuum mechanics that studies deformation in solids that are in contact with each other (i.e., are touching each other at one or more points involving elastic, plastic, and viscous deformation). The problems of contact mechanics can be broadly divided into static and dynamic problems. The classical and best known problem of contact mechanics is the problem of contact of two curved elastic surfaces. An isotropic linear elastic material is characterized by two elastic parameters, the Young modulus E and the Poisson's ratio v. The contact problem for the deformable elastic sphere in contact with a flat rigid surface was solved in 1882 by Heinrich Hertz, who found that, for the circular contact area, the area is proportional to the power of two thirds of the normal force, W, pressing the bodies together

$$A = \pi \left(\frac{3WR}{4E} \right)^{2/3} \tag{6.1}$$

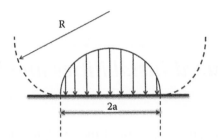

FIGURE 6.1
Schematic illustration of Hertzian contact: a sphere of radius R in contact with a flat surface.

where E is the elastic modulus of the sphere of radius R in contact with a flat surface (Figure 6.1). The normal stress (contact pressure) distribution as a function of the radius r inside the contact zone is given by

$$p(r) = p_0\sqrt{1 - \frac{r^2}{a^2}} \tag{6.2}$$

where

$$a = \left(\frac{3WR}{4E}\right)^{1/3} \tag{6.3}$$

a is the radius of contact (so that $A = \pi a^2$) with the assumption that $a \ll R$ and

$$p_0 = \frac{1}{\pi}\left(\frac{6WE^2}{R^2}\right)^{1/3} \tag{6.4}$$

is the maximum normal pressure (at the center of the contact zone). The result can be easily generalized for the case of an elliptic contact zone, two surfaces with elastic modules of E_1 and E_2, Poisson's ratios of v_1 and v_2, contact radii of R_1 and R_2, by substituting the effective properties into Equations (6.1)–(6.4):

$$\frac{1}{E} = \frac{1 - v_1^2}{E_1} + \frac{1 - v_1^2}{E_1} \tag{6.5}$$

$$\frac{1}{R} = \frac{1}{R_1} + \frac{1}{R_2} \tag{6.6}$$

Two important consequences of Hertz's result are that the contact area is proportional to the power two thirds of the normal force (rather than, say, a linear proportionality) and that the stress distribution given by Equation (6.2) is elliptical so that the derivative of $p(r)$ is discontinuous at the border of the contact zone $r = a$. The first consequence indicates that the linear Coulomb's law cannot be deduced directly from the Hertzian contact by simply assuming that the asperities have circular tops. The second consequences will result in the singularity of contact problem formulation, which will be discussed later.

Hertz's result was generalized for the case of the contact of two cylinders with parallel axes

$$a = \left(\frac{4WR}{\pi E} \right)^{1/2} \tag{6.7}$$

$$p_0 = \left(\frac{WE}{\pi R} \right)^{1/2} \tag{6.8}$$

for the conical indenter of angle α:

$$W = \frac{\pi}{2} a^2 E \cot \alpha \tag{6.9}$$

for the cylindrical indenter of radius a:

$$W = 2aEd = 2a^2 p\pi \tag{6.10}$$

$$p(r) = p_0 \sqrt{1 - \frac{r^2}{a^2}} \tag{6.11}$$

$$p_0 = \frac{Ed}{\pi a} \tag{6.12}$$

where d is the indentation depth.

Finally, for the point load with the normal and tangential components of W and F, the x- and y-components of the displacement (u and v) at the surface are given by

$$
\begin{pmatrix} u \\ v \end{pmatrix} = \begin{pmatrix} \dfrac{\kappa+1}{4\pi G}\ln|x| & \dfrac{\kappa-1}{8G}\dfrac{x}{|x|} \\[2ex] \dfrac{\kappa+1}{4\pi G}\ln|x| & \dfrac{\kappa-1}{8G}\dfrac{x}{|x|} \end{pmatrix} \begin{pmatrix} F \\ W \end{pmatrix} \qquad (6.13)
$$

where $\kappa = 3-4\nu$ for plain strain 2D elasticity and $\kappa = (3-\nu)/(1+\nu)$ for plain stress 2D elasticity.

Many contact problems are formulated in the form of an integral equation, which can be obtained using Green's function method and, for a 2D problem, has the form

$$
\delta_n(x) = \int_A [K_{nn}(x,\xi)\tau_{yy}(\xi) + K_{nt}(x,\xi)\tau_{xy}(\xi)]d\xi \qquad (6.14)
$$

$$
\delta_t(x) = \int_A [K_{tn}(x,\xi)\tau_{yy}(\xi) + K_{tt}(x,\xi)\tau_{xy}(\xi)]d\xi \qquad (6.15)
$$

where τ_{yy} and τ_{xy} are the normal and tangential stresses, δ_n and δ_t are the normal and tangential displacements, K are the "coefficients of influence" which relate the stress at a point ξ with the displacement at the point x, and the integration is performed through the entire contact area A.

Typically, the stresses are unknown, whereas the displacements are specified by the boundary conditions. The coefficients of influence can often be determined from the expression for the concentrated force, such as Equation (6.13), although in many cases (e.g., for a problem with periodic boundary conditions), various specific techniques can be used to obtain the integral equation form. Note also that the contact area is not necessarily constant and can be a function of the unknown stresses, which makes the contact problem non-linear in the sense that the sum of deformations of two stress fields is not equal to the deformation caused by the sum of these two stress fields because the relevant area of contact is different.

Other static problems of contact mechanics include problems with tangential loading, plastic yield, and adhesion. Tangential loading with friction may result in the formation of stick and slip zones. For the plane strain contact of two cylinders, it was shown independently by Cattaneo (1938) and Mindlin (1949) that there is a central stick region surrounded by two slip zones. As the tangential force increases, the size of the stick region decreases until overall sliding of the asperity begins. This sliding occurs with Coulomb's law of sliding friction satisfied by ($F = \mu W$), where μ is the coefficient of friction.

For normal loading of an elastic–plastic material, the Hertz solutions remain valid until the applied load is sufficiently large so as to initiate plastic deformation (Hill 1950). The Tresca or Van Mises maximum shear stress criterion can be applied. For the Hertz contact of two spheres, the maximum shear stress for $v = 0.3$ occurs at a depth of 0.48 a (and not at the surface) and has a value of 0.31 p_0. Yielding will initiate in the material with the lower yield strength (Adams and Nosonovsky 2000). Note also that the yield stress is known to depend on the size of contact. Plastic yield occurs due to the presence of dislocations and defects in the materials. It was argued that in small volumes of a material, there may not be a sufficient number of dislocations and thus the hardness and yield stress are smaller than for bulk material. Conventional plasticity theories lack a length scale, so they are incapable of predicting this effect. Recently, several strain-gradient theories of plasticity have been developed that provide the needed length scale (Nosonovsky and Bhushan 2004; Nosonovsky and Esche 2008; Nix and Gao 1998; Hutchinson 2000).

At a scale of many nanometers, the solid bodies can still be treated as a continuum, but the effects of surface forces and adhesion in the immediate vicinity of the contact region can become important. Subsequently, two different models were proposed for the contact of elastic spheres by Johnson, Kendall, and Roberts (JKR; 1971) and Derjaguin, Muller, and Toporov (DMT; 1975). Both theories predict a force needed to separate a sphere from a surface: the so-called pull-off force. The JKR theory assumes that the adhesive forces are confined to inside the contact area and thus give a pull-off force of 1.5 πwR, where w is the work of adhesion. The DMT model assumes that the adhesive forces act outside the contact area and yield the pull-off force of 2 πwR. Inelastic deformation can lead to adhesion hysteresis. For inelastic unloading, the energy released must overcome dissipation as well as the work of adhesion; consequently, additional work is needed to separate the deformed surfaces (Adams and Nosonovsky 2000).

6.1.2 Elastodynamic Problems

Dynamic problems of elasticity are significantly more diverse than static problems. The inertial term $\rho(\partial^2 \mathbf{u}/\partial t^2)$ should be included in the equations of elasticity in order to incorporate the dynamic effects. The plane-strain equations of motion for a linear isotropic elastic solid (the 2D Navier–Cauchy equations of the plain-strain linear elasticity) are

$$(\lambda + G)\nabla(\nabla \cdot \mathbf{u}) + G\mathbf{u}^2 = \rho \frac{\partial^2 \mathbf{u}}{\partial t^2} \tag{6.16}$$

where
$\lambda = vE/[(1 + v)(1 + 2v)]$ is the elastic Lame parameter
$G = E/(2 - 2v)$ is the elastic shear modulus
ρ is the density of the material
$\mathbf{u}$ is the vector of displacement

The component notation of this equation is

$$(\lambda+G)\hat{u}_{j,jk}+G\hat{u}_{k,jj} = \rho\frac{\partial\hat{u}_k}{\partial t} \qquad (k=1,2) \tag{6.17}$$

The well-known solution of Equation (6.17) constitutes propagating elastic waves in the materials. Two types of elastic waves are possible in the body of an infinite elastic space: the so-called P-waves, or dilatational waves, and the so-called S-waves, or shear waves (Figure 6.2a and b). For the dilatational waves, the displacement of the vibrating particles of the material occurs in the same direction as the direction of wave propagation. Thus, for a wave propagating in the x-direction in an infinite elastic medium, the displacements are given by

$$u(x,y,t) = A\cos[k(x-V_p t)]$$
$$v(x,y,t) = 0 \tag{6.18}$$

where A and k are the amplitude and the wave-number (both dependent on the initial or boundary conditions of the problem), whereas V_p is the speed of the dilatational waves, which is a material parameter related to the elastic properties as

$$V_p = \sqrt{\frac{\lambda+2G}{\rho}} \tag{6.19}$$

For the shear waves, the displacement of the vibrating particles of the material occurs in a direction perpendicular to the direction of wave propagation. Thus, for the wave propagating in the x-direction in an infinite elastic medium, the displacements are given by

$$u(x,y,t) = 0$$
$$v(x,y,t) = A\cos[k(x-V_s t)] \tag{6.20}$$

where V_s is the speed of the shear waves, which is a material parameter related to the elastic properties as

$$V_p = \sqrt{\frac{G}{\rho}} \tag{6.21}$$

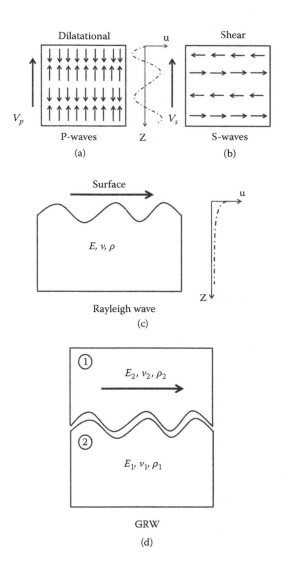

FIGURE 6.2
Elastic waves: (a) P-wave or dilatational wave (displacement in the direction of wave propagation); (b) S-wave or shear wave (displacement in the perpendicular direction to wave propagation); (c) surface or Rayleigh wave (amplitude decreases exponentially with the distance from the surface); (d) generalized Rayleigh waves (GRWs) at the interface (two coupled Rayleigh waves) for slightly dissimilar materials.

It is noted that the speed of the shear wave is almost smaller than that of a dilatational wave.

The velocities of the dilatational and shear waves given by Equations (6.19) and (6.21) can be directly deduced from Equation (6.16). For that end, one has to present the displacement vector as a sum of two components given by the gradient of a scalar potential ϕ and the rotor of a vector potential $\boldsymbol{\psi}$:

$$\mathbf{u} = \nabla\phi + \nabla \times \boldsymbol{\psi} \tag{6.22}$$

Substituting Equation (6.22) into Equation (6.16) yields two independent equations:

$$(\lambda + 2G)\Delta(\nabla\phi) = \rho\frac{\partial^2(\nabla\phi)}{\partial t^2} \tag{6.23}$$

$$G\Delta(\nabla \times \boldsymbol{\psi}) = \rho\frac{\partial^2(\nabla \times \boldsymbol{\psi})}{\partial t^2} \tag{6.24}$$

that describe the propagation of the two types of waves.

At a plane surface of an elastic half-space, a so-called surface wave or Rayleigh wave (Figure 6.2c) can propagate along the surface. The amplitude of the Rayleigh wave decreases exponentially with the distance from the surface:

$$u(x, y, t) = F(y)\cos[k(x - V_R t)]$$
$$v(x, y, t) = G(y)\sin[k(x - V_R t)] \tag{6.25}$$

where $F(y) = A\exp(-\alpha y) + B\exp(-\beta y)$ and $F(y) = C\exp(-\gamma y) + D\exp(-\delta y)$ and are certain functions decreasing exponentially with a distance from the surface. The constants A, B, C, D, α, β, γ, and δ can be determined from the boundary and initial conditions and from Equation (6.16). The speed of the Rayleigh wave V_R is smaller than that of the shear wave.

When two elastic half-spaces with identical mechanical properties are pressed together and form a flat interface between them, they have the same Rayleigh wave speed. Therefore, a pair of Rayleigh waves with identical wavelength and amplitude can propagate in each body along the interface without interfering with each other. For two bodies with slightly different mechanical properties, the Rayleigh wave speeds are close. In that case, a so-called generalized Rayleigh wave (GRW; Figure 6.2d) can propagate along the interface, with the speed lying in between the Rayleigh wave speeds of the two bodies and the amplitude decreasing exponentially with the distance from the interface. For significantly different mechanical properties of the two bodies, the GRW does not exist.

Another type of wave that is worth mentioning is the Stoneley wave. These waves propagate along the interface between two elastic solids bound to each other.

Many dynamic problems of contact mechanics involve stability analysis. In general, Lyapunov's method, which has been described in the preceding chapters, can apply. The basis of this method is to construct a Lyapunov function $V(t)$ that is an arbitrary function (but in many cases it can be the energy of the system) defined in such a manner that it is positive definite and reaches zero at the equilibrium state. The Lyapunov function should decrease with time so that its derivative is negative definitive in order for the equilibrium to be stable.

Suppose that the state of the system is described by the variables $\mathbf{u}(x,y,t)$, which may denote the deformation vector. Using the variable separation method, the functions can be presented as a superposition of modes:

$$\mathbf{u}(x,y,t) = \sum_k \mathbf{U}_k(x,y)T_k(t)$$ (6.26)

where $\mathbf{U}_k(x,y)$ is the spatial mode and $T_k(t)$ is the temporal mode. The time derivative is then given by

$$\dot{\mathbf{u}}(x,y,t) = \sum_k \mathbf{U_k}(x,u)\frac{\partial T_k(t)}{\partial t}$$ (6.27)

Then the time derivative of the Lyapunov function can be presented as

$$\frac{dV}{dt} = \sum_j \frac{\partial V}{\partial u_j} \sum_k U_{kj}(x,u)\frac{dT_{kj}(t)}{dt}$$ (6.28)

The stability depends on whether all the derivatives $dT_{kj}(t)/dt$ are negative. It is clear that if one of the modes is unstable, the entire solution is unstable.

6.1.3 Elastodynamics with Friction: Paradoxes and Instabilities

When contact elastodynamics is combined with friction, a number of important effects can emerge. First is the behavior of the GRWs. For a small constant coefficient of friction at the sliding interface, the amplitude of the GRW is not constant anymore. Quite oppositely, it can grow with time in an exponential manner, making the sliding unstable. This is the so-called Adams–Martins instability that will be discussed in detail in subconsequent sections of this chapter. The growing GRWs correspond to unstable modes of vibration at the interface and signify the fact that frictional sliding of two ideally elastic infinite bodies (half-spaces) is dynamically unstable for a large range of material parameters (the Adams–Martins instabilities exist only for slightly dissimilar material combinations in terms of their elastic properties, for which the GRWs exist). The fact that such a basic problem as the frictional sliding of two elastic surfaces is dynamically unstable indicates that there are significant issues of compatibility between elasticity and the Coulomb friction.

An even more striking problem is the so-called frictional paradoxes. The simplest example is the Painlevé paradoxes, a family of paradoxes with rigid (non-deformable) bodies and Coulomb friction (Genot and Brogliato 1999; Stewart 2000; Leine, Brogliato, and Nijmeijer 2002; Anh 2003). The systems with the Painleve paradoxes have non-unique solutions or no solutions because the friction force $F = \mu W$ is directed in the same direction as the velocity of the material point. Usually, in order to solve the equations of statics, an assumption should be made about the direction of the friction force. However, after the solution is obtained, it may turn out that the initial assumption about the direction of the friction force contradicts the direction of the velocities or other forces in the system.

For example, in the system shown in Figure 6.3, there are two sliders with the masses $M_1 = M_2 = m$, connected by a link with a constant length l forming the angle of α with the sliding surface. The upper slider is frictionless while the lower slider is frictional with the coefficient of friction μ. An external force P is applied to the upper slider. The motion of such a system is governed by the equation

$$2m\ddot{x} = P - \mu|R|\,sign(\dot{x}) \qquad (6.29)$$

where R is the normal force acting at the slider M_1. From the balance of forces acting on the slider M_2 one can conclude

$$m\ddot{x} = P + \frac{R}{\tan(\varphi)} \qquad (6.30)$$

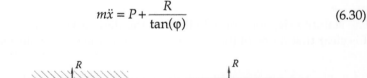

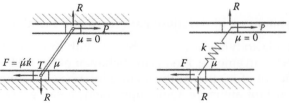

FIGURE 6.3
A system with the Painleve paradoxes (rigid link) is resolved for elastically deformable link.

Note that the force $T = R \cos \phi$ is the compression force in the link. Equations (6.29) and (6.30) form a system for the unknown $\ddot{x}$ and R. In order to solve the system, one should assume the value of $sign(\dot{x})$. However, if $\mu \tan(\varphi) > 2$, then two solutions exists for $\dot{x} > 0$:

$$m\ddot{x} = \frac{P(1 \pm \mu \tan(\varphi))}{2 \pm \mu \tan(\varphi)} \tag{6.31}$$

and no solution exists for $\dot{x} < 0$.

The Painleve paradoxes indicate that the Coulomb friction is not always compatible with the rest of the equations of mechanics. However, there is an interesting relationship between the Painlevé paradoxes and the friction-induced instabilities. If an elastically deformable link is considered instead of the rigid link, the sliding system has an additional degree of freedom (Leine et al. 2002). In that case, the paradoxes correspond to the unstable solution with the reaction force growing until the value of φ decreases so that the paradox condition $\mu \tan(\varphi) > 2$ will not be satisfied anymore (Figure 6.3b). Genot and Brogliato (1999) showed that the Painleve paradoxes are related to the so-called linear complementary problem (LCP) of the mathematical optimization theory.

The important conclusion is that the Painleve paradoxes are the result of over-simplification of the frictional system due to the reduction of the number of degrees of freedom and ignoring the elastic deformations. When these degrees of freedom and deformation are not ignored, frictional instabilities can occur, which themselves constitute a problem of compatibility of the Coulomb friction with the laws of mechanics. The instabilities lead to the limit cycles corresponding to the stick–slip motion, self-excited vibrations, and other self-organized phenomena. Therefore, the potential for self-organization is embedded in the very nature of the Coulomb friction.

Speaking of elastodynamic phenomena with friction, the so-called Schallamach waves should be mentioned as well. These are the detachment (separation) waves that were experimentally observed during the sliding of rubber against glass. Since Schallamach's discovery, there have been many attempts to explain Schallamach waves, but there exists no general relationship between Schallamach wave features and interfacial material properties.

6.2 Velocity Dependency of Coefficient of Friction and Stability Criterion

The standard and, in a sense, the model example of frictional instabilities is when the coefficient of friction decreases with increasing sliding velocity

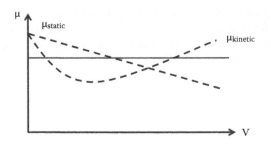

FIGURE 6.4
Different typical types of the velocity dependency of the coefficient of friction: constant, decreasing, the Stribeck curve.

(Figure 6.4). A positive feedback loop is established in that case: An increased sliding velocity (for example, due to a random fluctuation) leads to decreased frictional resistance to the motion and thus to acceleration and a further increase of the sliding velocity. Note that such a decrease, called "negative viscosity," is typical for dry friction. This is because the strength of individual asperity contacts increases with a growing age of contact due to non-elastic deformation involving plastic relaxation and creep. As a result, it is not only the static coefficient of friction that is greater than the kinetic coefficient of friction and tends to be higher for a longer age of contact. Furthermore, for higher sliding velocity, the average age of contact decreases, and therefore, the real area of contact and the coefficient of friction tend to decrease slightly. In addition, the normal degree of freedom plays a role in kinetic friction (both dry and wet) in such a manner that the sliding body tends to "elevate" (maybe for a fraction of a millimeter) with increasing velocity, so that the real area of contact decreases and the coefficient of friction decreases.

The decrease of dry friction with the increasing sliding velocity as a cause of self-excited vibrations was suggested by Lord Rayleigh as an explanation of why a violin string vibrates when excited by the bow. The bow slides along the string with an almost constant velocity, V_{bow}, that is much smaller than the amplitude of the string velocity at the point where it meets the bow, V_{string}. However, as the string vibrates, the relative velocity is larger during the half-period when the string moves in the direction opposite to the bow, $V_{string} + V_{bow}$, than when the string moves in the same direction, $V_{string} - V_{bow}$. As a consequence, the friction force between the string and the bow is larger when they move in the same direction and more energy is pumped from the bow into the string during that half-period than the amount of energy dissipated during the opposite half-period. Therefore, the self-excited vibrations of the string are maintained.

Since here we investigate the contact from a pure mechanical point of view (i.e., without considering the temperature (T) effect), the entropic stability criterion effectively becomes the energy criterion. It states that the second variation of the produced heat, $\delta^2 Q$, should be positive or, simply, the variation of the applied force δP and the sliding velocity δV should be of the same sign:

$$\delta^2 \dot{S} = \frac{1}{T} \delta^2 \dot{Q} = -\frac{1}{T} \delta PV = -\frac{1}{T} \frac{\partial P}{\partial V} (\delta V)^2 = \frac{W}{T} \frac{\partial \mu}{\partial V} (\delta V)^2 > 0 \qquad (6.32)$$

where the applied force during the stationary sliding is the negative of the friction force $P = -F = -\mu W$. In other words, since $W/T(\delta V)^2$ is positive, the stability criterion yields immediately that if the coefficient of friction decreases with increasing sliding velocity, the motion is unstable.

There are several approaches to the analysis of frictional self-excited vibrations dealing with various types of the dependency of the friction force on the sliding velocity. For example, non-linear dependency was suggested by Panovko and Gubanova (1979) and Anh, Angulo, and Ruiz-Medina (1999):

$$F = 3F_0 \left(1 - \frac{\dot{x}}{\gamma} + \frac{\dot{x}^3}{3\gamma^3} \right)$$

$$(6.33)$$

where F_0 and γ are parameters, resulting in limit cycles (attractors). On the other hand, the simple difference between the static and dynamic friction force

$$F = F_{static}, \qquad \text{for } \dot{x} = 0$$

$$F = F_{kinetic}, \qquad \text{for } |\dot{x}| > 0$$

$$(6.34)$$

can result in a stick–slip motion.

Note that the sliding velocity is coupled with the coefficient of friction in these models in the sense that the coefficient of friction depends on the velocity, while velocity (or, more exactly, the change of velocity or acceleration) depends on the coefficient of friction. Thus, the feedback loop is formed. It is interesting that any other parameter coupled with friction can play the same role. For example, when the coefficient of friction depends on the thickness of the tribofilm h formed in situ at the interface, there can exist a stationary state such that the first variation is zero:

$$\delta \dot{S} = \frac{1}{T} WV \delta \mu = \frac{WV}{T} \frac{\partial \mu}{\partial h} \delta h = 0, \quad \text{or simply} \quad \frac{\partial \mu}{\partial h} = 0 \qquad (6.35)$$

In other words, friction does not change with changing thickness of the tribofilm. The stability of the stationary state depends on the sign of the second variation

$$\delta^2 \dot{S} = \frac{1}{T} \delta^2 \dot{Q} = \frac{WV}{T} \delta^2 \mu = \frac{WV}{T} \frac{\partial^2 \mu}{\partial h^2} (\delta h)^2 > 0, \quad \text{or simply} \quad \frac{\partial^2 \mu}{\partial h^2} > 0 \quad (6.36)$$

The temperature at the sliding contact can be coupled with the coefficient of friction in such a manner that the coefficient of friction increases with increasing temperature, whereas higher friction leads to rates of heat production:

$$\delta^2 \dot{S} = \delta \frac{1}{T} \delta \dot{Q} = -\frac{1}{T^2} \delta T W V \frac{\partial \mu}{\partial T} \delta T = -\frac{WV}{T^2} \frac{\partial \mu}{\partial T} (\delta T)^2 > 0,$$

$$\text{or simply} \quad \frac{\partial \mu}{\partial T} < 0 \tag{6.37}$$

In other words, if the coefficient of friction increases with increasing temperature, sliding is unstable and understandably so, since additional friction will result in a feedback loop and further growth of heat production and temperature. These examples will be discussed in the next chapters along with another example.

To conclude this section, for frictional sliding of a material point, the most general variational entropic criterion of the stability of the stationary state ($\delta \dot{S} = 0$) should be applied, $\delta^2 \dot{S} > 0$. For many cases of constant temperature ($T = \text{const}$), the entropic criterion is simplified to the pure mechanical form, $\delta^2 \dot{Q} > 0$. For continuum systems, the stability criterion should be applied to the modes of vibration.

6.3 Thermoelastic Instabilities

A large family of frictional instabilities, which has been studied extensively, is the so-called thermoelastic instabilities (TEIs), the result of the interaction between frictional heating, hermoelastic distortion, and contact pressure (Figure 6.5). As the interface temperature grows, the near-surface volumes of the contacting bodies expand, so the contact pressure grows as well. As a result, the friction force increases, resulting in excess heat generation and the further growth of the temperature, thus creating another type of positive feedback. The TEI leads to the formation of "hot spots" or localized high-temperature regions at the interface (Barber 1969).

The formation of such localized hot spots is indicative of high local stresses that can lead to material degradation and eventual failure. The problem of the TEI is particularly prevalent in energy dissipation systems such as brakes and clutches. Hot spots can cause material damage and wear and are also a source of undesirable frictional vibrations, known in the automotive disk

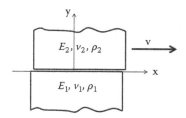

FIGURE 6.5
Two elastic half-spaces with the elastic modulus E, Poison's ratio v, and density p sliding relative to each other with the velocity V. Frictional sliding can be destabilized due to thermoelastic instability or destabilization of a generalized Rayleigh wave (Adams–Martins instability). GRW exists for slightly dissimilar materials.

brake community as "hot roughness" or "hot judder" (Anderson and Knapp 1990, Zagrodzki 1990).

Current research dealing with TEI began with the classical work of Barber (1969). Before this work, it was accepted in general that owing to the inevitable roughness of the surfaces, the contact at any particular time is restricted to a few small areas that are distributed over the nominal contact area. And considering the fact that dissipated energy in sliding is almost entirely converted into heat at or near the contact surface, one can conclude that high temperatures will be reached at those actual contact areas (Barber 1969). There were some experimental works (Bowden and Tabor 1950; Bowden and Thomas 1954) to establish values for these "flash" temperatures.

Evidence of localized heating on a large scale was first observed by Parker and Marshall (1948) during the sliding of brake blocks on railway wheels. This work was largely neglected for many years, but it became the subject of increased interest after discovery of supporting metallurgical evidence (e.g., Hundy, 1957, who observed how heated regions exceeded a temperature of 700°C and were rapidly cooled). This means that while the flash temperatures are high enough to transform the surface layer, their duration is too short to affect the bulk of the solid (Barber 1969).

6.3.1 Barber's Experiment

Barber (1969) set up an experiment to investigate the phenomenon of thermoelastic instabilities. In this experiment, a cast iron block slid against the circumference of a mild steel wheel. Normal load was applied through a lever mechanism and the sliding speed was continuously variable. It was immediately apparent that the heat input to the block was non-uniform since red hot areas could be seen on the sides of the block adjacent to the interface. These areas were transient, lasting for a few seconds. Through thermocouples that were fitted below the surface of the block at different positions, temperatures were recorded (Barber).

Barber (1969) observed that the temperature remained constant at a particular point for a relatively long period and then increased considerably for a period of a few seconds. A large number of these fluctuations were recorded. This experiment also showed that the normal load was not uniformly distributed over the nominal contact area, but rather was concentrated into a relatively small area whose location changed every few seconds. Barber concluded that any non-uniform loading over the surface causes a corresponding non-uniform generation of frictional heat and the temperature field associated with this heat input will have maxima in the regions of concentrated loading. These regions of high temperature will therefore expand more than their surroundings, causing the surface near to them to move outward. Thus, it is conceivable that, during a contact cycle, the loaded area is sustained above the predominant surface level by different thermal expansion. When the load is transferred to a new area, the heat input at the old one is discontinued and the solid will contract to its original position, leaving a depression in the surface that corresponds to the volume of material worn away during the cycle.

This led Barber (1969) to the formulation of the concept of the TEI. Let us consider a rough solid surface sliding against a smooth surface. The pressure will vary over the interface, and to a first approximation, the heat input per unit area and the linear wear rate will be proportional to the local pressure. The initial expansion will be a linear function of the heat input so that, after a short period of sliding, the surface profile will be altered at each point by an amount that depends on the local pressure. This change in the surface profile will itself change the distribution of pressure so that the series of effects forms a closed loop or feedback system. Thermal expansion tends to increase the amplitude of the surface roughness and therefore contributes positive feedback to the system. However, the wear process introduces a negative feedback term, as it removes material from the highest parts of the surface. Thus, the stability of the system depends on the relation between the rates of wear and the thermal expansion; if the expansion is greater than the wear, any initial irregularity in load distribution will be exaggerated and the load may eventually be concentrated into a relatively small area.

According to Barber (1969), two conditions must be satisfied if the thermal expansion of the solid is to have a significant effect on the sliding process: the initial expansion of a contact must exceed the wear rate and the normal expansion produced must be significant in comparison with the compliance of the surface. If the wear is always greater than the thermal expansion, the system will be stable and will tend to distribute the load over the entire nominal contact area. For small values of the expansion, λ is given by

$$\lambda = \frac{2\alpha\mu\sigma LV(1+v)t}{\rho cA} \tag{6.38}$$

where the material has density ρ, thermal capacity c, coefficient of linear thermal expansion α, and Poisson's ratio νAL and V are also the cross-sectional area, normal load, and the sliding velocity.

The linear wear is given by

$$w = WLVt/A \tag{6.39}$$

where W is the wear coefficient. Thus, the system is stable if (Barber 1969)

$$\frac{\mu\sigma}{W} > \frac{\rho c}{2\alpha(1+\nu)} \tag{6.40}$$

Paradoxically, a very low wear rate can also minimize the effect of the instability by increasing the duration of the cycle until the load is carried at one point throughout a complete test. If the wear is negligible on the time scale under consideration, the thermoelastic distortion will still alter the load distribution, but the latter will achieve a steady state.

6.3.2 Extension of Barber's Work

The research by Barber (1969) also indicated that the division of frictional heat between sliding solids should be sensitive to the interfacial boundary conditions. Thus, it should provide a guide to the nature of asperity interactions and a more accurate statement of the boundary conditions for any subsequent analysis (Berry and Barber 1984). Measurement of the division of heat in a symmetric "cylinder on cylinder" system was frustrated by the widespread occurrence of thermoelastic instability. Berry and Barber developed an alternative specimen geometry that has permitted the division of heat between sliding solids of various materials to be investigated experimentally. The characteristics of microscopic thermal resistance at the surface of a sliding solid have been investigated theoretically for several types of asperity interaction. An approximate method was employed to estimate the thermal resistance of an oxidized surface. Berry and Barber (1984) interpreted the observed division of heat with reference to the characteristic behavior associated with the various types of asperity interaction. They showed that in a mild wear regime, oxide films have an appreciable effect on microscopic thermal resistance and hence on the thermal behavior of the sliding solids, particularly the division of heat between them.

Burton, Nerlikar, and Kilaparti (1973) used a perturbation method to investigate the stability of contact between two sliding half planes. The system was linearized about the uniform pressure state and perturbations were sought that can grow exponentially with time. Their results provided useful insight into the nature of the phenomenon, but there is no inherent length scale in the problem as defined and it was found that sufficiently long wavelengths are always unstable. A length scale can be artificially introduced into the analysis by restricting attention to perturbations below a certain wavelength, estimated as being comparable with the linear dimensions of the practical system, but the resulting predictions for critical speed did not generally show good agreement with those observed experimentally (Dow and Stockwell 1977, Banerjee and Burton 1979).

Lee and Barber (1993) developed a model in which the brake disc was of finite thickness. This research was the first solution of a TEI problem involving a geometric length scale. They used Burton's method to analyze the stability of a layer sliding between two half planes. They pointed out that the metal thickness is important in determining a realistic instability threshold. The threshold of instability is often translated into the critical speed beyond which TEI is likely to occur.

According to Lee and Barber (1993), Burton's semi-infinite analysis with a plane-strain hypothesis yields a critical speed beyond that observed experimentally in automotive disk brake systems. They attributed this phenomenon in part to neglecting the finite thickness of the conducting body. Another important factor is the consideration of the antisymmetric mode predicted by Lee and Barber, giving rise to alternating hot spots. Yi, Barber, and Zagrodzki (2000) developed a finite-element model for determining the critical speed for TEI of an axisymmetric clutch or brake and compared it with the analytical models.

This geometry provided a first step toward that of the typical disk brake assembly, where a disk slides between two pads of a friction material. Using typical material properties from automotive applications, it was found that stability is governed by a deformation mode that is anti-symmetric with respect to the mid-plane of the layer and that has a wavelength proportional to the layer thickness.

Despite the considerable idealizations involved in the theory proposed by Lee and Barber (1993), it provides plausible predictions for the critical speed and the mode shape in typical brake assemblies and is therefore quite widely used in the brake and clutch industry for TEI analysis. However, there is a clear need for a method that will account for other features of the system geometry, such as the finite width of the sliding surface, the axi-symmetric geometry of the disk, and the "hat" section used to attach the disk to its support. One approach is to use the finite element method to solve the coupled transient thermoelastic contact problem in time (Zagrodzki 1990; Johannson 1993; Zagrodzki et al. 1999). This method is extremely flexible in that it can accommodate non-linear or temperature-dependent constitutive behavior, more realistic friction laws, and practical loading cycles. However, it is also extremely computer intensive and appears unlikely to be a practical design tool for three-dimensional (3D) problems in the foreseeable future.

A promising alternative approach was to implement Burton's perturbation method numerically, leading to an eigenvalue problem to determine the stability boundary. If the exponential growth rate of the dominant perturbation can be assumed to be real, the critical sliding speed is defined by the condition that there exists a steady-state equilibrium perturbation—that is, one with zero growth rate. Du et al. (1997) used the finite element method to develop the matrix defining this eigenvalue problem. Yi et al. (2000) used Du's method to explore the effect of disk geometry in an idealized disk brake in which the brake pads are assumed to be rigid and non-conducting. Their results showed that the critical speed is in many cases quite close to that predicted by the considerably simpler analysis of Lee and Barber (1993), which probably explains the success of that analysis in practical applications.

Du's method rests on the assumption that the dominant perturbation has a real growth rate. The limited range of problems that have been solved analytically suggest that this assumption is justified if one of the two sliding bodies is a thermal insulator, or if the dominant perturbation is independent of the coordinate in the sliding direction, as in "banding" instabilities in axi-symmetric systems, where the frictional heating is concentrated in an axi-symmetric annular band within the contact area. However, a rigorous proof of this result has never been advanced. When both materials are thermally conducting, the stability boundary is generally determined by a disturbance that migrates with respect to both bodies in, or opposed to, the direction of sliding (Burton et al. 1973). In a stationary frame of reference, the perturbation will then appear to oscillate in time, corresponding to a complex exponential growth rate. The migration speed is smaller relative to the better thermal conductor, and this relative motion approaches zero when the other body tends to the limit of thermal insulation. Practical systems such as brakes and clutches usually involve a steel or cast iron disk sliding against a composite friction material of significantly lower conductivity (typically two orders of magnitude lower than that of steel). As a result, the dominant perturbation migrates only very slowly relative to the metal disk. However, this migration plays an important part in the process, because it reduces the thermal expansion due to a given perturbation in heat input and hence increases the critical speed.

Burton's method can be implemented numerically for systems of two thermal conductors by defining the eigenvalue problem for the exponential growth rate. This method was first suggested by Yeo and Barber (1996), who developed it in the context of the related static thermoelastic contact problem, where instability results from the pressure dependence of an interfacial contact resistance.

Yeo and Barber (1996) first assumed that the temperature, stress, and displacement fields could be written as the product of a function of the spatial coordinates and an exponential function of time when these expressions were substituted into the governing equations and boundary conditions of the problem, the exponential factor canceled, and a modified system of equations in which the growth rate appears as a linear parameter was retained. A finite element discretization of this modified problem then yielded a linear eigenvalue problem for the growth rate.

In order to adapt this method to the sliding contact problem, Yeo and Barber (1996) needed to choose a suitable frame of reference relative to which at least one of the bodies would necessarily be moving. This introduced convective terms into the heat conduction equation and could present numerical problems when the convective term was large (Christie et al. 1976). The relative magnitude of convective and diffusive terms could be assessed by calculating the Peclet number $Pe = Va/k$, where V is the convective velocity, k is the thermal diffusivity, and a is a representative length scale. Peclet numbers in tribological applications are typically very large. For example, a steel clutch disk of mean diameter 0.2 m rotating at 2000 rpm corresponds to a Peclet number of about 35,000, using the mean diameter for a, and even the element Peclet number will be large compared with unity for a realistic discretization. Thus, the convective terms will tend to dominate the finite element solution.

Fortunately, difficulties with convective terms can be avoided by using Fourier reduction in the sliding direction as long as (1) no material points on either sliding body experience intermittent contact, and (2) periodic boundary conditions apply in the sliding direction. These conditions are satisfied for multi-disk brakes and clutches, which have an axi-symmetric geometry, but which often exhibit signs of damage attributable to TEI with a non-axi-symmetric eigenmode, as shown in preceding figures. In this case, orthogonality arguments show that all the eigenmodes must have Fourier form in the circumferential direction and the sinusoidal function can be factored out of the equations, leading to an eigenvalue problem in radial and axial coordinates only, for given values of wave number and rotational speed. A finite element description of this problem leads to a linear eigenvalue problem for the growth rate.

6.3.3 TEI in a Generalized System with Liquid Lubricant

Extensive work of Barber (1969) with application to braking systems is reported by several researchers, such as Berry and Barber (1984), Barber et al. (1985), Lee and Barber (1993), and Du et al. (1997). Applications of TEI to mechanical seals include some of the pioneering analytical work by Burton et al. (1973), Kilaparti and Burton (1977), Banerjee and Burton (1976), and Lebeck (1980). Burton (1980) gives an excellent overview of the subject, with a clear physical interpretation of the results. These papers are restricted to application of nominally flat and smooth surfaces in dry contact. The implications of these limitations are discussed by Jang and Khonsari (1999).

Jang and Khonsari (2003) developed a model for treating TEI in a conductor–insulator system in the presence of a liquid lubricant. Their model consisted of a body of finite thickness with high thermal conductivity (referred to as the conductor) and a rough frictional surface made of a material with very low thermal conductivity (i.e., an insulator). The gap between the two bodies is filled with a lubricant of known viscosity forming a nominal film thickness between the two bodies. One surface is stationary and the other undergoes a sliding motion at a constant speed. The system is operating under a steady-state condition. To determine the susceptibility of this system to TEI, a small surface wave, representing a disturbance, is imposed. The imposed surface wave, even though it is very small, will affect the velocity profile in the fluid, the hydrodynamic pressure distribution, and the heat balance within the film. As a result, the entire deformation field—both thermal and mechanical—will be affected.

Jang and Khonsari (2003) studied whether at a given operating speed the combination of these deformations grow unbounded, thus resulting in a failure, or die out, implying that the system is thermoelastically stable. Both the symmetric and anti-symmetric modes were considered in their analysis. A general set of governing equations was derived and solved for the critical speed. They considered surface-roughness characteristics and fluid–solid interaction with a conducting body of finite thickness. They showed that the moving-wave solution provides a complete description of TEI.

However, a much simpler solution is obtained by using the stationary-wave assumption. It is shown that the stationary-wave solution provides a lower limit on the critical speed, beyond which TEI is predicted to occur. As such, the stationary-wave results for the insulator-conductor system may be viewed as conservative and hence useful for design purposes. Jang and Khonsari's theory showed that the critical speed is governed by five independent dimensionless parameters and that their interaction determines the susceptibility of a system to TEI. Of particular interest are the thickness of the conducting member and its surface roughness, the influence of the lubricant viscosity and film thickness, and the physical properties of the contacting bodies. Extensive parametric simulations reveal that surface roughness tends to initiate TEI at a much lower operating speed than that of a "smooth" surface. However, the roughness effect vanishes beyond $h_0 = 3\sigma$, where h_0 is forming a nominal film thickness and σ is roughness root mean square (RMS) parameter.

Jang and Khonsari (2003) also showed that the number of surface disturbances is directly related to the number of hot spots. Comparison with independent experimental results revealed a satisfactory prediction of the number of hot spots. For the symmetric mode, the maximum number of hot spots in an infinitely thick conductor is found to be smaller than that of a conductor of finite thickness. Based on the stationary-wave solution, the critical speed for the anti-symmetric mode is higher than that for the symmetric mode due to the rigidity of the insulator. By this approach, the number of hot spots can also be estimated.

6.3.4 Sliding Thermoelastodynamic Instability

Several distinct categories of instability in sliding of elastic bodies are known to result from the interaction between relatively simple physical processes—notably, the elastic deformation of the contacting bodies, the development of frictional forces at the interface opposing the motion, and the consequent generation of frictional heat. It has long been known that frictional vibrations can result if the coefficient of friction is a decreasing function of speed or, in the case of stick–slip vibrations, if the static coefficient exceeds the dynamic coefficient. However, Martins, Guimaraes, and Faria (1995) and Adams (1995) have shown that the steady sliding of two elastic half-planes can be unstable, even assuming the elementary Coulomb friction law in which the frictional traction is proportional only to the local normal contact pressure.

Both frictional instabilities and TEI can be analyzed by using linear perturbation methods, leading to a characteristic equation for the exponential growth rate of an initial perturbation. In the discrete (e.g., finite element) formulation, this takes the form of a linear eigenvalue problem for the growth rate (Yi et al. 2000; Yi and Barber 2001). However, there has been little attention to the possible interaction between the two mechanisms of instability. Thermoelastic deformations are neglected in the analysis of frictional instabilities and a quasi-static approximation is used in the analysis of TEI. Some justification for this "decoupling" of the two phenomena is provided by the widely divergent timescales involved. Thermoelastic instabilities occur on the rather slow timescale of thermal diffusion, whereas elastodynamic processes are governed by the elastic wave speeds in the materials (Afferrante, Ciavarella, and Barber 2006).

In caliper disk brakes, the typical unstable TEI mode involves a set of equally spaced hot spots around the disk and, as these pass through the brake pads, they can cause mechanical vibration known as "hot judder." However, this is analyzed as a "one-sided" interaction. The TEI problem is assumed to be quasi-static, which then merely serves to define the excitation for a dynamic analysis. It would clearly be much more satisfactory to develop an analysis of the coupled problem including both elastodynamic and thermoelastic effects in the same perturbation analysis. If the conventional uncoupled wisdom is justified, we should then be able to classify the resulting eigenfunctions into TEI modes and elastodynamic modes.

Afferrante et al. (2006) initiated this investigation by considering a simple possible system comprising a one-dimensional elastic layer bonded to a rigid half-space and sliding against a second rigid half-space. They found that although the modes can in fact be classified in the described way, the extremely weak coupling between the two mechanisms destabilized the elastodynamic natural vibration modes, causing the system to be unstable at arbitrarily low speeds.

Afferrante et al. (2006) found a simple mechanism for the occurrence of frictional vibrations in which thermo-mechanical coupling destabilizes the lowest mode of natural vibration. The transient behavior is characterized by a flutter instability at a frequency close to the first natural frequency of the elastodynamic system and it leads ultimately to a limit cycle with alternating periods of contact and separation also at this frequency. The instability occurs regardless of the wide difference in timescales of the thermal and elastodynamic processes as characterized by the smallness of the coupling parameter γ. In fact, the system is predicted to be unstable for arbitrarily small values of γ. The mechanism leads to vibrations normal to the sliding interface and provides a plausible explanation for observations of such vibrations.

6.4 Adams–Martins Instabilities

The stability of sliding two pure elastic half-spaces with a constant coefficient of friction between them is a relatively simple problem. However, it was not thoroughly studied until the 1990s when Adams (1995) and Martins et al. (1995) discovered that, for a broad range of material parameters, the motion is dynamically unstable. In particular, the instability occurs when the elastic properties of two materials are slightly dissimilar, as opposed to the case of significantly dissimilar materials that does not lead to instability. The instability is in the form of interface waves (generalized Rayleigh waves) with the amplitude growing exponentially with time. Furthermore, the solution is *ill posed* in the sense that the rate of instability becomes infinite in the limit of the short-wavelength interface waves. Dynamic instability was found also in the case of a simplified model (a beam on an elastic foundation instead of the elastic half-space) and in the more complex case of a wavy (rather than smooth) surface with a periodic set of contact and separation zones.

The Adams–Martins instability indicates that the Coulomb friction model has significant problems when combined with linear elasticity. One approach to regularize the frictional sliding of elastic half-spaces is to consider a dynamic state-and-rate friction law with the coefficient of friction dependent on both the sliding velocity and the age of contact (Rice, Lapusta, and Ranjith 2001). It is noted also that the highest rates of instabilities occur at the low wavelength limit, where the continuum elasticity breaks due to the wavelength approaching the molecular scale length, so the regularization can also occur naturally due to the discrete molecular structure of matter in the limit of small length.

6.4.1 Stability Analysis

The influence of elastodynamic phenomena on friction has been intensely investigated in the past several decades. These studies concentrated on the possibility of relative sliding of two elastic bodies with propagating slip and separation zones and, later, on the study of the stability of friction sliding with a reduced coefficient of friction. The possibility of two elastic bodies sliding relative to each other without slipping, due to a separation pulse, has been investigated by Comninou and Dundurs (1978) for identical materials and by Adams (1999) for different materials. In these studies, the coefficient of friction, µ, was assumed to be speed independent; therefore, no distinction is made between static and kinetic friction. The two semi-infinite isotropic elastic bodies, of different material properties, satisfied Coulomb's friction inequality at their common interface, $F < \mu W$, where W and F were the normal and shear forces applied to the half-spaces (Figure 6.5). Therefore, applied normal and shear stresses are insufficient to produce global slipping. Although Coulomb's inequality is satisfied at the interface, the separation or slip can occur locally. Therefore, the force necessary to produce relative motion is less than μW as would be expected from Coulomb's law.

Comninou and Dundurs (1977) suggested that separation zones can form at the interface. However, these separation zones are similar to penny-shaped cracks and involve square-root singularities that would require energy sources and sinks (Freund 1978). This makes the existence of traveling separation zones questionable, if not completely unrealistic. However, the slip zones (Comninou and Dundurs 1977) do not involve any singularities and thus can exist.

Further studies were concentrated on the stability of sliding without any separation. Renardy (1992) investigated frictional sliding of an elastic solid against a rigid substrate and found that such sliding can be unstable. Martins et al. (1995) investigated the sliding of elastic and viscoelastic half-spaces against a rigid surface. Dynamic instabilities were found for cases in which the friction coefficient and the Poisson's ratio were large. These instabilities were thought to play a role in Schallamach waves (Schallamach 1971). Adams (1995) showed that the steady sliding of two elastic half-spaces is also dynamically unstable, even at low sliding speeds. The instability mechanism is essentially one of slip-wave destabilization.

Steady-state sliding is shown to give rise to a dynamic instability in the form of self-excited motion. These self-excited oscillations are generally confined to a region near the sliding interface and can eventually lead either to partial loss of contact or to propagating regions of stick–slip motion. These waves can contribute to the formation of friction-induced vibrations. The rate of amplitude increase is proportional to the wave number, which causes infinitesimal wavelengths to increase at an unlimited rate. Thus, the propagation of an arbitrary pulse is *ill posed,* as was found by Renardy (1992) for sliding of an elastic solid against a rigid substrate with a sufficiently high friction coefficient. Simoes and Martins (1998) investigated the effect of introducing an intrinsic length scale into the problem and found a method of regularization of the ill posedness by using a non-local friction law.

The existence of these instabilities does not depend upon a friction coefficient that decreases with increasing speed, nor does it require a non-linear contact model, as with Martins and Oden (1990). These analytical results are consistent with the numerical simulations of Andrews and Ben-Zion (1997) for tectonic plate friction during earthquakes.

In a different investigation, Adams (1996) used a simple beam-on-elastic-foundation model in order to investigate instabilities caused by sliding of a rough surface on a smooth surface. The mechanism of instability in that investigation is due to the interaction of a complex mode of vibration with the sliding friction force. Adams (1998) then investigated the sliding of two dissimilar elastic bodies due to periodic regions of slip and stick propagating along the interface. It was found that such motion, which results from a self-excited instability, allows for the interface sliding conditions to differ from the observed sliding conditions. In particular, an interface coefficient of friction (the ratio of interface shear stress to normal stress) and an apparent coefficient of friction (ratio of remote shear to normal stress) were defined.

The interface friction coefficient can be constant or an increasing/decreasing function of slip velocity. However, the apparent coefficient of friction is less than the interface friction coefficient. Furthermore, the apparent coefficient of friction can decrease with sliding speed even though the interface friction coefficient is constant. Thus, the measured coefficient of friction does not necessarily represent the behavior of the sliding interface. Also, the presence of slip waves may make it possible for two frictional bodies to slide without a resisting shear stress and without any interface separation. In the limit as the slip region becomes very small compared with the stick region, the results of Adams (1998) become that of a slip pulse traveling through a region that otherwise sticks. Nosonovsky and Adams (2000, 2004) showed that dynamic instabilities can occur also for the contact of an elastic periodic rough versus flat elastic surface.

Ranjith and Rice (2001) extended the work of Adams (1995) and analyzed frictional sliding of two elastic half-spaces in the small velocity limit. If a generalized Rayleigh wave exists at the interface of the two bodies for frictionless contact, steady sliding with arbitrary small friction becomes unstable. Generalized Rayleigh waves, also known as slip waves or smooth contact Stoneley waves, exist when the material mismatch is not very high and they become equivalent to Rayleigh surface waves in the limiting case of identical materials (Achenbach and Epstein 1967). Ranjith and Rice (2001) showed that if a memory-dependent rate-and-state friction law is considered instead of the instantaneous Coulomb's law, the pulse-propagation problem becomes well posed, although the solution is still unstable.

In the following decade, a number of studies on the origin of slip pulse have been conducted with molecular dynamics (MD) and other simulation techniques (Rottler and Robbins 2005; Ben-David, Rubinstein, and Fineberg 2010; Li et al. 2011), especially in application to the study of earthquakes.

Adams (1995) investigated sliding of two perfectly flat elastic half-spaces sliding with respect to each other (Figure 6.5) with velocity $\hat{V}_0$ and coefficient of sliding friction μ. A summary of his analysis follows. The plane-strain equations of motion for a linear isotropic elastic solid are

$$(\lambda + G)\frac{\partial^2 \hat{u}_j}{\partial \hat{x}_j \partial \hat{x}_k} + G\frac{\partial^2 \hat{u}_k}{\partial \hat{x}_j^2} = \rho\frac{\partial^2 \hat{u}_k}{\partial \hat{t}^2}(k = 1, 2) \tag{6.41}$$

where

$\hat{u}_k = \hat{u}_k(\hat{x}_1, \hat{x}_2, \hat{t})$ are the components of displacement in the $\hat{x}_k$ direction

$\hat{t}$ is the time

G is the shear modulus

Λ is the Lame's constant

P is the density

Letters with hats on top are used for dimensional quantities. Equation (6.41) pertains to the lower body, whereas the corresponding quantities for the upper body are denoted with a prime ('). Equation (6.41) is transformed into a dimensionless coordinate system moving to the right with constant speed $\hat{V}$ ($\hat{V}$ not necessarily equal to $\hat{V}_0$, which yields, Adams 1995)

$$(\beta^2 - V^2)\frac{\partial^2 u_1}{\partial x_1^2} + \frac{\partial^2 u_1}{\partial x_2^2} + (\beta^2 - 1)\frac{\partial^2 u_2}{\partial x_2 \partial x_1} = \frac{\partial^2 u_1}{\partial t^2} - 2V\frac{\partial^2 u_1}{\partial x_1 \partial t}$$

$$\beta^2 \frac{\partial^2 u_2}{\partial x_2^2} + (1 - V^2)\frac{\partial^2 u_2}{\partial x_1^2} + (\beta^2 - 1)\frac{\partial^2 u_1}{\partial x_1 \partial x_2} = \frac{\partial^2 u_2}{\partial t^2} - 2V\frac{\partial^2 u_2}{\partial x_1 \partial t}$$

(6.42)

where

$$x_1 = (\hat{x}_1 - \hat{v}\hat{t})/l, \quad x_2 = \hat{x}_2/l, \quad \hat{t} = tc_2/l$$

$$u_k(x_1, x_2, t) = \hat{u}_k(\hat{x}_1, \hat{x}_2, \hat{t})/l \quad \text{(for } k = 1, 2)$$

$$\beta = \frac{c_1}{c_2} = \sqrt{\frac{2(1-v)}{(1-2v)}}, \quad c_1 = \sqrt{\frac{\lambda + 2G}{\rho}}, \quad c_2 = \sqrt{\frac{G}{\rho}}, \quad V = \hat{V}/c_2$$

v is Poisson's ratio, c_1 and c_2 and are the dilatational and shear wave speeds, respectively. In this equation, l is the characteristic length. The dimensionless stresses and the stress-displacement laws are defined by

$$\tau_{jk}(x_1, x_2, t) = \frac{\hat{\tau}_{jk}(\hat{x}_1, \hat{x}_2, \hat{t})}{G} = \frac{\partial u_k}{\partial x_j} + \frac{\partial u_j}{\partial x_k} + (\beta^2 - 2)\left(\frac{\partial u_1}{\partial x_1} + \frac{\partial u_2}{\partial x_2}\right)\delta_{jk}$$

(6.43)

$$(j, k = 1, 2)$$

In the next step, the solution of Equation (6.41) is decomposed into two parts (Adams 1995):

$$u_k(x_1, x_2, t) = u_k^*(x_1, x_2) + Real\{\bar{u}_k(x_1, x_2, t)\}$$

$$u_k(x_1, x_2, t) = U_k(x_2)e^{2\pi i x_1 + \Lambda t}$$

(6.44)

The nominally steady-state solution ($u_k{}^*$) is for a moving surface load distribution that is invariant in time and space. The corresponding dimensionless stresses are

$$\tau_{12}{}^* = -\mu p^*, \quad \tau_{22}{}^* = -p^* \tag{6.45}$$

The second term ($\bar{u}_k$) in Equation (6.44) is the unsteady part. Adams (1995) noted that Λ and $U_k(x_2)$ appearing in Equation (6.44) are, in general, complex numbers, and thus, the unsteady solution represents a spatially periodic propagating motion. Furthermore, if the real part of Λ does not vanish, the unsteady solution possesses a time-varying amplitude of oscillation. Moreover, the characteristic length (l) has been taken to be the dimensioned wavelength. The displacements given by Equation (6.44) are substituted into Equation (6.41) and yield (Adams 1995)

$$\frac{\partial^2 U_1}{\partial x_2^2} - Q_1 U_1 + 2\pi i(\beta^2 - 1)\frac{\partial U_2}{\partial x_2} = 0$$

$$\beta^2 \frac{\partial^2 U_2}{\partial x_2^2} - Q_2 U_2 + 2\pi i(\beta^2 - 1)\frac{\partial U_1}{\partial x_2} = 0 \tag{6.46}$$

where

$$Q_1 = 4\pi i V \Lambda - \Lambda^2 - 4\pi^2(\beta^2 - V^2)$$

$$Q_2 = 4\pi i V \Lambda - \Lambda^2 - 4\pi^2(1 - V^2)$$

Arriving at Equation (6.46), the steady-state solution has been subtracted out (Adams 1995). Then, the solution of Equation (6.46) can be written as

$$U_1(x_2) = Ae^{sx_2}, \quad U_2(x_2) = Be^{sx_2} \tag{6.47}$$

Upon substitution into Equation (6.46), this leads to the homogeneous characteristic equations

$$(s^2 + Q_1)A + 2\pi i(\beta^2 - 1)sB = 0$$

$$2\pi i(\beta^2 - 1)sA + (\beta^2 s^2 + Q_2)B = 0 \tag{6.48}$$

Non-trivial solutions of Equation (6.48) require vanishing of the determinant—that is,

$$\beta^2 s^4 + [Q_1\beta^2 + Q_2 + 4\pi^2(\beta^2 - 1)]s^2 + Q_1 Q_2 = 0 \tag{6.49}$$

which is the characteristic equation and is quadratic in s^2. The four roots of Equation (6.49) are denoted with $\pm s_1$ and $\pm s_2$. The condition that stresses vanishes at infinity leading to proper selection of the signs of s_1 and s_2—that is (Adams 1995),

$$Real\{s_1\} < 0, \quad Real\{s_2\} < 0 \tag{6.50}$$

for the half-space. Furthermore, Equation (6.48) gives

$$B_k = \alpha_k A_k, \quad \alpha_k = -\frac{s_k^2 Q_1}{2\pi i(\beta^2 - 1)s_k} \tag{6.51}$$

which finally results in (Adams 1995)

$$U_1(x_2) = A_1 e^{s_1 x_2} + A_2 e^{s_2 x_2}$$
$$U_2(x_2) = \alpha_1 A_1 e^{s_1 x_2} + \alpha_2 A_2 e^{s_2 x_2} \tag{6.52}$$

For sliding friction, the shear stress is proportional to the normal stress—that is (Adams 1995),

$$\bar{\tau}_{12}(x_1, 0, t) = \mu \bar{\tau}_{22}(x_1, 0, t) \tag{6.53}$$

From Equations (6.43), (6.44), (6.52), and (6.53),

$$A_2 = \theta A_1$$
$$\theta = -\frac{s_1 + 2\pi i\alpha_1 - \mu[2\pi i(\beta^2 - 2) + \beta^2 s_1\alpha_1]}{s_2 + 2\pi i\alpha_2 - \mu[2\pi i(\beta^2 - 2) + \beta^2 s_2\alpha_2]} \tag{6.54}$$

A similar procedure was used by Adams (1995) for the upper half-space in which involving quantities appear with the prime. Furthermore,

$$V - V'\kappa = V_0, \quad Real\{s_1'\} > 0, \quad Real\{s_2'\} > 0$$
$$\Lambda' = \Lambda/\kappa, \quad \kappa = \frac{c_2'}{c_2}, \quad \Lambda' t' = \Lambda t \tag{6.55}$$

It was noted (Adams 1995) that although u_k' is a function of (x_1, x_2, t'), the solution for u_k' was expressed as a function of (x_1, x_2, t) by using Equation (6.55). Finally, the normal displacements and normal stresses on two half-spaces are continued across their boundary—that is,

$$\overline{\tau}_{22}(x_1, 0, t) = (G'/G)\overline{\tau}_{22}'(x_1, 0, t)$$
$$\overline{u}_2(x_1 0, t) = \overline{u}_2'(x_1, 0, t) \tag{6.56}$$

which yields

$$\left(\frac{G'}{G}\right)(\alpha_1 + \theta\alpha_2)\{2\pi i[\beta'^2 - 2] + \beta'^2(s_1'\alpha_1' + s_2'\alpha_2'\theta') + 2\pi i\theta'[\beta'^2 - 2]\}$$
$$-(\alpha_1' + \theta'\alpha_2')[2\pi i[\beta^2 - 2 + \beta^2(s_1\alpha_1 + s_2\alpha_2\theta) + 2\pi i\theta[\beta^2 - 2]] \tag{6.57}$$

and

$$A_1' = A_1(\alpha_1 + a_2\theta)/(\alpha_1' + \alpha_2'\theta') \tag{6.58}$$

For a given sliding velocity, a coefficient of friction, and material properties, all of the terms in Equation (6.57) can be determined for a chosen value of Λ (Adams 1995). Thus, Equation (6.57) represents one complex non-linear equation with one complex unknown (Λ) and can be solved by standard methods.

With Λ known, the dimensionless contact pressure (p) and the dimensionless slip velocity (V_S) at the sliding interface can be found according to (Adams 1995)

$$p = p^* - Ce^{2\pi i x_1 + \Lambda t}$$
$$C = \{2\pi i[\beta^2 - 2](1 + \theta) + \beta^2(s_1\alpha_1 + s_2\alpha_2\theta)\}A_1 \tag{6.59}$$

$$V_S = V_0 + (D' - D)e^{2\pi i x_1 + \Lambda t}$$
$$D' = \Lambda(1 + \theta')A_1', \quad D = \Lambda(1 + \theta)A_1 \tag{6.60}$$

Finally, the solutions obtained as described earlier are subject to the inequality constraints that the contact pressure and the slip velocity both be non-negative—that is,

$$p \geq 0, \quad V_S \geq 0, \quad -\infty < x_1 < \infty \tag{6.61}$$

Obtaining the results using the method described previously, Adams (1995) analyzed stability for different conditions. The damping ratio ζ was defined by

$$\zeta = -\Lambda^R / \left| \Lambda^I \right|, \quad \Lambda^R = Real\{\Lambda\}, \quad \Lambda^I = Imag\{\Lambda\} \tag{6.62}$$

analogously to a lightly damped one-degree-of-freedom system. Thus, positive damping ($\zeta > 0$) corresponded to stable motion ($\Lambda^R < 0$), whereas negative damping ($\zeta < 0$) represented unstable (self-excited) motion ($\Lambda^R > 0$). The damping ratio, rather than the real part of the eigenvalue, was taken (Adams) as the measure of the severity of the instability for the same reason that it was used for positively damped systems.

Adams (1995) considered the case in which identical materials slide at very low velocities ($V_0 \approx 0$). This configuration allows for two Rayleigh waves of equal amplitude and equal wavelength but with a 180° phase difference to propagate in the same direction without interaction. Thus, $\zeta = 0$ for all values of the friction coefficient. For very low sliding velocities and different material combinations, the results for $\mu = 0$ have Stoneley waves for smooth contact (Achenbach and Epstein 1967). The results obtained by Adams (1995) showed that such slip waves existed for cases in which the material mismatch was small enough. As μ was increased from zero, the value of $-\zeta$ was also seen to increase from zero.

He observed the same trend for the material combinations that corresponded to larger mismatches in materials. For these cases, smooth contact Stoneley waves did not exist for $\mu = 0$, but analogous waves appeared at finite values of the friction coefficient. For all of the cases in which $V_0 = 0$, if a given value of Λ was a solution, then $-\Lambda$ was also a solution. Since the imaginary parts of these two solutions had opposite signs, the $-\Lambda$ solution moved in the opposite direction as the Λ solution. Also, these motions had the reverse stability properties since their real parts were of different signs. Adams (1995) noted that the solutions were all for negative values of Λ^I that corresponded to waves traveling to the right. Thus, only waves traveling in the same direction as sliding were destabilized. It was also observed that the material combinations with the lower κ were more susceptible to dynamic instabilities than the higher κ cases.

Adams (1995) also discussed the physical mechanism that was responsible for the self-excited motion described previously. The energy dissipated (due to friction) per unit surface area over one period of oscillation was given by

$$\varepsilon = f \, Real\{p\} \, Real\{V_s\} dt \tag{6.63}$$

He considered the difference between the phases of the variable parts of V_s and of p (Equations 6.59 and 6.60). The results showed that, for small friction, this phase difference was approximately 0° and 180° for left- and right-traveling waves, respectively. Thus, for small friction, ε was somewhat less than $\mu p^* V_0$ for a right-traveling wave, and hence, the energy dissipated at the interface was less than the work done in moving the bodies. This extra energy was the contributor to the self-excited motion that continues to grow with time. For a left-traveling wave $\varepsilon > \mu p^* V_0$, the energy dissipated at the interface was greater than the work done in moving the bodies, forcing the left-traveling wave to decay. For larger values of friction, the same general trends were true (Adams 1995); ε vanished for zero friction.

For cases in which dynamic instability existed, the amplitude of the motion continued to increase until either of the equations in Equation (6.61) ceased to hold. Regions of loss of contact were initiated when the contact pressure became zero, whereas areas of stick–slip commenced when the slip velocity vanished. From Equations (6.59)–(6.61), the critical value of v_0/p^* is defined as (Adams 1995)

$$\left(\frac{v_0}{p^*}\right)_C = \frac{|D - D'|}{|C|} \tag{6.64}$$

and

$$\left(\frac{v_0}{p^*}\right) > \left(\frac{v_0}{p^*}\right)_C = \text{loss-of-contact occurs first}$$

$$\left(\frac{v_0}{p^*}\right) < \left(\frac{v_0}{p^*}\right)_C = \text{stick–slip occurs first}$$

$$\tag{6.65}$$

Adams (1995) concluded that self-excited oscillations exist for a wide range of material combinations, friction coefficients, and sliding speeds (including very low speeds). The magnitude of these dynamic instabilities tends to increase with the extent of mismatch in the shear wave speeds of the two materials. These self-excited vibrations are usually restricted to the region close to the sliding interface. Their existence is due to destabilization of interface waves that travel in the same direction as sliding. These dynamic instabilities can eventually give rise either to partial separation or to regions of stick–slip. The greater the mismatch in shear wave speeds is, the less likely is stick–slip as opposed to loss of contact. Furthermore, for a given material pair, high contact pressure and slow sliding speeds make stick–slip more likely to occur.

Ranjith and Rice (1999) proposed different solutions of the stability problem and showed that the problem is not just unstable, but ill posed, which constitutes a higher level of instability. The ill posedness is manifest of the unstable growth of the interfacial disturbances of all wavelengths, with the growth rate inversely proportional to the wavelength. They established the connection between the ill posedness and the existence of certain interfacial waves in friction contact, called the generalized Rayleigh wave. They also showed that, for material combinations where the generalized Rayleigh wave exists, steady sliding with Coulomb friction is ill posed for arbitrarily small values of friction. Then, they studied regularization of the problem by an experimentally motivated friction law and showed that a friction law with no instantaneous dependence on normal stress but a simple fading memory of prior history of normal stresses makes the problem well posed. We will discuss this problem more in Section 6.4.4.

In general, this unexpected result shows that Coulomb friction is incompatible with elasticity. As discussed, the mathematical formulation of quasi-static sliding of two elastic bodies (half-spaces) with a frictional interface, governed by Amontons–Coulomb's rule, is a classical contact mechanics problem. Interestingly, the stability of such sliding was not investigated until the 1990s, when Adams (1995) showed that the steady sliding of two elastic half-spaces is dynamically unstable, even allowing for sliding speeds.

The instability mechanism is essentially one of slip-wave destabilization. Steady-state sliding was shown to give rise to a dynamic instability in the form of self-excited motion. These self-excited oscillations are confined to a region near the sliding interface and can eventually to lead either partial loss of contact or to propagating regions of stick–slip motion (slip waves). The existence of these instabilities depends upon the surfaces' elastic properties; however, it does not depend upon the friction coefficient or require a non-linear contact model. The same effect was predicted theoretically by Nosonovsky and Adams (2004) for the contact of rough periodic elastic surfaces.

As discussed earlier in this chapter, two types of elastic waves can propagate in an elastic medium: shear and dilatational waves. In addition, surface elastic waves may exist, and their amplitude decreases exponentially with the distance from the surface. For two slightly dissimilar elastic materials in contact, the interface waves (Rayleigh waves) may exist at the interface zone. Their amplitude decreases exponentially with the distance from the interface.

The previously mentioned instabilities are a consequence of energy being pumped into the interface as a result of the positive work of the driving force (that balances the friction force). As a result, the amplitude of the interface waves grows with time. In a real system, of course, the growth is restricted by the limits of applicability of linear elasticity and linear vibration theory. This type of friction-induced vibration may be at least partially responsible for noise and other undesirable effects during friction (Nosonovsky and Adams 2004). These instabilities are a consequence of the inherent non-linearity of the boundary conditions with Coulombian friction. Whereas the interface waves occur for slightly dissimilar (in the sense of their elastic properties) materials, for very dissimilar materials, waves would be radiated along the interfaces. This provides a different mechanism of pumping energy away from the interface (Nosonovsky and Adams 2004).

Adams et al. (2005) also demonstrated that dynamic effects lead to new types of frictional paradoxes, in the sense that the assumed direction of sliding used for Coulomb friction is opposite to that of the resulting slip velocity. In a strict mathematical sense, Coulomb friction is inconsistent not only with rigid body dynamics (the Painlevé paradoxes) but also with the dynamics of elastically deformable bodies.

6.4.2 Adams's Simple Dynamic Model of a Beam on an Elastic Foundation

After frictional dynamic instabilities were discovered by Adams (1995) and Martins et al. (1995), they were thoroughly investigated. For example, Adams (1998) investigated the effect of these instabilities on layered surfaces. In addition, Adams (1996) developed a simple model that uses a beam on elastic foundation (modeled mathematically by a simple differential equation) to represent an elastic half-space, which is useful for illustrative purposes.

His model consists of a beam on an elastic foundation acted upon by a series of moving linear springs, where the springs represent the asperities on one of the surfaces. The aim was to examine the possibility of instabilities in the nominally steady sliding of a rough surface on a flat body with a constant coefficient of friction. Although sliding contact is non-linear, the contact behavior can be linearized about the nominally steady position. Thus, the rough surface is modeled by a series of linear springs. The problem is decomposed into a steady-state part and an unsteady portion. An eigenvalue analysis of the dynamic part shows that dynamic instabilities in the form of self-excited motion exist for any finite speed. The mechanism of instability is due to the interaction of a complex mood of vibration with friction force exerted by the moving loads.

Adams's model (1996), shown schematically in Figure 6.6, consists of a tensioned beam on an elastic foundation, acted upon by a series of moving linear springs and used as a qualitative model of two bodies in sliding contact. The action of the springs represents the asperities on the sliding surface

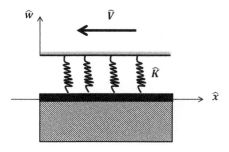

FIGURE 6.6
Simplified model of the elastic contact: a series of moving springs acting on a tensioned beam on an elastic foundation representing the elastic medium.

and includes a constant coefficient of friction (μ). This model intended to capture some of the essential features of the interaction model between two sliding bodies.

The dimensionless partial differential equation of motion for the traverse deflection $w(x,t)$ of a tensioned beam on a damped elastic foundation is given by (Adams 1996)

$$\frac{\partial^4 w}{\partial x^4} - (T - v^4)\frac{\partial^2 w}{\partial x^2} + kw + c\frac{\partial w}{\partial t} + cv\frac{\partial w}{\partial x} +$$

$$2v\frac{\partial^2 w}{\partial x \partial y} + \frac{\partial^2 w}{\partial t^2} = p(x,t) \tag{6.66}$$

$$-\infty < x < \infty$$

This is a coordinate system that moves to the left with constant dimensionless speed v. The dimensionless parameters used in Equation (6.66) are

$$\hat{w}/L, \ x = (\hat{x} + \hat{v}\hat{t})/L, \ t = (\hat{t}/L^2)\sqrt{EI/(\rho A)},$$

$$T = \hat{T}L^2/(EI), \ k = \hat{k}L^4/(EI), \ c = \hat{c}L^2/\sqrt{EI\rho A},$$

$$v = \hat{v}L\sqrt{\rho A/(EI)}, \ p = \hat{p}L^3/(EI)$$

where
   x is a dimensionless coordinate moving to the left with the springs
   t is the dimensionless time
   L is the spring spacing
   EI is the beam flexural rigidity
   ρA is the mass per unit length of the beam
   $\hat{T}$ is the beam tension
   $\hat{k}$ and $\hat{c}$ are the foundation stiffness and damping coefficients, respectively
   $\hat{p}(\hat{x},\hat{t})$ is the external force per unit length applied to the beam

At a typical frictional spring located at the point $x = n(n = 0, \pm1, \pm2, ...)$, there is continuity of displacement and slope (Adams 1996):

$$w(n^-,t) = w(n^+,t)$$

$$\frac{\partial w}{\partial x}(n^-,t) = \frac{\partial w}{\partial x}(n^+,t) \tag{6.67}$$

The spring of stiffness $(\hat{k})$ causes a discontinuity in the internal shear force, while the friction force produces a discontinuity in the internal bending moment (Adams 1996). The jump in the moment is due to the combination of the friction force and the finite height ($\hat{h}$) of the beam. According to Adams (1996), these conditions appear as

$$\frac{\partial^2 w}{\partial x^2}(n^-,t) = \frac{\partial^2 w}{\partial x^2}(n^+,t) + \alpha K w(n,t)$$

$$\frac{\partial^3 w}{\partial x^3}(n^-,t) = \frac{\partial^3 w}{\partial x^3}(n^+,t) + \alpha K w(n,t)$$

(6.68)

At each support, ($n = 0, \pm1, \pm2,\ldots$), where

$$K = \frac{\hat{K}L^3}{EI}, \quad \alpha = \frac{\mu\hat{h}}{2L}$$

(6.69)

are the dimensionless spring stiffness and friction parameter, respectively.

Due to the periodic spacing of the supports, the solution is taken to be spatially periodic. The periodicity of the solution must be an arbitrary inter-multiple (N) of the spacing—that is (Adams 1996),

$$w(x+N,t) = w(x,t)$$

(6.70a)

which gives

$$w(x+1,t) = e^{2\pi i/N} w(x,t)$$

(6.70b)

Thus, as Adams (1996) showed with applications of Equations (6.67)–(6.70), the partial differential Equation (6.66) needs only to be solved in the interval between two adjacent springs ($0 < x < 1$) for any value of N.

Finally, Adams (1996) wrote the complete solution of Equation (6.66) as the sum of two solutions. The first part represents the steady-state behavior due to a series of moving spring loads of constant magnitude and the second describes vibrations about the steady configuration. The unsteady solution is characterized by free vibrations of the system in the form of

$$w(x,t) = \text{Re}\{W(x)e^{\lambda t}\}$$

(6.71)

Adams (1996) also showed that the complex form of the displacement function, for a steady solution, is in the form of

$$w(x) = C_1 e^{i\gamma_1 x} + C_2 e^{i\gamma_2 x} + C_3 e^{i\gamma_3 x} + C_4 e^{i\gamma_4 x}$$

(6.72)

where γ_1, γ_2, γ_3, and γ_4 are determined from the characteristic equation of the system.

Adams (1996) presented results by applying the mathematical model to several cases. He obtained the maximum ratios of the real part of λ to the absolute value of the imaginary pan of λ as a function of dimensionless speed v and presented results for various values of the friction parameter (α) for different values of the foundation stiffness (k) for various values of beam tension (T) and for different values of the asperity spring stiffness (K). Of particular importance were the existences of eigenvalues with positive real parts that indicate a dynamic instability. In these results, as the speed was increased from zero, the values of the dynamic instabilities also increased. The results for different friction parameters showed that as this value increased, so did the dynamic instability. The foundation parameter (k) was a measure of the interaction of two points on the beam separated by a distance equal to the spring asperity spacing. Large values of k indicated less interaction than did small values of k. Adams's (1996) results indicate that $\lambda^R/|\lambda^I|$ decreases with increasing k. He also obtained that increased beam tension decreases the magnitude of the dynamic instability. This effect was similar to the effect of large k in localizing the deflections.

In order to understand the reason for the unstable response, Adams (1996) discussed first the case in which friction and damping were absent and the springs were stationary ($v = 0$). In this case the eigenvalues λ are purely imaginary and the eigenvectors are real, resulting in the well-known normal modes of vibration. Adams (1996) assumed that the springs were no longer stationary, but that friction and damping were still absent. Such a system could be shown to be gyroscopic, in which case the eigenvalues λ were purely imaginary and the eigenvectors yielded complex modes of vibration (Meirovitch 1974; Wickert and Mote 1990). In a complex mode, there was a fixed relation between the real and imaginary parts of the mode that were vibrating 90° out of phase with each other. However, with friction included, the eigenvalue problem became general, rather than gyroscopic, and the eigenvalues as well as the eigenvectors were complex. Hence, even without damping, the system was non-conservative as a consequence of the lack of symmetry introduced by friction. Furthermore, the eigenvalues can have positive or negative real parts. An eigenvalue with a negative real part corresponds to dissipation, whereas an eigenvalue with a positive real part corresponds to an increase in energy (Adams).

Adams's results (1996) showed that, if the speed of the moving springs were reduced to zero, a complex mode would tend toward a real mode corresponding to a "pinned" configuration (i.e., one with zero beam deflection and moment at the spring). Similarly, a complex mode with $v = 0$ would approach a "spring" mode (i.e., one in which the beam deflection was symmetric about the spring). Adams also considered the case that the value of the friction coefficient was small. The displacement can be written as

$$w(x,t) = \text{Re}\{W(x)e^{\lambda t}\} = [W^R(x)\cos\lambda^I t - W^I(x)\sin\lambda^I t]e^{\lambda^k t} \qquad (6.73)$$

where

$$\lambda = \lambda^R + i\lambda^I, \quad w(x) = W^R(x) + iW^I(x) \qquad (6.74)$$

Adams (1996) considered the case in which the slope of the beam at the spring was governed essentially by the real part of the mode, whereas the displacement was dictated by the imaginary part. For those values of time for which $\sin\lambda^I t = 1$, it can be seen from Equation (6.73) that the spring displacement (at $x = 0$) was a maximum. Hence, at those instants of time, the spring force (compressive), friction force (acting to the left), and the frictional moment exerted by the spring (counterclockwise) were at a maximum. However, since $\cos\lambda^I t = 0$ at those times, the beam slope at the spring was approximately zero and its angular velocity was negative (clockwise). Hence, the rate of work done on the beam by the frictional moment was negative. It can also be shown that the integral of this rate of work over a complete cycle is negative and hence this described mode dissipates energy.

Adams (1996) also considered the case in which the slope of the beam at the spring was governed by the complex mode. In this case, the real part primarily determined the spring deflection (hence, the spring force, friction force, and friction moment), whereas the imaginary part principally determined the beam slope. Arguments similar to those used for the first case demonstrated that this mode absorbed energy from the applied friction force and hence led to unstable self-excited motion, as indicated in Equations (6.67)–(6.70).

The question arises as to what happens as the dynamic instability causes the displacements to increase slowly with time. There are two distinct possibilities, according to Adams (1996). First, the rotational motion of the beam cross sections may cause the tangential velocity of points on the upper surface of the beam to have a speed equal to that of the sliding surface. This is the onset of asperity-level stick–slip and thus microscopically non-uniform motion. Alternatively, the transverse beam displacements can become large enough to cause loss of contact between a spring and the beam. This corresponds to repeated impacts in which energy is dissipated. Thus, whether the result is microscopic stick–slip or partial loss of contact is apt to have a profound effect on the observed macro-scale friction.

Finally, as Adams (1996) concluded, self-excited motion exists for a range of friction coefficient and material parameters at any finite sliding speed. The magnitude of these instabilities tends to increase with increases in either the friction coefficient or the asperity spring stiffnesses, or with decreases in either the foundation stiffness or the beam tension. The mechanism of instability is the result of an interaction between a complex mode of vibration and the frictional force from the sliding body. The self-excited motion will give rise either to partial loss of contact (repeated impacts with some energy dissipation) or to non-uniform asperity-level slick–slip motion. It is anticipated that these vibrations play a role in dry frictional sliding.

6.4.3 Dynamic Instabilities during Sliding of Rough Surfaces

As discussed in the preceding section, dynamic instabilities can occur during the contact of flat elastic surfaces. In the present section we will show that these instabilities also exist in the case of non-flat surfaces. Nosonovsky and Adams (2000, 2004) investigated the vibration and stability of dry frictional sliding of two elastic bodies with a wavy contact interface, and this section is based on their work. For the limiting case of complete contact with friction, the problem reduces to the one investigated by Adams (1995) and discussed in the preceding sections of this book. That instability is in the form of traveling interfacial waves, growing in amplitude with time. For incomplete contact, a solution in the form of complex modes of vibration is sought. In the frictionless quasi-static limit, this solution, if it exists, describes normal modes of vibration that are localized near the interface. With increasing sliding velocity and the coefficient of friction, these vibration modes become complex and the solution can become unstable (Nosonovsky and Adams 2004).

6.4.3.1 Formulation of the Problem

An important parameter of the frictional contact is the real area of contact between two sliding rough surfaces. Virtually all surfaces, including those usually considered nominally flat, are rough due to the asperities. Therefore, the contact of two bodies occurs only at the tops of the asperities, so the real area of contact constitutes only a small part of the nominal contact area. The real area of contact depends on both the form of the asperities and the way they are deformed. Some contact models for rough surfaces, especially those developed following the seminal work by Greenwood and Williamson (1966), deal with particular statistical distributions of random asperity heights and radii: Gaussian, exponential, stochastic, or fractal.

Since complete analytical solutions of the equations of elasticity combined with these statistical models are very difficult because of their complexity, these models usually deal with local zones of elastic deformation. Therefore, the asperities are considered independently of each other, and their effect consists of the summation of the actions of many single asperities. These are the so-called uncoupled contact models (Adams and Nosonovsky 2000). In many cases the uncoupled models give a good approximation of an actual deformed state of the bodies. These types of models can deal with a pure elastic deformation or with an elastic–plastic deformation law. According to many theories of this kind, the true area of contact is linearly proportional to the loading due to a statistical distribution of asperity heights. As the load increases, the number of asperities involved in the contact also increases. This is true for both elastic–plastic and pure elastic materials. The force of friction is, in turn, proportional to the area of contact due to adhesion. In this manner the linear dependence of the area of true contact upon the normal loading leads to the linear Coulomb law (Adams and Nosonovsky 2000).

Another approach deals with the elastic deformation of the body as a whole, or the so-called coupled contact models of rough surfaces. The contact surface profile must be represented here by a rather simple function, in order to make possible analytical or numerical solutions of the equations of elasticity. The simplest model of the rough contact surface is a wavy sinusoidal surface. This model represents the first term in a Fourier decomposition of the actual rough surface. Mathematically, this problem leads to mixed periodical boundary conditions that represent a combination of contact and separation zones. It is noted, however, that the boundary conditions cause this problem to be non-linear and results for different Fourier components cannot be superimposed (Nosonovsky and Adams 2000).

The frictionless two-dimensional elastic contact problem for a surface loaded by a periodic system of rigid flat punches is a classical problem of contact mechanics that was solved by Sadowski (1928). Later, Westergaard (1939) used a complex stress function and found a closed form solution for the 2D contact problem of an elastic half-space with a wavy surface. He obtained an expression for contact stresses as well as the dependence of the contact area on the applied pressure. Independently, Shtaerman (1949) obtained the same result, using Green's function method in order to formulate the contact problem as an integral equation for normal contact stress. He found a general form of the mathematical solution for an arbitrary periodic contact profile and a particular solution for a sinusoidal profile. He also obtained an integral equation formulation for the non-periodic frictional contact problem and a general form of the solution.

Dundurs, Tsai, and Keer (1973) used Fourier analysis in a stress function approach and produced a series solution to the frictionless problem with a wavy surface, which reduces to the form obtained by Westergaard (1939). Kuznetsov (1985) obtained a solution of the same problem, with one rigid body, by using the complex potential method of Muskhelishvili. He also solved the frictional low-velocity problem (Kuznetsov 1976). Johnson, Greenwood, and Higginson (1985) obtained a numerical solution, as well as asymptotic solutions for small and large zones of contact, for the frictionless case of 2D waviness (using, however, 3D elasticity). Their method was based on Fourier analysis. Manners (1998) extended Westergaard's problem to contact surfaces with more complicated periodic profiles. Since a lot of heat is generated during friction, thermoelastic analysis is also of interest. Brock (1996) presented a formulation of the thermoelasticity problem with a general non-periodic contact profile, making approximations for the coupling of contact regions.

Nosonovsky and Adams (2000, 2004) investigated the steady-state solution and the stability of small vibrations' relative sliding of a flat elastic half-space against an elastic body with a wavy surface. The effects of friction and of the sliding velocity were both included in their analysis. The pressure needed to close the gap between the two bodies was determined and found to depend on both the coefficient of the friction coefficient and the sliding speed. Furthermore, the contact area was determined as a function of the applied pressure for various material combinations, coefficients of friction, and sliding velocities. Their results are summarized next.

Similarly to the preceding sections, Nosonovsky and Adams (2000, 2004) considered the plane-strain sliding frictional contact of two semi-infinite elastic bodies. However, they assumed that one of the bodies is flat and that the other one has a slightly wavy surface. The bodies are pressed together by a uniform normal traction $\hat{p}$ and slide under a uniform tangential traction $\mu\hat{p}$ applied at infinity. The coefficient of kinetic friction between the two bodies is μ. The lower surface is assumed to have a purely sinusoidal boundary at $\hat{y} = 0$ with a period of $2l'$ and peak-to-valley amplitude $\hat{g}$—that is,

$$\hat{y} = \frac{\hat{g}}{2}\left(1 - \cos\frac{\pi(\hat{x}+\hat{E})}{\hat{i}}\right) \tag{6.75}$$

where $\hat{x}$ and $\hat{y}$ are the coordinates attached to the lower body and the eccentricity $\hat{E}$ will be defined later. The upper body is moving to the right with the constant velocity $\hat{V}$. Now the dimensionless coordinates and parameters

$$x = \frac{\pi\hat{x}}{\hat{i}}, \quad y = \frac{\pi\hat{y}}{\hat{i}}, \quad g = \frac{\pi\hat{g}}{\hat{i}}, \quad E = \frac{\pi\hat{E}}{\hat{i}}, \quad V = \hat{V}\sqrt{\frac{\rho_1}{G_1}}, \quad t = \frac{\pi\hat{t}}{\hat{i}}\sqrt{\frac{G_1}{\rho_1}} \tag{6.76}$$

are defined, where ρ_1 and ρ_2 are the densities of the two bodies, and G_1 and G_2 are their elastic shear moduli.

The contact regions are determined from the solution of the steady-state problem, whereas small vibrations near the steady-state solution should be investigated in order to analyze the stability of the steady-state solution.

6.4.3.2 The Steady-State Solution

First, Nosonovsky and Adams (2000) investigated steady sliding. For the upper body, the Navier equations for the dimensionless displacements $u(x,y)$, $v(x,y)$ in the x- and y-directions are given by

$$(\beta_1^2 - V^2)\frac{\partial^2 u}{\partial x^2} + \frac{\partial^2 u}{\partial y^2} + (\beta_1^2 - 1)\frac{\partial^2 v}{\partial y\partial x} = 0$$

$$(\beta_1^2)\frac{\partial^2 v}{\partial y^2} + (1 - V^2)\frac{\partial^2 v}{\partial x^2} + (\beta_1^2 - 1)\frac{\partial^2 u}{\partial y\partial x} = 0 \tag{6.77}$$

Here the ratio of longitudinal (C_L) and shear (C_S) wave speeds are

$$\beta_k = \left(\frac{C_L}{C_S}\right)_k = \sqrt{\frac{2(1-v_k)}{1-2v_k}} \tag{6.78}$$

in which $k = 1$ for the upper body and $k = 2$ for the lower body.

The deformed state of the bodies was considered as the superposition of uniform stresses (caused by the dimensionless applied pressure $p = \hat{p}/G_1$ and dimensionless tangential traction μp and the dimensionless residual stresses ($\sigma_{ij}^{(k)} = \partial_{ij}^{(k)}/\mu_k$). The boundary conditions at infinity ($|y| = \infty$) state that stresses and strains should be bounded. On the contact surface ($y = 0$); here, are two kinds of boundary conditions: non-mixed and mixed boundary conditions. The non-mixed conditions are valid on the entire surface (Nosonovsky and Adams 2000):

$$\sigma_{xy}^{(1)} = \mu\sigma_{yy}^{(1)} \quad -\infty < x < \infty \tag{6.79}$$

$$\sigma_{xy}^{(2)} = \mu\sigma_{yy}^{(2)} \quad -\infty < x < \infty \tag{6.80}$$

$$\mu_1\sigma_{yy}^{(1)} = \mu_2\sigma_{yy}^{(1)} \quad -\infty < x < \infty \tag{6.81}$$

The shear moduli appear in Equation (6.81) due to the manner in which the stresses were non-dimensionalized. The mixed boundary conditions are satisfied only in the separation zone:

$$\sigma_{yy}^{(1)} = p \quad c < |x| < \pi \tag{6.82}$$

and in the contact zone:

$$\frac{\partial v^{(1)}}{\partial x} - \frac{\partial v^{(2)}}{\partial x} = \frac{g}{2}\sin(x+E) \quad 0 < x < c \tag{6.83}$$

Here, x is equal to zero at the center of the contact zone. Thus, the parameter E is the coordinate distance representing the peak of the surface waviness relative to the center of the contact zone. Furthermore the solution is periodic with period 2π. Finally, any solution of the preceding equations must be such that the contact pressure remains positive everywhere in the contact region. Satisfaction of that condition also guarantees that the shear stress opposes slip throughout the contact region.

6.4.3.3 Steady State: Analytical Solution

The Navier equations (6.77) yield a solution for the upper body in the form of (only the real parts of displacements and stresses are considered here and in the subsequent formulas)

$$u^{(1)}(x,y) = \operatorname{Re} \sum_{n=1}^{\infty} A_n(e^{s_1 y} + \delta_3 e^{s_2 y})e^{inx}$$

$$v^{(1)}(x,y) = \operatorname{Re} \sum_{n=1}^{\infty} A_n(e^{s_1 y} + \delta_2\delta_3 e^{s_2 y})e^{inx}$$

(6.84)

Here, s_1 and s_2 are roots of the corresponding characteristic equation. Since the solution must be bounded at infinity, among the four roots only the two roots (s_1, s_2) with positive real parts are considered. The Navier equations further require (Nosonovsky and Adams 2000)

$$\delta_m = i\frac{s_m^2 - n^2(\beta_1^2 - V^2)}{(n)(\beta_1^2 - 1)s_m} \quad m = 1,2$$

(6.85)

where δ_3 can be determined from the Coulomb friction boundary conditions (Equation 6.79), namely,

$$\delta_3 = -\frac{\mu in(\beta_1^2 - 2) + \mu\beta_1^2 s_1\delta_1 - (s_1 + \delta_1 in)}{\mu in(\beta_1^2 - 2) + \mu\beta_1^2 s_2\delta_2 - (s_2 + \delta_2 in)}$$

(6.86)

For the lower body, which is subject to a stationary load distribution, there are repeated roots of the characteristic equation, and a solution is sought in the following:

$$u^{(2)}(x,y) = \operatorname{Re} \sum_{n=1}^{\infty} iB_n\left(-1 + \delta_4\frac{\beta_1^2 + 1}{\beta_1^2 - 1}e^{s_1 y} - \delta_4 ny\right)e^{-ny + inx}$$

$$v^{(2)}(x,y) = \operatorname{Re} \sum_{n=1}^{\infty} B_n(1 + \delta_4 ny)e^{-ny + inx}$$

(6.87)

that satisfies the Navier equation (6.77) with zero sliding velocity. In order to satisfy the friction law for the lower body, the coefficient δ_4 is given by (Nosonovsky and Adams 2000)

$$\delta_4 = (\beta_1^2 - 1)\frac{i+\mu}{i\beta_2^2+\mu} \tag{6.88}$$

The condition of continuity of σ_{yy} on the surface gives

$$B_n = \delta_5 A_n, \quad \delta_5 = \left(\frac{\mu_1}{\mu_2}\right)\frac{i\beta_1^2+\mu}{2in(1-\beta_2^2)}[(1+\delta_3)(\beta_1^2-2)in+\beta_1^2(s_1\delta_1+s_2\delta_3\delta_2)] \tag{6.89}$$

where δ_5 is independent of n. Summarizing, the non-mixed boundary conditions have been used in order to determine all the unknown coefficients except for A_n.

The two mixed boundary conditions Equations (6.82) and (6.83) now yield

$$\sum_{n=1}^{\infty}[\delta_1+\delta_2\delta_3-\delta_5]inA_ne^{inx/m} = \frac{g}{2}\sin(x+E) \quad |x|<c \tag{6.90}$$

$$\sum_{n=1}^{\infty}[(1+\delta_3)(\beta_2^2-2)in+\beta_1^2(s_1\delta_1+s_2\delta_3\delta_2)]A_ne^{inx} = p \quad c<|x|<\pi \tag{6.91}$$

In order to satisfy Equation (6.91), the unknown function $\phi(\xi)$ is introduced such that

$$[(1+\delta_3)(\beta_1^2-2)in+\beta_1^2(s_1\delta_1+s_2\delta_3\delta_2)]A_n = \int_{-c}^{c}\phi(\xi)e^{-in\xi}\,d\xi \tag{6.92}$$

Then, Equation (6.91), with the use of the identities for generalized functions (Gelfand and Shilov 1964),

$$\sum_{n=1}^{\infty}e^{inx} = -\frac{1}{2}+\pi\delta(x)+\frac{i}{2}\cot\frac{x}{2} \quad |x|<\pi \tag{6.93}$$

becomes

$$\frac{1}{2}\int_{-c}^{c}\phi(\xi)\,d\xi = -p \tag{6.94}$$

Now the mixed boundary condition Equation (6.90), along with Equation (6.93), can be written as (Nosonovsky and Adams 2000)

$$K_p^r - K^r \pi \phi(x) - \frac{1}{2} K^i \int_{-c}^{c} \phi(\xi) \cot\left(\frac{x-\xi}{2}\right) d\xi = \frac{g}{2} \sin(x+E) \quad |x| < c \qquad (6.95)$$

where

$$K = K^r + iK^i = \frac{\delta_1 + \delta_2 \delta_3 - \delta_5}{(1+\delta_3)(\beta_1^2 - 2) + \beta_1^2 (s_1 \delta_1 + s_2 \delta_3 \delta_2) \frac{1}{in}} \qquad (6.96)$$

does not depend on n.

Nosonovsky and Adams (2000) introduced the coordinate transformations

$$\tan\frac{\xi}{2} = au, \quad \tan\frac{x}{2} = as, \quad \tan\frac{E}{2} = ae, \quad \tan\frac{c}{2} = a, \quad \psi(u) = \frac{\phi(\xi)}{1+a^2 u^2} \qquad (6.97)$$

that reduce integral Equation (6.95) to

$$K^2 \psi(s) + \frac{K^i}{\pi} \int_{-1}^{1} \frac{\psi(u)}{u-s} du = f(s) \qquad (6.98)$$

where

$$f(s) = gf_1(s) + pf_2(s) = g \frac{1}{\pi} \frac{a(s+\varepsilon)(1+\varepsilon s a^2)}{(1+a^2 s^2)^2 (1+a^2 e^2)}$$

The resultant condition (6.94), after the coordinate transformation (6.97), yields

$$\int_{-1}^{1} \psi(u) du = -\frac{p}{a} \qquad (6.99)$$

To solve Equation (6.98), which is subject to Equation (6.99), the procedure developed by Erdogan and Gupta (1972) was used by Nosonovsky and Adams (2000). The unknown $\psi(u)$ was taken as

$$\psi(u) = (1-u)^\alpha (1+u)^{1-\alpha} \sum_{n=0}^{\infty} c_n P_n^{(\alpha,1-\alpha)}(u) \qquad (6.100)$$

where $P_n^{(\alpha,1-\alpha)}(u)$ is the Jacobi polynomial of the *n*th order and

$$\alpha = \frac{1}{2\pi i} \log \frac{K^r - iK^i}{K^r + iK^i} = -\frac{1}{\pi} \arctan \frac{K^i}{K^r} \qquad (6.101)$$

characterizes the degree of the singularity of the unknown function at the leading edge of the contact region ($u = 1$). The parameter α was to be chosen such that $0 < \alpha < 1$ in order that the solution be bounded at both ends.

Nosonovsky and Adams (2000) obtained the coefficients C_n of Equation (6.100):

$$-\frac{2K^i}{\sin \pi \alpha} \theta_k^{(-\alpha,-1+\alpha)} C_{k-1} = \int_{-1}^{1} P_k^{(\alpha,1-\alpha)}(s)(1-s)^{-\alpha} \times (1+s)^{-1+\alpha} f(s)\,ds \qquad (6.102)$$

$$k = 0,1,2,\ldots$$

where

$$\theta_k^{(-\alpha,-1+\alpha)} = \frac{2^{\alpha+(1-\alpha)+1}}{2k+\alpha+(1-\alpha)+1} \frac{\Gamma(k+\alpha+1)\Gamma(k+(1-\alpha)+1)}{k!\Gamma(k+\alpha+(1-\alpha)+1)} \qquad (6.103)$$

and $\Gamma(.)$ is the gamma function.

By substituting Equation (6.100) into Equation (6.101) and using Equations (6.102) and (6.103), the resultant condition Equation (6.104) was reduced to (Nosonovsky and Adams 2000)

$$-\frac{2\sin \pi \alpha}{K^i} \int_{-1}^{1} P_1^{(-\alpha,-1+\alpha)}(s)(1-s)^{-\alpha} \times (1+s)^{-1+\alpha}[gf_1(s) + pf_2(s)]\,ds = -\frac{p}{a} \qquad (6.104)$$

In order for the solution to be bounded at the ends, $f(s)$ must satisfy the consistency condition (Erdogan and Gupta 1972):

$$\int_{-1}^{1} (1-s)^{-\alpha} \times (1+s)^{-1+\alpha}[gf_1(s) + pf_2(s)]\,ds = 0 \qquad (6.105)$$

Equations (6.104) and (6.105) are solved for p/g with the use of Equations (6.106) and (6.107):

$$\frac{p}{g} =$$

$$-\frac{a}{1+a^2\varepsilon^2}\frac{\dfrac{2}{2\pi}(\varepsilon I_1+I_2-a^2\varepsilon^2 I_2-a^2\varepsilon I_3)+T\dfrac{1-2\alpha}{2\pi}(\varepsilon I_0+I_1-a^2\varepsilon^2 I_1-a^2\varepsilon I_2)}{-\left(\dfrac{T}{\pi}\right)K^r\left(\dfrac{1}{2}J_1+\dfrac{1-2\alpha}{2}J_0\right)+\left(\dfrac{T}{\pi}\right)K^i a\left(\dfrac{1}{2}J_2+\dfrac{1-2\alpha}{2}J_1\right)+\dfrac{1}{a}} \quad (6.106)$$

$$\frac{p}{g}=-\frac{a}{1+a^2\varepsilon^2}\frac{\varepsilon I_0+I_1-a^2\varepsilon^2 I_1-a^2\varepsilon I_2}{-J_0 K^r+aK^i J_1} \quad (6.107)$$

where

$$I_m=\int_{-1}^{1}\frac{s^m(1-s)^{-\alpha}(1+s)^{-1+\alpha}}{(1+a^2 s^2)^2}ds \quad m=0,1,2,3$$

$$ (6.108)$$

$$I_m=\int_{-1}^{1}\frac{s^m(1-s)^{-\alpha}(1+s)^{-1+\alpha}}{1+a^2 s^2}ds \quad m=0,1,2$$

When combined together, Equations (6.106) and (6.107) constitute a quadratic equation for ε that has two roots, ε_1 and ε_2, such that their product is equal to $-1/a^2$. These two roots correspond to two values of eccentricity, E_1 and E_2, shifted by π relative to each other (i.e., $E_1 = E_2 - \pi$). From Equations (6.106) and (6.107) note also that the shifted second root gives the same absolute value of p/g as the first root, but with a negative sign (Nosonovsky and Adams 2000). This can be interpreted as shifting of the wavy surface (Equation 6.81) by π so that the second root corresponds to the same solution as the first one, but translated by π in the x-direction due to the change of sign of g.

For a given value of material parameters, coefficient of friction, and p/g, Equations (6.106) and (6.107) can be solved for a and ε. However, it is simpler mathematically to prescribe a and calculate ε; then p/g can be determined from Equations (6.106) and (6.107).

In order to determine the minimum value of pressure (p^*) that gives complete contact and the corresponding value of eccentricity E^*, Equations (6.90) and (6.91) (in which $A_n = 0$ or $n > 1$) can be reduced for $c = \pi$ to

$$Kp^* e^{ix} = \frac{g}{2}\sin(x+E) \quad |x| < \pi \quad (6.109)$$

Finally, the complex Equation (6.100) along with Equation (6.101) gives two equations for the real numbers $p*$ and $E*$ coefficient:

$$E* = -\arctan\left(\frac{K^r}{K^I}\right) = \pi\left(\frac{1}{2} - \alpha\right), \quad p* = \frac{g}{2}\frac{1}{|K|} \tag{6.110}$$

6.4.3.4 Steady State: Results and Discussion

For the small velocity case $V \to 0$, the dependence of the parameter K on the material parameters and the friction coefficient agree numerically with the result by Shtaerman (1949)—that is,

$$K^I = 1 - v_1 + \frac{\mu_1}{\mu_2}(1 - v_2) \tag{6.111}$$

$$K^r = \mu\left[(1 - 2v_2)\frac{\mu_1}{\mu_2} - (1 - 2v_1)\right] \tag{6.112}$$

Note that by an appropriate adjustment of v_1 in Equations (6.111) and (6.112) and p/g in the governing Equation (6.95), the problem with two elastic bodies can be reduced to a problem with one elastic and one rigid body ($\mu_1/\mu_2 = 0$). Thus, for the dynamic case with friction, a steady-state solution has been obtained by Nosonovsky and Adams (2000) for the dependence of the contact zone length and the contact pressure on the applied pressure, the friction coefficient, and the velocity of sliding. The solution obtained demonstrates a decrease of the minimum pressure required to close the gap between the two bodies with an increase of the frictional coefficient and/or the sliding velocity. A resonance exists for the value of the sliding velocity equal to the surface wave velocity of the upper body. The solution is valid for both small-velocity and high-velocity sliding. These results have relevance in furthering our understanding of sliding friction as well as in the design and analysis of seals. By properly controlling the surface topography, it is possible to control the contact area and the extent of separation.

6.4.3.5 Small Vibrations near the Steady-State Solution

In the preceding sections, steady-state sliding was investigated. Now we will study the stability of small perturbations near the steady-state solution. Mathematically, the formulation is quite similar; however, the stability analysis involves dynamic (time-dependent) terms. The following solution follows Nosonovsky and Adams (2004).

The Navier equations for the dimensionless displacements $u(x,y,t)v(x,y,t)$ in the x- and y-directions, respectively, are given by

$$(\beta_1^2 - V^2)\frac{\partial^2 u^{(1)}}{\partial x^2} + \frac{\partial^2 u^{(1)}}{\partial y^2} + (\beta_1^2 - 1)\frac{\partial^2 v^{(1)}}{\partial x \partial y} = \frac{\partial^2 u^{(1)}}{\partial t^2} - 2V\frac{\partial^2 u^{(1)}}{\partial x \partial t}$$

$$(\beta_1^2)\frac{\partial^2 v^{(1)}}{\partial y^2} + (1 - V^2)\frac{\partial^2 v^{(1)}}{\partial x^2} + (\beta_1^2 - 1)\frac{\partial^2 u^{(1)}}{\partial x \partial y} = \frac{\partial^2 v^{(1)}}{\partial t^2} - 2V\frac{\partial^2 v^{(1)}}{\partial x \partial t}$$

$$(\beta_1^2)\frac{\partial^2 u^{(2)}}{\partial x^2} + \frac{\partial^2 u^{(2)}}{\partial y^2} + (\beta_1^2 - 1)\frac{\partial^2 v^{(2)}}{\partial x \partial y} = \kappa^2\frac{\partial^2 u^{(2)}}{\partial t^2}$$ (6.113)

$$(\beta_2^2)\frac{\partial^2 v^{(2)}}{\partial y^2} + \frac{\partial^2 v^{(2)}}{\partial x^2} + (\beta_2^2 - 1)\frac{\partial^2 v^{(2)}}{\partial x \partial y} = \kappa^2\frac{\partial^2 u^{(2)}}{\partial t^2}$$

where the shear wave speed ratio is k and the ratios of longitudinal (C_L) and shear (C_S) wave speeds are related to the Poisson's ratios of the two materials v_1 and v_2 according to

$$\beta_k = \left(\frac{C_L}{C_S}\right)_k = \sqrt{\frac{2(1 - v_k)}{1 - 2v_k}}, \quad k^2 = \left(\frac{C_L^{(1)}}{C_S^{(2)}}\right)^2 = \frac{G_1\rho_2}{G_2\rho_1}$$ (6.114)

in which $k = 1$ for the upper body and $k = 2$ for the lower body. Also, note that indices (1) and (2) are related to the upper and the lower bodies, respectively.

The deformed state of the bodies is considered as the superposition of uniform stresses (caused by the dimensionless applied pressure $p = \hat{p}/G_1$, dimensionless tangential traction μ_p, and the dimensionless residual stresses $(\sigma_{xx}, \sigma_{xy}, \sigma_{yy})$. The dimensionless residual stresses are related to the dimensional residual stresses according to (Nosonovsky and Adams 2004)

$$\sigma_{ij}^{(k)} = \frac{\partial_{ij}^{(k)}}{\mu_k}$$ (6.115)

and vanish at infinity. The dimensionless stresses are related to the dimensionless deformations by

$$\sigma_{yy}^{(k)} = (\beta_{12}^2 - 1)\frac{\partial u}{\partial x}u_x^{(k)} + \beta_k^2\frac{\partial_v^{(k)}}{\partial y}$$

$$\sigma_{xy}^{(k)} = \frac{\partial u^{(k)}}{\partial y} + \beta_k^2\frac{\partial v^{(k)}}{\partial x}$$ (6.116)

The boundary conditions at infinity, ($|y| = \infty$), state that stresses should be equal to the applied tractions. On the contact surface ($y = 0$), there are two kinds of boundary conditions: non-mixed and mixed boundary conditions. The non-mixed conditions are valid on the entire surface (Nosonovsky and Adams 2004):

$$\sigma_{xy}^{(1)} = \mu\sigma_{yy}^{(1)} \qquad -\infty < x < \infty \tag{6.117}$$

$$\sigma_{xy}^{(2)} = \mu\sigma_{xy}^{(2)} \qquad -\infty < x < \infty \tag{6.118}$$

$$G_1\sigma_{yy}^{(1)} = G_2\sigma_{yy}^{(1)} \qquad -\infty < x < \infty \tag{6.119}$$

The shear moduli appear in Equation (6.119) due to the manner in which the stresses were non-dimensionalized in Equation (6.115). The mixed boundary conditions are satisfied only in the separation zone:

$$\sigma_{yy}^{(1)} = p - c < |\pi| < c \tag{6.120}$$

and in the contact zone (Nosonovsky and Adams 2004):

$$\frac{\partial v^{(1)}}{\partial v} - \frac{\partial v^{(2)}}{\partial x} = \frac{g}{2}\sin(x + E) \tag{6.121}$$

Here, x is equal to zero at the center of the contact zone. Thus, the parameter E is the coordinate distance representing the peak of the surface waviness relative to the center of the contact zone.

Suppose that the steady-state solution is given by $u_0^{(1)}(x, y)$, $v_0^{(1)}(x, y)$, $u_0^{(2)}(x, y)$, $v_0^{(2)}(x, y)$, with the length of the contact zone $2c$ and the eccentricity E. In order to analyze the stability of the steady-state solution, small vibrations near the steady-state solution are considered. The complete solution $\bar{u}^{(k)}$, $\bar{v}^{(k)}$ of Equation (6.113) is a superposition of the steady-state solution and the small vibrations—that is,

$$\bar{u}^{(k)}(x, y, t) = u_0^{(k)}(x, y, t) + u^{(k)}(x, y, t)$$

$$\bar{v}^{(k)}(x, y, t) = v_0^{(k)}(x, y, t) + v^{(k)}(x, y, t)$$

In order to satisfy the Coulomb friction inequalities, it is required that the sliding velocity V always be greater than the local x-component of the small vibration relative velocity at the interface—that is,

$$\frac{\partial u^{(2)}(x,0,t)}{\partial t} - \frac{\partial u^{(1)}(x,0,t)}{\partial t} < V \qquad (6.122)$$

For small vibrations, the non-mixed boundary conditions (Equations 6.117–6.119) remain in the same form, whereas the mixed boundary conditions Equations (6.120) and (6.122) are reduced to

$$\sigma_{yy}^{(1)} = 0 \qquad c < |x| < \pi \qquad (6.123)$$

$$\frac{\partial v^{(1)}}{\partial x} - \frac{\partial v^{(2)}}{\partial x} = 0 \qquad |x| < c \qquad (6.124)$$

6.4.3.6 Formulation in Integral Equation Form

Any initial small perturbation can be represented as a superposition of the modes $u_k(x,y)$, $v_k(x,y)$ that correspond to the eigenvalues Λ_k—that is,

$$u(x,y,t) = \operatorname{Re} \sum_{k=1}^{\infty} C_k u_k(x,y) e^{\Lambda_k t}$$

$$(6.125)$$

$$v(x,y,t) = \operatorname{Re} \sum_{k=1}^{\infty} D_k v_k(x,y) e^{\Lambda_k t}$$

A stable solution must have the real parts of all Λ_k negative or zero.
The modes are sought in the form of

$$u^{(1)}(x,y) = \sum_{k=-\infty}^{\infty} A_n (e^{s_1 y} + \delta^3 e^{s_2 y}) e^{inx/m}$$

$$(6.126)$$

$$v^{(1)}(x,y) = \sum_{k=-\infty}^{\infty} A_n (e^{s_1 y} + \delta_3' e^{s_2 y}) e^{inx/m}$$

for the upper body and

$$u^{(2)}(x,y) = \sum_{k=-\infty}^{\infty} A_n (e^{s_1 y} + \delta_3' e^{s_2 y}) e^{inx/m}$$

$$(6.127)$$

$$v^{(1)}(x,y) = \sum_{k=-\infty}^{\infty} A_n (e^{s_1 y} + \delta_2' \delta_3' e^{s_2 y}) e^{inx/m}$$

for the lower body, where $m = 1$ for a mode with a period of 2π and $m = 2$ for an anti-symmetric mode $v(x + 2\pi, y) = -v(x, y)$ with a period of 4π. The terms that represent the translation of the system as a rigid body can be set to zero, ($A_0 = 0$ and $B_0 = 0$), without loss of generality. Note that the subscript k has been omitted for conciseness.

In order to satisfy the Navier equation (6.113)—s_1 and s_2 in Equation (6.126)—the latter must be roots of the characteristic equation

$$\beta_1^2 s^4 + \left[Q_1 \beta_1^2 + Q_2 + \left(\frac{n}{m} \right)^2 (\beta_1^2 - 1) \right] s^2 + Q_1 Q_2 = 0 \tag{6.128}$$

The solution of Equation (6.128) is

$$s_1 = \pm \sqrt{-Q_2}, \quad s_2 = \pm \sqrt{-Q_1 / \beta_1^2} \tag{6.129}$$

where

$$Q_1 = 2 \left(\frac{n}{m} \right) iVA - \Lambda^2 - \left(\frac{n}{m} \right)^2 (\beta_1^2 - V^2)$$

$$Q_2 = 2 \left(\frac{n}{m} \right) iVA - \Lambda^2 - \left(\frac{n}{m} \right)^2 (1 - V^2) \tag{6.130}$$

Since the solution must be bounded at infinity, among the four roots (Equation 6.129) only the two roots with positive real parts are considered for the upper body. The Navier equations further require that

$$\delta_1 = i \frac{s_1^2 + Q_1}{\left(\dfrac{n}{m} \right)(\beta_1^2 - 1) S_1} = -i \frac{n}{m S_1} \tag{6.131}$$

$$\delta_2 = i \frac{s_2^2 + Q_1}{\left(\dfrac{n}{m} \right)(\beta_2^2 - 1) S_2} = -i \frac{m S_2}{n} \tag{6.132}$$

while Λ_n and δ_3 are unknown coefficients that will be determined from the boundary conditions. The displacement field (Equation 6.126) produces stresses on the contact surface given by

$$\sigma_{yy}^{(1)} = \sum_{-\infty}^{\infty} A_n \left[(1+\delta_3)\left(\frac{n}{m}\right)(\beta_1^2 - 1) + \beta_1^2(s_2\delta_1 + s_2\delta_2\delta_3) \right] e^{inx/m} \quad -\alpha < x < \alpha$$

$$\sigma_{xy}^{(1)} = \sum_{-\infty}^{\infty} A_n \left[\left(s_1 + \delta_1 i\left(\frac{n}{m}\right) \right) + (s_2\delta_3 + s_2\delta_2\delta_3) \right] e^{inx/m} \quad -\alpha < x < \alpha$$

(6.133)

Coulomb friction's law (Equation 6.117) on the surface requires

$$\delta_3 = -\frac{\mu i\left(\dfrac{n}{m}\right)(\beta_1^2 - 2) + \mu\beta_1^2 s_1\delta_1 - \left(s_1 + \delta_1 i\left(\dfrac{n}{m}\right) \right)}{\mu i\left(\dfrac{n}{m}\right)(\beta_1^2 - 2) + \mu\beta_1^2 s_2\delta_2 - \left(s_2 + \delta_2 i\left(\dfrac{n}{m}\right) \right)}$$

(6.134)

Similar relations can also be written for the lower body; the corresponding variables are marked with a prime (′)—that is, Q_1', Q_1', δ_1', δ_2', and δ_3'. The real part of s_1' and s_2' must be negative for the solution to be bounded at infinity. The condition of continuity (Equation 6.119) of σ_{yy} on the surface gives

$$B_n = \delta_4 A_n$$

$$\delta_4 = \left(\frac{\mu_1}{\mu_2}\right)\frac{(1+\delta_3')(\beta_1^2 - 2)i\left(\dfrac{n}{m}\right) + \beta_1^2(s_1\delta_1 + s_2\delta_3\delta_2)}{(1+\delta_3)(\beta_2^2 - 2)i\left(\dfrac{n}{m}\right) + \beta_2^2(s_1'\delta_1' + s_2'\delta_3'\delta_2')}$$

(6.135)

Summarizing, the non-mixed boundary conditions have been used in order to determine all the unknown coefficients except for A_n. The two mixed boundary conditions (Equations 6.123 and 6.124) now yield

$$\sum_{n=-\infty}^{\infty} [\delta_1 + \delta_2\delta_3 - \delta_4(\delta_1 + \delta_2\delta_3)]i\left(\frac{n}{m}\right)A_n e^{inx/m} = 0 \qquad |x| < 0$$

(6.136)

$$\sum_{n=-\infty}^{\infty} \left[(1+\delta_3)(\beta_2^2 - 2)i\left(\frac{n}{m}\right) + \beta_1^2(s_1\delta_1 + s_2\delta_3\delta_2) \right] A_n e^{inx/m} = 0 \qquad c < |x| < \pi$$

(6.137)

Note that for $n = 1$, the vanishing of the term in brackets of Equation (6.138) constitutes the slip wave equation for a wavelength of $2m\pi$, while the vanishing of the term in brackets of Equation (6.137) constitutes the Rayleigh wave equation for the upper body. The parameter δ_4 is undefined at the Rayleigh wave speed of the upper body. Now in order to satisfy Equations (6.136) and (6.137), the unknown function $\phi(\xi)$ is introduced such that

$$\left[(1+\delta_3)(\beta_1^2 - 2)i\left(\frac{n}{m}\right) + \beta_1^2(s_1\delta_1 + s_2\delta_3\delta_2)\right] = \int_C \phi(\xi)e^{-in\xi}\,d\xi \qquad (6.138)$$

The integration is performed in the contact zone C ($-c < \xi < c$, for $m = 1$), ($-c < \xi + 2\pi < c$, for $m = 1$). Then, with the use of the identities for generalized functions (Gelfand and Shilov 1964)

$$\sum_{n=1}^{\infty} \cos(nx) = -\frac{1}{2} + \pi\delta(x) \quad |x| < \pi \qquad (6.139)$$

$$\sum_{n=1}^{\infty} \sin(nx) = \frac{1}{2}\cot\left(\frac{x}{2}\right) \qquad (6.140)$$

Equation (6.139) becomes

$$\int_C \phi(\xi)\sum_{n=-\infty}^{\infty} e^{in(x-\xi)/m}d\xi = \int_C \phi(\xi)2m\pi\delta(x-\xi)d\xi = 0 \qquad x \notin C \qquad (6.141)$$

which is satisfied automatically in the separation zone. Furthermore, Equation (6.136) yields a homogeneous integral equation for $\phi(\xi)$ given by

$$\int_C \phi(\xi)\sum_{n=-\infty}^{\infty} K_n e^{in(x-\xi)/m}d\xi = 0 \qquad x \in C \qquad (6.142)$$

where

$$K_n = \frac{\delta_1 + \delta_2\delta_3 - \delta_4(\delta_1' + \delta_2'\delta_3')}{(1+\delta_3)(\beta_1^2 - 2) + \beta_1^2(s_1\delta_1 + s_2\delta_3\delta_2)\dfrac{m}{in}} \qquad (6.143)$$

Integral Equation (6.142) is singular because the summation in its kernel diverges as $x \to j$. In order to solve Equation (6.142), K_n is decomposed into two terms:

$$K_n = K^\infty + K_n^{rem} \qquad n > 0$$

$$K_n = K^{-\infty} + K_n^{rem} \qquad n < 0 \qquad\qquad (6.144)$$

$$K_n = K_0^{rem} = 0$$

where

$$K^\infty = K_r^\infty + iK_i^\infty = \lim_{n \to \infty} K_n$$

$$K^{-\infty} = K_r^\infty - iK_i^\infty = \lim_{n \to -\infty} K_n$$

in which K_r^∞ and K_i^∞ are the real and imaginary parts of K^∞ and K^{rem} is the remaining part. Equation (6.142) now can be written for $m = 1$ as

$$2\pi\phi(x)K_r^\infty + \int_{-C}^{C} \phi(\xi)\left\{-K_r^\infty - K_i^\infty \cot\left(\frac{x-\xi}{2}\right) + k(x-\xi)\right\}d\xi = 0 \qquad |x| < C \qquad (6.145)$$

where

$$k(x-\xi) = \sum_{n=-\infty}^{\infty} K_n^{rem} e^{in(x-\xi)} \quad (m=1) \qquad\qquad (6.146)$$

for $m = 2q$. Equation (6.142) yields

$$4\pi\phi(x)K_r^\infty + \int_{-C}^{C} \phi(\xi)\left\{-K_i^\infty \cot\left(\frac{x-\xi}{4}\right) - K_i^\infty \tan\left(\frac{x-\xi}{4}\right) + k(x-\xi)\right\}d\xi = 0$$

$$|x| < C \qquad\qquad (6.147)$$

where

$$k(x-\xi) = \sum_{n=-\infty}^{\infty} K_n^{rem} e^{in(x-\xi)/2}(1-(-1)^n) \quad (m=2) \qquad\qquad (6.148)$$

In the derivation of Equations (6.145)–(6.148), the properties of $\phi(x)$—namely, the anti-symmetry $\phi(x + 2\pi) = -\phi(x)$ for $m = 2$, and the periodicity $\phi(x + 2\pi) = \phi(x)$ for $m = 1$, have been used, as well as double angle formulas for the trigonometric functions. Note that the natural frequencies of vibration for complete contact are given by the roots of

$$K_n = 0 \quad m = 1, 2, \quad n = 1, 2, 3 \tag{6.149}$$

where $m = 1$ corresponds to the symmetric modes and $m = 2$ corresponds to the anti-symmetric modes. Making the change of variables (Nosonovsky and Adams 2004)

$$x \to cx, \quad \xi \to c\xi, \quad \phi(x) = \phi(cx) \tag{6.150}$$

yields, for $m = 1$,

$$\frac{2\pi}{c}\phi(x)K_r^\infty + \int_{-1}^{1} \phi(\xi)\left\{-K_r^\infty - K_i^\infty \cot\frac{c(x-\xi)}{2} + k(c(x-\xi))\right\}d\xi = 0 \qquad |x| < 1 \tag{6.151}$$

while, for $m = 2$,

$$\frac{4\pi}{c}\phi(x)K_r^\infty + \int_{-1}^{1} \phi(\xi)\left\{-K_i^\infty \cot\frac{c(x-\xi)}{4} - K_i^\infty \tan\frac{c(x-\xi)}{4} + k(c(x-\xi))\right\}d\xi = 0 \tag{6.152}$$

$$|x| < 1$$

Note that Equations (6.151) and (6.152) govern the stability of the steady-state solution, which, in turn, was obtained from the equation

$$\frac{2\pi}{c}\phi(x)K_r^\infty + \int_{-1}^{1} \phi(\xi)\left\{-K_i^\infty \cot\frac{c(x-\xi)}{2}\right\}d\xi = f(x) \qquad |x| < 1 \tag{6.153}$$

where $f(x)$ describes the slope of the interface profile, which is sinusoidal for this problem (Nosonovsky and Adams 2000).

The stability analysis is independent of the particular form of the profile $f(x)$ and uses only the length of the contact zone c obtained from the steady-state solution of Equation (6.153). Equation (6.153), which was solved by Kuznetsov (1976) for the particular case of one rigid and one elastic body and low sliding velocities, governs several important problems of contact elasticity, such as propagating inter-sonic stick–slip regions (Adams 2001) and ultra-sonic motors (Zhari 1996). The algorithm of stability analysis that is considered in this work can be applied to the stability analysis of steady-state problems governed by Equation (6.153), whereas the particular form of the function $k(c(x - \xi))$ is different for each problem.

6.4.3.7 Special Solutions for Limiting and Resonance Cases

6.4.3.7.1 Frictionless Case

Simplifications can be made for a number of special cases. Let us first consider the frictionless static case ($\mu = 0$, $V = 0$). It is possible to show that the roots are purely imaginary (no energy is dissipated; $\Lambda = i\lambda$ where l is real) for this case and that the eigenvalue problem is self-adjoint. For purely imaginary roots, Equations (6.129–6.135) simplify to

$$Q_1 = \lambda^2 - \left(\frac{n}{m}\right)^2 \beta_1^2, \quad Q_2 = \lambda^2 - \left(\frac{n}{m}\right)^2 \tag{6.154}$$

$$s_1 = -\sqrt{\left(\frac{n}{m}\right)^2 - \lambda^2}, \quad s_2 = -\sqrt{\left(\frac{n}{m}\right)^2 - \lambda^2 / \beta_1^2}$$

$$s_1' = -\sqrt{\left(\frac{n}{m}\right)^2 - \lambda^2 / \kappa^2}, \quad s_2' = -\sqrt{\left(\frac{n}{m}\right)^2 - \lambda^2 / (\kappa^2 \beta_1^2)} \tag{6.155}$$

where

$$\Lambda = \pm i\lambda \tag{6.156}$$

For δ_3', we will have:

$$\delta_3 = \frac{s_1^2 + \left(\frac{n}{m}\right)^2}{2s_1 s_2} \tag{6.157}$$

$$\delta_3' = \frac{s_1'^2 + \left(\frac{n}{m}\right)^2}{2s_1' s_2'} \tag{6.158}$$

for the case of $n \to \infty$

$$K_\infty = K_i^\infty, \quad K_\gamma = 0 \tag{6.159}$$

and

$$k(x - \xi) = 2i \sum_{n=1}^{\infty} (K_n - K_i^{\infty}) \sin[n(x - \xi)] \quad (m = 2) \qquad (6.160)$$

Note that are Q_1, Q_2, Q_1', Q_2', s_1, s_2, s_1', s_2', δ_3, δ_3', and δ_4 purely real, while δ_1, δ_1', δ_2, and δ_2' are purely imaginary. All the quantities remain the same when the sign of λ is reversed.

6.4.3.7.2 Small Separation Zone

For the case of a vanishingly small separation zone ($c \rightarrow \pi$), Equation (6.137) must be satisfied only at a single separation point $|x| = \pi$, whereas Equation (6.136) has a solution in the form of

$$A_{\pm 1} \neq 0$$
$$A_{\pm n} = 0 \quad n > 1 \qquad (6.161)$$

This is the solution for slip waves (generalized Rayleigh waves) with two waves of the same amplitude traveling in opposite directions and forming a standing slip wave. To satisfy Equation (6.137) at $|x| = \pi$, the phase of the standing slip wave must be chosen that gives zero normal stress at the separation points.

6.4.3.7.3 Small Contact Zone

For a small contact zone ($c \rightarrow 0$), Equation (6.136) must be satisfied only for the contact points $x = 2\pi n$. A resonance type of solution (Equation 6.161) in the form of standing Rayleigh waves in each body, which are independent of each other, exists and satisfies Equation (6.137) only if the Rayleigh wave speeds of the two bodies $C_R^{(1)}$ and $C_R^{(2)}$ are related in the following manner:

$$pC_R^{(1)} = qC_R^{(2)} \qquad p, q = 1, 2, 3 \qquad (6.162)$$

In this case, standing Rayleigh waves with wavelengths of $2\pi/p$ and $2\pi/q$ can exist in the two bodies and their nodes will coincide with each other and with the contact points. In the general case when the Rayleigh wave speed ratio does not permit such a resonance, a standing Rayleigh wave can exist in only one body with node points at the points of contact. It must be stressed that although, mathematically, it is possible to consider any small value of c, physically, the displacements caused by the small vibrations must be much smaller than those of the steady-state solution and should not cause any sufficient change of the contact zone length c. In the limiting case of $c \rightarrow 0$ this condition cannot be satisfied.

6.4.3.7.4 Equal Rayleigh Wave Speeds

In the case of equal Rayleigh wave speeds for the two materials in contact, the bracketed term on the left-hand side of Equation (6.138) vanishes and thus the trivial solution of Equations (6.145) and (6.147) given by

$$\phi(x) = 0 \tag{6.163}$$

does not correspond to each A_n vanishing. Note that in this case, Equations (6.138) and (6.139) have a solution simultaneously as the slip wave speed and Rayleigh wave speeds coincide. Therefore, a solution of the form Equation (6.161) exists. This solution implies that the contact pressure is equal to zero, while the normal displacements are continuous but not zero. It corresponds to the case of two Rayleigh waves (traveling or standing) of the same amplitude, wavelength, and phase, which exist simultaneously in the two bodies without interaction. This solution exists for any value of the friction coefficient. It exists also for the case of high sliding velocity if the Rayleigh wave speeds are related to the sliding velocity by

$$C_R^{(1)} - C_R^{(2)} = V \tag{6.164}$$

6.4.3.8 Numerical Analysis of the Integral Equation

6.4.3.8.1 Reduction to a System of Linear Algebraic Equation

The lowest frequency of vibration, which corresponds to $m = 2n = 1$ in Equation (6.151), is the most important, so we concentrate on a numerical investigation of this case. Applying the method developed by Erdogan and Gupta (1972), we write Equation (6.152) in the form of

$$4\pi\phi(x)K_r^\infty - 4K_i^\infty \int_{-1}^{1} \frac{\phi(\xi)d\xi}{x - \xi} +$$

$$\int_{-1}^{1} \phi(\xi)\left\{ K_i^\infty \frac{\phi(\xi)}{(x-\xi)} - cK_i^\infty \cot\frac{c(x-\xi)}{4} - cK_i^\infty \tan\frac{c(x-\xi)}{4} + ck(c(x-\xi)) \right\}d\xi = 0$$

$$|x| < 1$$

$$\tag{6.155}$$

The singular term is in the first integral, whereas the second integrand is bounded. Following Erdogan and Gupta (1972), we define the degree of singularity at the leading edge as

$$\alpha = \frac{2}{2\pi i} \log\left(\frac{K_r^\infty - iK_i^\infty}{K_r^\infty + iK_i^\infty}\right) = -\frac{1}{2\pi i}\arctan\left(\frac{K_i^\infty}{K_r^\infty}\right) \qquad 0 < \alpha < 1 \qquad (6.166)$$

and the weighting function as

$$w(x) = (1-x)^\alpha (1+x)^{1-\alpha} \qquad (6.167)$$

Note that the stresses are bounded at the transitions between separation and contact zones. The unknown function $f(x)$ is sought in the form of

$$\phi(x) = \sum_{n=0}^{\infty} c_n w(x) P_n^{(\alpha, 1-\alpha)}(x) \qquad (6.168)$$

where $P_n^{(\alpha, 1-\alpha)}(x)$ is a Jacobi polynomial of order n. The coefficients c_n can be found from the system of linear equations:

$$D(x_i) = \sum_{n=0}^{\infty} c_n D_{ni}$$

$$= \sum_{n=0}^{\infty} c_n \{-8K_i^\infty \pi\} P_{n+1}^{(-\alpha, -\alpha)}(x_i) + \int_{-1}^{1} w(\xi) P_n^{(\alpha, \alpha)}(\xi) \qquad (6.169)$$

$$\left(\frac{4K_i^\infty}{x-\xi} - cK_i^\infty \cot\frac{c(x_i - \xi)}{4} - cK_i^\infty \tan\frac{c(x_i - \xi)}{4} + ck(c(x_i - \xi))\right) d\xi = 0$$

where D_{ni} denotes the term in figure brackets. In order to solve this system of equations, the summation can be truncated at $n = N$ with a finite value of N. A collocation method can be applied to Equation (6.169). The collocation points in this bounded integrand are taken as evenly spaced in the interval $(-1, 1)$—that is, x_i ($i = 0, 1, 2, \ldots, N$)—and the integration points are defined according to the Jacobi–Gauss method:

$$\xi_i = \cos\frac{\pi(2j+1)}{(2N+1)} \qquad j = 0, 1, 2, \ldots, N \qquad (6.170)$$

In order for non-trivial solutions of Equation (6.170) to exist, the determinant of the $N + 1$ by $N + 1$ matrix is necessary:

$$D(\Lambda, \mu) = \left| D_{p-1, q-1} \right| = 0 \qquad (6.171)$$

6.4.3.8.2 Small Fraction and Sliding Velocity

Nosonovsky and Adams (2004) then considered the case of small friction $\mu \to 0$; the roots are expected to be close to those of the frictionless case. It was found in the analysis of the complete contact case (Ranjith and Rice 2001) that the real part of the root depends linearly on μ or small μ, so let us also assume such a linear dependence. For a small friction coefficient, the root is localized near the root for the frictionless case that lies on the imaginary axis—that is,

$$\Lambda = \pm i\lambda + \mu\Lambda_1 \tag{6.172}$$

where $i\lambda$ is purely imaginary. The determinant (Equation 6.171), which is a function of both μ and Λ, can be represented as

$$D(\Lambda,\mu) = \frac{\partial D}{\partial \mu}\mu + \frac{\partial D}{\partial \mu}\mu\Lambda_1 \tag{6.173}$$

which yields

$$\Lambda_1 = \frac{\dfrac{\partial D}{\partial \mu}}{\dfrac{\partial D}{\partial \Lambda}} \tag{6.174}$$

The determinant $D(i\lambda,\mu)$ is a purely real function when its first argument is purely imaginary, which results in a purely imaginary derivative $\partial D/\partial \Lambda$. For the case of zero sliding velocity ($V = 0$), it can be shown that the function $k(c(x - \xi))$ given by Equations (6.146) and (6.148) is purely real for non-zero μ, which results in a pure imaginary $\partial D/\partial \mu$ and yields a pure imaginary, thus, for zero sliding velocity and small friction, the roots of $D(\Lambda)$ remain purely imaginary and no energy dissipation or instability occurs.

The numerical investigation shows that for a small non-zero V, the imaginary part of $\partial D/\partial \mu$ is proportional to V—that is,

$$\text{Im}\left(\frac{\partial D}{\partial \mu}\right) = \gamma V \tag{6.175}$$

The degree of stability η can be defined as

$$\eta = \Lambda_1 / V \tag{6.176}$$

6.4.3.9 Discussion

As described in the previous section, Nosonovsky and Adams (2004) modeled roughness of the surfaces by a periodic wavy surface on one of the bodies. The contact occurs at periodically spaced contact regions on the peaks of the wavy asperities. The length of the contact regions is known from the solution of the steady-state problem; then the stability of sliding was analyzed. The overall solution was sought in the form of a superposition of the steady-state solution and small vibrations. Using a Fourier series representation for the vibration modes, the dynamic problem was reduced to a singular integral equation for determining the eigenvectors (modes) and eigenvalues (frequencies). The singular integral equation was analyzed by decomposing the unknown function as a Jacobi polynomial series multiplied by a weighting function and applying a collocation method. This procedure allowed us to replace the singular integral equation with a system of linear algebraic equations that was solved numerically.

For the frictionless case it was found that normal vibration modes exist for a wide range of material combinations. These modes correspond, in the case of complete contact, to generalized Rayleigh waves and can be interpreted as standing generalized Rayleigh waves. In the case of incomplete contact, and for the wavelength equal to twice the waviness period, the lowest frequency of vibration lies in the range between the lowest Rayleigh wave frequency of the two bodies and the generalized Rayleigh wave frequency. For the case of a vanishing small contact zone, the vibration mode can be interpreted as a standing Rayleigh wave in one of the bodies. For the general case of incomplete contact, vibration has been found for a wider range of material parameters than that for which generalized Rayleigh waves exist.

For the case of non-zero friction, complex vibration modes and eigenvalues were found. There are eigenvalues with a positive real part, which means that the vibrations have an amplitude that grows exponentially with time and thus steady sliding is unstable. In the case of incomplete contact, the degree of instability for small friction and velocity is proportional to both the friction coefficient and the sliding velocity. The degree of instability may be high enough to contribute to the formation of friction-induced vibrations for low-damping and moderate- or high-sliding velocities (of the order of 0.001 of the shear wave speed or higher). In the limiting case of complete contact, the results reduce to the destabilizing generalized Rayleigh wave analyzed by Adams (1995) and subsequently by Ranjith and Rice (1998). This result demonstrates that the effect of the dynamic sliding instability of Coulomb friction exists not only in the case of complete contact, but also in the more realistic case when the true contact area is much less than the nominal contact area. Finally, note that this analysis considers the cases for which the wavelength of vibration is equal to one or two times the waviness period. For wavelengths much greater than the waviness period, it might be anticipated that the effect of surface roughness would be less significant than was found here and thus the results of Adams (1995) and Ranjith and Rice (1998) would be more directly applicable.

These results have relevance in furthering our understanding of sliding friction as well as in the design and analysis of lip seals, such as the synthetic rubber seals used extensively in automobiles and other devices. Some models of lip seals include sinusoidal micro-undulations and micro-asperities (Salant and Flaherty 1995). A typical value of Young's modulus is 10 MPa, which gives a Rayleigh wave speed of about 100 m/s. Hence, the sliding velocity need not be extremely high in order to be a significant fraction of the Rayleigh wave speed.

6.4.4 Ill Posedness and Regularization

The ill posedness is manifest of the unstable growth of the interfacial disturbances of all wavelengths, with growth rate inversely proportional to the wavelength. Ranjith and Rice (1999) established the connection between the ill posedness and the existence of certain interfacial waves in friction contact, called the generalized Rayleigh wave. They also showed that for material combinations where a generalized Rayleigh wave exists, steady sliding with Coulomb friction is ill posed for arbitrarily small values of friction. Then, they studied regularization of the problem by an experimentally motivated friction law and showed that a friction law with no instantaneous dependence on normal stress but a simple fading memory of prior history of normal stresses makes the problem well posed.

To describe ill posedness, let V denote the velocity of steady sliding, the same at every point along the interface, and τ the shear stress at the interface. When the shear stress is perturbed in a single spatial mode of wave number k,

$$\Delta\tau = Q(t)e^{ikx_1} \tag{6.177}$$

where x_1 is the coordinate axis along the interface and $Q(t)$ is an arbitrary function of time, t, propagating slip-rate modes of the form

$$\Delta V = A(k)e^{ik(x_1 - ct)}e^{a|k|t} \tag{6.178}$$

are found, $A(k)$ is the amplitude of the mode, a and c are independent of the wavelength, and where $a > 0$ or a broad range of friction coefficients and material pairs. For such $a > 0$ cases, all wavelengths in the slip response are unstable and the growth rate of the instability is inversely proportional to the wavelength. An observer traveling with the speed c of the instability sees a perturbation velocity field that is the sum of an infinite number of modes, namely,

$$\Delta V(x+ct,t) = \int_{-\infty}^{\infty} A(k)e^{ik(x_1-ct)}e^{a|k|t}\,dk \qquad (6.179)$$

where $x = x_1 - ct$. Clearly, this integral fails to exist (diverges by oscillation for $x \neq 0$ in an arbitrarily small time after the perturbation is turned on, unless $A(k)$ decays exponentially or faster with $|k|$. Such a problem is said to be ill posed.

Renardy (1992) studied the sliding of a neo-Hookean elastic solid against a rigid substrate. In the limit of linear elasticity, he showed that ill posedness exists when the friction coefficient is sufficiently high—namely, greater than unity. This limiting case was studied independently by Martins et al. (1995). On the other hand, Adams (1995) showed that when the two solids on either side of the interface are linear elastic and not very dissimilar, the problem can be ill posed for arbitrarily small values of friction.

Earlier, Weertman (1963) and Achenbach and Epstein (1967) had shown that, in frictionless sliding of dissimilar elastic half-spaces constrained against formation of opening gaps, an interfacial wave solution can exist when the material mismatch is not very high. It is called the generalized Rayleigh wave since its speed of propagation reduces to that of the Rayleigh surface wave when the two materials are identical. The numerical results of Adams (1995) suggested a connection between the existence of the generalized Rayleigh wave and the ill posedness. Ranjith and Rice (1999) showed that for conditions under which the generalized Rayleigh wave exists in frictionless contact, the stability problem with Coulomb friction is ill posed for arbitrarily small values of friction.

Weertman (1980) argued that when such a wave exists, a self-healing slip pulse can propagate along the frictional interface between dissimilar elastic solids, even when the remote shear stress is less than the frictional strength of the interface, and a family of such pulse solutions has been constructed by Adams (1998). The velocity of propagation of the slip pulse is precisely that of the generalized Rayleigh wave. Numerical studies of the nucleation and propagation of such slip pulses with a Coulomb friction law at the interface, by Andrews and Ben-Zion (1997), Ben-Zion and Andrews (1998), and Harris and Day (1997), have difficulties that seem to originate in the instability and ill posedness results cited earlier. They observe that their simulations depend on mesh size and that a nucleated slip pulse splits into a number of pulses. Using the spectral numerical methodology for bi-materials, Breitenfeld and Geubelle (1998) and Cochard and Rice (1999) illustrate the ill posedness by showing that the more terms are in their spectral basis set, the more the pulse splitting is for a case that is ill posed in the sense discussed previously. They showed that the same method gives results that converge with enlargement of the basis set for parameter choices in the well-posed range.

Ranjith and Rice (1999) showed that an experimentally motivated friction law incorporating a memory dependence rather than instantaneous dependence on normal stress regularizes the problem of steady sliding along an interface between dissimilar elastic solids. A friction law for this problem must necessarily incorporate the response to varying normal stress, since slip along a dissimilar material interface alters normal stress. Creep-slippage experiments by Linker and Dieterich (1992) at sliding rates of order 1 μm/s suggest that friction has an instantaneous as well as a memory dependence on normal stress.

That is, in response to a step change $\Delta\sigma$ in compressive normal stress at approximately constant slip rate, which will ultimately lead to shear strength increase of $f\Delta\sigma$, a partial strength increase of $(f - \alpha)\Delta\sigma$ occurs at the time of the step and further memory-like increases $\alpha\Delta\sigma$ occur with continuing slip over a few μm distance. They found $\alpha \approx 0.30\ 0.8f$, depending on how their data fit.

But subsequent work by Prakash and Clifton (1993) and Prakash (1998) on high-speed sliding (1 to 10 m/s) induced by oblique shock impact, in which normal stress was altered over a much shorter timescale by a wave reflection, showed no instantaneous effect, but just a fading memory of a prior history of normal stress (i.e., $\alpha = f$). Ranjith and Rice (1999) used a friction law suggested by the latter experiments. Furthermore, studies by Martins and Simoes (1995) and Simoes and Martins (1998) showed that regularization can also be achieved by using a friction law in which the usual instantaneous dependence on normal stress is replaced by a dependence on normal stress averaged over some small finite area, although such a law seems not to be directly motivated by experiments and would not, for example, be consistent with memory effects in those just mentioned, for which normal stress was altered uniformly over a macroscopic sliding surface.

6.5 Radiation of Elastic Waves by Friction

As discussed in preceding sections of this chapter, starting with the classic discovery of elastic dilatational and shear waves conducted in the nineteenth century, many different types of elastic waves have been investigated. A Rayleigh wave can propagate along the free surface of a semi-infinite elastic body and has amplitude that decays exponentially with distance from the free surface. Similar waves can travel along the interface of two contacting elastic bodies. Such waves were investigated by Stoneley (1924) for bonded contact and are known as Stoneley waves. Stoneley waves exist only if the shear wave speeds of the two materials do not differ greatly. Achenbach and Epstein (1967) investigated interface waves in unbonded frictionless contact in which separation does not occur. These "smooth contact Stoneley waves" (also known as slip waves or generalized Rayleigh waves) are qualitatively similar to those for bonded contact and occur for a wider range of material combinations.

Comninou and Dundurs (1977) investigated slip waves with periodic regions of separation along a frictionless interface. The possibility of two identical half-spaces sliding with friction due to the presence of separation waves and/or stick–slip waves was also studied by Comninou and Dundurs (1978). Both of these analyses showed that such waves could exist only with square-root singularities at the tips of the slip zones. Freund (1978) pointed out that the singularities encountered by Comninou and Dundurs would require energy sources and sinks.

The notion that certain observed friction behavior is not a property of the interface, but rather a consequence of system dynamics, was suggested by Martins et al. (1990). Adams (1998) investigated the sliding of two dissimilar elastic bodies due to periodic regions of slip and stick propagating along the interface. He found that such motion allows for interface sliding conditions to differ from observed sliding conditions. In particular, the interface coefficient of friction can be constant or an increasing/decreasing function of slip velocity. However, the apparent coefficient of friction can be less than the interface friction coefficient.

Furthermore, the apparent coefficient of friction can decrease with sliding speed even though the interface friction coefficient is constant. Thus, the measured coefficient of friction does not necessarily represent the behavior of the sliding interface. In the limit, as the slip region becomes very small compared with the stick region, the results of Adams (1998) become that of a slip pulse traveling through a region that otherwise sticks. Rice (1997) derived that result in a simpler manner than the periodic solution of Adams by using the moving dislocation formulation of Weertman (1980). Andrews and Ben-Zion (1997) obtained a numerical solution for a slip pulse, the amplitude of which increases and the width of which decreases as the pulse continues to propagate. This self-sharpening effect is consistent with the analytical solution of Adams (1995) for sliding.

Adams (2000) investigated the sliding of an elastic half-space against a rigid surface and showed that steady sliding is compatible with the formation of a pair of body waves (a plane dilatational wave and a plane shear wave) radiated from the sliding interface. He also showed that a rectangular wave train, or a rectangular pulse, can allow for motion of the two bodies with a ratio of remote shear to normal stress (defined as the apparent friction coefficient) that is less than the ratio of shear to normal stress at the interface (defined as the interface coefficient of friction). Thus, the apparent coefficient of friction is less than the interface coefficient of friction.

Furthermore, the apparent friction coefficient decreases with increasing speed even if the interface friction coefficient is speed independent. These results support the interpretation of certain friction behavior as a consequence of the dynamics of the system, rather than as strictly an interface property (Martins et al. 1990).

Nosonovsky and Adams (2001) investigated the interaction of elastic body waves with frictional sliding. They showed that the phenomenon of elastic waves radiated from a sliding interface also exists in the general case of two elastic bodies in sliding contact. Each wave moves at a different angle with respect to the interface such that the trace velocities along the interface are equal and supersonic with respect to both media. This supersonicity does not violate causality as it is only the trace velocity that is supersonic; the waves move at the dilatational and shear wave speeds in their respective bodies. Nosonovsky and Adams (2001) also showed that a slip pulse can allow for motion of the two bodies with an apparent friction coefficient that is less than the interface coefficient of friction. In their analysis, which will be presented in the following section, both speed-independent and speed-dependent interface frictional laws are considered.

6.5.1 Formulation of the Problem

Nosonovsky and Adams (2001) considered two perfectly flat elastic half-spaces, moving relative to each other with constant speed V_0. The system is subjected to a remotely applied compressive normal traction (p^*) and shearing traction (q^*) as shown in Figure 6.7.

A well-known solution to the plane strain equations of motion is in the form of plane body waves in an infinite medium (e.g., Graf 1975). Consider plane dilatational and plane shear waves that move away from the interface at angles θ_1 and θ_2 in the lower body and θ_1' and θ_2' in the upper body, respectively. The displacement components in the x-y coordinate system for the lower body are

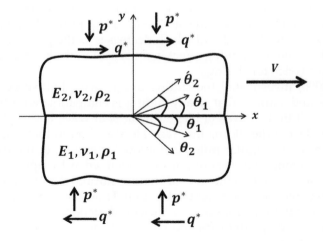

FIGURE 6.7
Frictional sliding of two elastic dissimilar semi-infinite bodies with dilatational and shear waves radiated from the interface in each body.

$$u(x, y, t) = A_1 \cos\theta_1 \exp[ik_1(x\cos\theta_1 - y\sin\theta_1 - c_1 t)]$$
$$+ A_2 \sin\theta_2 \exp[ik_2(x\cos\theta_2 - y\sin\theta_2 - c_2 t)]$$
$$v(x, y, t) = -A_1 \sin\theta_1 \exp[ik_1(x\cos\theta_1 - y\sin\theta_1 - c_1 t)]$$
$$A_2 \cos\theta_2 \exp[ik_2(x\cos\theta_2 - y\sin\theta_2 - c_2 t)]$$

$$(6.180)$$

Areas for the upper body displacements are

$$u'(x, y, t) = A_1' \cos\theta_1' \exp[ik_1'(x\cos\theta_1' - y\sin\theta_1' - c_1' t)]$$
$$+ A_2' \sin\theta_2' \exp[ik_2'(x\cos\theta_2' - y\sin\theta_2' - c_2' t)]$$
$$v'(x, y, t) = -A_1' \sin\theta_1' \exp[ik_1'(x\cos\theta_1' - y\sin\theta_1' - c_1' t)]$$
$$+ A_2' \cos\theta_2' \exp[ik_2'(x\cos\theta_2' - y\sin\theta_2' - c_2' t)]$$

$$(6.181)$$

as were determined by standard coordinate transformations.

In Equation (6.180), $u(x,y,t)$ and $v(x,y,t)$ are the components of displacement in the x- and y-directions, respectively; A_1 and A_2 are the wave amplitudes; k_1 and k_2 are the wave numbers for dilatational and shear waves, respectively; c_1 and c_2 are the dilatational and shear wave speeds, respectively. In Equation (6.181) the same parameters for the upper body are marked with primes ('). In order for the boundary conditions at the sliding interface to be satisfied, it is necessary that the trace velocity (c) and the wave number along the interface (k) of both waves in each body be identical—that is (Nosonovsky and Adams 2001),

$$c \equiv c_1 / \cos\theta_1 = c_2 / \cos\theta_2 = c_1' / \cos\theta_1 = c_2' / \cos\theta_2 \qquad (6.182)$$

$$k \equiv k_1 \cos\theta_1 = k_2 \cos\theta_2 = k_1' \cos\theta_1' = k_2' \cos\theta_2' \rightarrow k_2 = k_1 c_1 / c_2 = k_1 \beta,$$
$$k_1' = k_1 c_1 / c_1' = k_1 \beta / (\beta' \kappa), \quad k_2' = k_1 c_1 / c_2' = k_1 \beta / \kappa \qquad (6.183)$$

$$\theta_2 \cos^{-1}((1/\beta)\cos\theta_1), \ \theta_1' = \cos^{-1}(\kappa\beta'/\beta)\cos\theta_1, \ \theta_2' = \cos^{-1}((\kappa/\beta)\cos\theta_1)) \quad (6.184)$$

where

$$\beta = \frac{c_1}{c_2} = \sqrt{\frac{2(1-v)}{1-2v}}, \quad \beta' = \frac{c_1'}{c_2'} = \sqrt{\frac{2(1-v')}{1-2v'}}, \quad \kappa = \frac{c_2'}{c_1'} = \sqrt{\frac{\rho G'}{\rho' G}} \qquad (6.185)$$

In Equation (6.185) v and v' are Poisson's ratios of the two bodies, G and G' are the shear moduli, and ρ and ρ' are their densities. Thus, Nosonovsky and

Adams (2001) observed from Equation (6.185) that the dilatational and shear waves yield wave motion along the interface that is supersonic with respect to both bodies. This behavior does not violate causality as it is only the trace velocity along the interface that is supersonic. Furthermore, Equation (6.175) provides constraints between the angles of propagation of the dilatational and shear waves in the two bodies.

Finally, the normal and shearing stresses in each body are given by Nosonovsky and Adams (2001):

$$\tau_{yy}(x,y,t) = A_1 G i k_1 [(c_1/c_2)^2 - 1 - \cos 2\theta_1] \exp[ik(x - y\tan\theta_1 - ct)]$$
$$- A_2 G i k_1 \sin 2\theta_2 \exp[ik(x - y\tan\theta_2 - ct)],$$

$$\tau_{xy}(x,y,t) = -A_1 G i k_1 \sin 2\theta_1 \exp[ik(x - y\tan\theta_1 - ct)]$$
$$+ A_2 G i k_2 \cos 2\theta_2 \exp[ik(x - y\tan\theta_2 - ct)]$$

$$\tau'_{yy}(x,y,t) = A'_1 G' i k_1 [(c'_1/c'_2)^2 - 1 - \cos 2\theta'_1] \exp[ik(x + y\tan\theta_1 - ct)]$$
$$+ A'_2 G' i k'_2 \sin 2\theta'_2 \exp[ik(x + y\tan\theta'_2 - ct)],$$

$$\tau'_{xy}(x,y,t) = A'_1 G' i k'_1 \sin 2\theta'_1 \exp[ik(x + y\tan\theta'_1 - ct)]$$
$$+ A'_2 G' i k'_2 \cos 2\theta'_2 \exp[ik(x + y\tan\theta'_2 - ct)]$$

(6.186)

6.5.2 Steady Sliding with Radiated Waves

For the frictional sliding contact of two elastic bodies, there are four boundary conditions (Nosonovsky and Adams 2001):

$$v(x,0,t) = v'(x,0,t) \qquad (6.187)$$

$$\tau_{yy}(x,0,t) = \tau'_{yy}(x,0,t) \qquad (6.188)$$

$$\tau_{xy}(x,0,t) = \tau'_{xy}(x,0,t) \qquad (6.189)$$

$$q*+\tau_{xy}(x,0,t) = \mu[p*-\tau_{yy}(x,0,t)] \qquad (6.190)$$

In Equation (6.190), μ is the interface coefficient of friction (i.e., the ratio of shear to normal stress at the interface). If steady motion occurs, then the ratio $q*/p*$ is defined as the apparent coefficient of friction ($\mu*$). It is this value that would be measured in a sliding experiment as the coefficient of friction. At this point of the analysis, Equation (6.181) requires that $\mu*$ and μ to be equal. As will be seen in the next section, $\mu*$ and μ may, in general, be different.

Because $q^* = \mu^* p^*$ and $\mu^* = \mu$, Equations (6.187–6.190) lead to a system of four equations:

$$-A_1 \sin \theta_1 + A_2 \cos \theta_2 - A_1' \sin \theta_1' - A_2' \cos \theta_2' = 0 \qquad (6.191)$$

$$-A_1 \sin 2\theta_1 + A_2 \beta \cos 2\theta_2 - A_1' \left(\frac{G}{G} \right) \left(\frac{\beta}{\kappa \beta'} \right) \sin 2\theta_1' - A_2' \left(\frac{G}{G} \right) \left(\frac{\beta}{\kappa} \right) \cos 2\theta_2' = 0 \quad (6.192)$$

$$A_1 (\beta^2 - 1 - \cos 2\theta_1) - A_2 \beta \sin 2\theta_2 - A_1' \left(\frac{G'}{G} \right) \left(\frac{\beta}{\kappa \beta'} \right) (\beta'^2 - 1 - \cos 2\theta_1')$$

$$- A_2' \left(\frac{G'}{G} \right) \left(\frac{\beta}{\kappa} \right) \sin 2\theta_2' = 0 \qquad (6.193)$$

$$M = -\tau_{xy}(x,0,t) / \tau_{yy}(x,0,t) = \frac{A_1 \sin 2\theta_1 - A_2 \beta \sin 2\theta_2}{A_1 (\beta^2 - 1 - \cos 2\theta_1) - A_2 \beta \sin 2\theta_2} \qquad (6.194)$$

Note that Equation (6.194) defines the parameter M, which is a ratio between the contributions of the wave solutions to the tangential and normal stresses, also known as the wave coefficient of friction (Graf 1975). At this point of the analysis, $M = \mu^* = \mu$, although it will be shown in the next section that these three quantities can differ. A summary of the definitions of these coefficients is provided in Table 6.1. In order to obtain solutions for sliding with a pair of body waves and for given material properties and coefficient of friction, it is necessary to solve Equations (6.191–6.194) for values of $M = \mu$ that give non-trivial solutions for A_1, A_2, A_1', and A_2'. It is, however, mathematically more convenient to vary θ_1, calculate θ_2, θ_1', and θ_2' from Equation (6.184), solve for A_1/A_2', A_2/A_2', and A_1'/A_2' from Equations (6.191–6.194), and then calculate M from Equation (6.194). Note that, if $\beta/\kappa\beta' < 1$ that corresponds to $c_1 < c_1'$ and

TABLE 6.1

Definitions of the Three Coefficients of Friction

Symbol	Name	Definition
μ	Interface coefficient of friction	Ratio of total shear to total normal stress at the interface
μ^*	Apparent coefficient of friction (can be less than μ due to the reduction of friction caused by self-organized stick–slip pattern)	Ratio of remote shear to remote normal stress
M	Wave parameter ("wave coefficient of friction" in Graf 1975)	Ratio of shear to normal stress due to wave motion

Source: Nosonovsky, M. and Adams, G. G. 2001.

then θ_1 can be varied only for $\theta_1 > \cos^{-1}(\beta/\kappa\beta')$. Because of the symmetry of the configuration, it is sufficient to consider the values of $\beta/\kappa\beta' > 1$ and $0 < \theta_1 < \pi/2$. Solutions with negative μ correspond to the opposite direction of sliding, which, because of the symmetry, is the same as $\pi/2 < \theta_1 < \pi$ with positive μ.

Equation (6.194) gives the ratio of shear stress to contact pressure, which is independent of wavelength, spatial position, and time. In a similar manner a parameter can be obtained that characterizes the ratio of shear stress to relative slip velocity and is defined according to Nosonovsky and Adams (2001) as

$$\alpha = [\tau_{xy}(x,0,t)/G]/(\dot{u}(x,0,t) - \dot{u}'(x,0,t))/c_2] \tag{6.195}$$

where

$$\dot{u}(x,0,t) = -ikc(A_1 \cos\theta_1 + A_2 \sin\theta_2)\exp[ik(x - ct)]$$
$$\dot{u}'(x,0,t) = -ikc(A_1' \cos\theta_1' + A_2' \sin\theta_2')\exp[ik(x - ct)] \tag{6.196}$$

Thus,

$$\alpha = \frac{2A_1 \sin\theta_1 \cos\theta_2 - A_2 \cos 2\theta_2}{(-A_1' \cos\theta_1' + A_2' \sin\theta_2' + A_1 \cos\theta_1 + A_2 \sin\theta_2)} \tag{6.197}$$

and α depends only on M and the material parameters of the two bodies; it is independent of wavelength, x and t.

If a more complicated frictional law than the one described by Equation (6.190) is considered (e.g., the one that includes a dependence on sliding velocity), Equations (6.191–6.193) still remain valid. However, Equation (6.194) must be replaced with a different relation. For example, the frictional law,

$$\tau_{xy} = -\mu\tau_{yy} + \eta(\dot{u}' - \dot{u}) \tag{6.198}$$

along with Equation (6.195), yields

$$M = \mu = -(1 + \eta c_2/G\alpha)\tau_{xy}/\tau_{yy} \tag{6.199}$$

Thus, the effect of η is to alter the relationship between the friction coefficient and the wave angles. Because α depends upon the wave angles, Equation (6.199) does not constitute a simple scaling of Equation (6.194). Also note that positive η corresponds to the coefficient of friction increasing with increasing velocity, whereas negative η corresponds to the dynamic coefficient of friction decreasing with increasing velocity. Each of these phenomena can be observed in tribological systems.

6.5.3 A Slip Pulse

Nosonovsky and Adams (2001) also investigated the possible relative motion of the two bodies due to the presence of a propagating stick–slip pulse. Equation (6.194) gives the ratio of shear and normal stresses (i.e., M), whereas Equation (6.197) gives the ratio of normalized shear stress to relative slip velocity (i.e., α). Each of these quantities is independent of position, time, wavelength, and wave amplitudes for given values of wave angles.

Nosonovsky and Adams (2001) considered a superposition of waves obtained in the previous section, the form of which is a finite length rectangular pulse that is invariant in the moving coordinates $x - ct$:

$$\begin{aligned}
\tau_{yy} &= \tau_{yy}^{slip} \qquad \text{if } |x - ct| < a \\
\tau_{yy} &= 0 \qquad\quad \text{if } |x - ct| > a
\end{aligned} \tag{6.200}$$

The shear stress and the relative slip velocity remain proportional to τ_{yy} according to $\tau_{xy} = -M\tau_{yy}$ and Equation (6.186), respectively, and thus τ_{xy} and $(\dot{u}' - \dot{u})$ also form rectangular pulses. In the slip region, Coulomb's law (Equation 6.190) yields

$$q* - M\tau_{yy}^{slip} = \mu(p* - \tau_{yy}^{slip}) \qquad |x - ct| < a \tag{6.201}$$

which results in the apparent coefficient of friction:

$$\mu* = \frac{q*}{p*} = \mu - \frac{\tau_{yy}^{slip}}{p*}(\mu - M) \tag{6.202}$$

This is different from the interface coefficient of friction μ. The relative slip velocity V_s is defined as

$$V_s = \dot{u}'(x,0,t) - \dot{u}(x,0,t) = M\tau_{yy}^{slip} \frac{c_2}{\alpha G}, \quad |x - ct| < a \tag{6.203}$$

and is constant in the slip zone and zero in the stick regions.

Nosonovsky and Adams (2001) noted that the slip velocity is discontinuous at the border between stick and slip. Thus, material points undergo an infinite acceleration as the leading/trailing edge of the pulse passes by. This behavior is consistent with the moving discontinuous stress field. Also, the pulse solution was obtained in the context of a speed-independent friction law. However, since the slip velocity is constant in the slip region, the same result would also hold with a speed-dependent friction law in which the value of μ corresponding to the speed V_s is used.

There are, however, some constraints that apply to this solution. In order for stick to occur, the ratio of remotely applied shear to normal stress must be less than the interface friction coefficient, or

$$q^* < \mu p^* \rightarrow \mu^* < \mu \tag{6.204}$$

To avoid separation between the bodies, the total normal stress must be positive:

$$p^* - \tau_{yy}^{slip} \geq 0 \tag{6.205}$$

For the range of parameters considered, it was found that $a > 0$ and $M > 0$. It can therefore be shown that $\tau_{yy}^{slip} > 0$ and that these constraints yield

$$\mu > \mu^* \geq M \tag{6.206}$$

Finally, the magnitude of the slip distance can be found by multiplying the relative slip velocity by the time required for the pulse to propagate a distance equal to its length. The result is

$$U^{Slip} = \left(\frac{2a}{c}\right) V_s = (\gamma \alpha \tau_{yy}^{slip} / p^*)(p^*/G) \quad \gamma \equiv 2(M/\alpha)\cos\theta_2 \tag{6.207}$$

Thus, the magnitude of the slip distance depends non-linearly upon M since M, α, and θ_2 are inter-related according to Equations (6.194) and (6.195) and linearly upon the magnitude and width of the pulse. Figure 6.8 shows schematically the slip parameter γ versus the parameter M for $\rho = \rho'$. The exact value of γ as a function of M can be obtained for various values of κ^2. The slip increases with increasing M; this tendency is more obvious with more similar materials ($\kappa^2 \rightarrow 1$).

6.5.4 Discussion

Nosonovsky and Adams (2001) showed that body waves can be induced by the frictional sliding of two elastic half-spaces. This phenomenon exists for the coefficient of friction greater than a certain minimum value μ_0. The waves occur with a speed-dependent or speed-independent friction law. The pairs of plane dilatational and shear waves are radiated from the interface and propagate at different angles away from the sliding interface. Nosonovsky

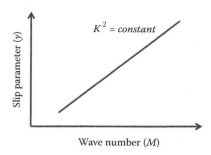

FIGURE 6.8
The slip parameter versus the wave number for constant k².

and Adams also investigated the possible existence of a slip pulse, formed by a superposition of these waves. Such a pulse constitutes a propagating slip zone in which the frictional slip condition is satisfied. The formation of such a slip zone, surrounded by regions of stick, allows for the apparent coefficient of friction to be less than the interface coefficient of friction.

6.6 Interaction of Elastic Waves with Friction

Nosonovsky and Adams (2002) investigated the interaction of incident elastic dilatational and shear waves with a frictional sliding interface. They used simple speed-independent Amontons–Coulomb friction without including rate-and-state dependent friction or any length scale effects associated with asperity interactions. Thus, they neglected any effect of frequency on the reflection and transmission coefficients. They determined the dependencies of the amplitudes and wave angles of the reflected and refracted waves on the material parameters, friction coefficient, and incoming wave angle. They also showed that, for an incoming wave that forms a rectangular train of pulses, a relative stick–slip motion of the two bodies can occur. Therefore, the vibrational motion of the elastic half-space can be transformed into a linear sliding motion of the half-spaces relative to each other.

This type of motion transformation is similar in principle to that of "ultrasonic propulsion" motors (Zhari 1996). Those devices are based on inducing propagating elastic surface waves in a piezoelectric stator, which cause the rotor to move over the stator. The subject of the study by Nosonovsky and Adams (2002) was motion transformation through the propagation of stick and slip zones induced by incident body waves, as opposed to propagation of contact and separation regions in conventional ultrasonic motors. The propagation of an isolated slip pulse can be used for position control, while the rectangular train of the pulses can serve for velocity control.

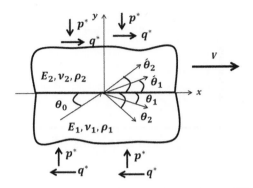

FIGURE 6.9
Effect of an incoming elastic wave on the frictional interface: two reflected and two refracted waves, both shear and dilatational.

6.6.1 Incoming Dilatational Wave

Nosonovsky and Adams (2002) considered two perfectly flat elastic half-spaces, moving relative to each other with constant speed V_0. The system is subjected to a remotely applied compressive normal traction (p^*) and shearing traction (q^*) as shown in Figure 6.9 (Nosonovsky and Adams 2002).

A well-known solution to the plane strain equations of motion is in the form of plane body waves in an infinite medium (e.g., Graf 1975). Consider a plane dilatational wave with wave number k_0 in the lower body that moves toward the interface at angle θ_0. After interacting with the interface, four plane waves will be produced—namely reflected dilational and shear waves at angles θ_1 and θ_2 in the lower body—and refracted dilatational and shear waves at angles θ_1' and θ_2' in the upper body, respectively. The displacement components in the x-y coordinate system for the lower body are

$$u(x,y,t) = A_0 \cos\theta_0 \exp[ik_0(x\cos\theta_0 + y\sin\theta_0 + y\sin\theta_0 - c_1t)]$$
$$+ A_1 \cos\theta_1 \exp[ik_1(x\cos\theta_1 + y\sin\theta_1 - c_1t)]$$
$$+ A_2 \sin\theta_2 \exp[ik_2(x\cos\theta_2 - y\sin\theta_2 - c_2t)] \tag{6.208}$$
$$v(x,y,t) = -A_1 \sin\theta_1 \exp[ik_1(x\cos\theta_1 - y\sin\theta_1 - c_1t)]$$
$$+ A_2 \cos\theta_2 \exp[ik_2(x\cos\theta_2 - y\sin\theta_2 - c_2t)]$$

whereas, for the upper body, the displacements are

$$u'(x,y,t) = A_1' \cos\theta_1' \exp[ik_1'(x\cos\theta_1' - y\sin\theta_1' - c_1't)]$$
$$+ A_2' \sin\theta_2' \exp[ik_2'(x\cos\theta_2' - y\sin\theta_2' - c_2't)] \tag{6.209}$$
$$v'(x,y,t) = -A_1' \sin\theta_1' \exp[ik_1'(x\cos\theta_1' - y\sin\theta_1' - c_1't)]$$
$$+ A_2' \cos\theta_2' \exp[ik_2'(x\cos\theta_2' - y\sin\theta_2' - c_2't)]$$

as were determined by standard coordinate transformations.

In Equation (6.208), $u(x,y,t)$ and $v(x,y,t)$ are the components of displacement in the x- and y-directions, respectively; A_0 is the amplitude of the incoming dilatational wave. A_1 and A_2 are the wave amplitudes; k_1 and k_2 are the wave numbers for dilatational and shear waves, respectively; c_1 and c_2 are the dilatational and shear wave speeds, respectively. In Equation (6.209), the same parameters for the upper body are marked with primes (′). In order for the boundary conditions at the sliding interface to be satisfied, it is necessary that the trace velocity (c) and the wave number along the interface (k) of both waves in each body be identical (Nosonovsky and Adams 2002)—that is,

$$c \equiv c_0 / \cos\theta_0 \equiv c_1 / \cos\theta_1 = c_2 / \cos\theta_2 = c_1' / \cos\theta_1 = c_2' / \cos\theta_2 \quad (6.210)$$

$$k \equiv k_0 \cos\theta_0 \equiv k_1 \cos\theta_1 = k_2 \cos\theta_2 = k_1' \cos\theta_1' = k_2' \cos\theta_2' \rightarrow k_1 = k_0, k_2 =$$
$$k_1 c_1 / c_2 = k_1 \beta, \; k_1' = k_1 c_1 / c_1' = k_1 \beta / (\beta' \kappa), \quad k_2' = k_1 c_1 / c_2' = k_1 \beta / \kappa \quad (6.211)$$

$$\theta_2 = \cos^{-1}((1/\beta)\cos\theta_1), \, \theta_1 = \cos^{-1}(\kappa\beta'/\beta)\cos\theta_1, \, \theta_2' = \cos((\kappa/\beta)\cos\theta_1)) \quad (6.212)$$

where

$$\beta = \frac{c_1}{c_2} = \sqrt{\frac{2(1-v)}{1-2v}}, \quad \beta' = \frac{c_1'}{c_2'} = \sqrt{\frac{2(1-v')}{1-2v'}}, \quad k = \frac{c_2'}{c_1'} = \frac{\sqrt{\rho G'}}{\rho' G} \quad (6.213)$$

In Equation (6.213) v and v' are Poisson's ratios of the two bodies, G and G' are the shear moduli, and ρ and ρ' are their densities. Thus, it is observed from Equation (6.210) that the dilatational and shear waves yield wave motion along the interface that is supersonic with respect to both bodies. This behavior does not violate causality as it is only the trace velocity along the interface that is supersonic. Furthermore, Equation (6.212) provides constraints between the angles of propagation of the dilatational and shear waves in the two bodies.

Finally, the normal and shearing stresses in each body are given by Nosonovsky and Adams (2002):

$$\tau_{yy}(x,y,t) = A_0 Gik_0[(c_1 / c_2)^2 - 1 - \cos 2\theta_0]\exp[ik(x + y\tan\theta_0 - ct)]$$
$$A_1 Gik_1[(c_1 / c_2)^2 - 1 - \cos 2\theta_1]\exp[ik(x - y\tan\theta_1 - ct)]$$
$$- A_2 Gik_1 \sin 2\theta_2 \exp[ik(x - y\tan\theta - ct)],$$
$$\tau_{xy}(x,y,t) = A_0 Gik_0 \sin 2\theta_0 \exp[ik(x + y\tan\theta_0 - (t)]$$
$$- A_1 Gik_1 \sin 2\theta_1 \exp[ik(x - y\tan\theta_1 - ct)]$$
$$+ A_2 Gik_2 \cos 2\theta_2 \exp[ik(x - y\tan\theta_2 - ct)],$$
$$\tau_{yy}'(x,y,t) = A_1'G'ik_1[(c_1' / c_2')^2 - 1 - \cos 2\theta_1]\exp[ik(x + y\tan\theta_1 - ct)]$$
$$+ A_2'G'ik_2 \sin 2\theta_2 \exp[ik(x + y\tan\theta_2 - ct)],$$
$$\tau_{xy}(x,y,t) = A_1'G'ik_1 \sin 2\theta_1 \exp[ik(x + y\tan\theta_1 - ct)]$$
$$+ A_2'G'ik_2' \cos 2\theta_2' \exp[ik(x + y\tan\theta_2' - ct)]$$

$$(6.214)$$

For the frictional sliding contact of two elastic bodies, there are four boundary conditions (Nosonovsky and Adams 2002):

$$v(x,0,t) = v'(x,0,t) \qquad (6.215)$$

$$\tau_{yy}(x,0,t) = \tau'_{yy}(x,0,t) \qquad (6.216)$$

$$\tau_{xy}(x,0,t) = \tau'_{xy}(x,0,t) \qquad (6.217)$$

$$q*+\tau_{xy}(x,0,t) = \mu[p*-\tau_{yy}(x,0,t)] \qquad (6.218)$$

In Equation (6.218), μ is the interface coefficient of friction (i.e., the ratio of shear to normal stress at the interface), as defined in Table 6.1. Moreover, if steady motion occurs, then the ratio $q*/p*$ is defined as the apparent coefficient of friction ($\mu*$). At this point of the analysis, Equation (6.218) requires $\mu*$ and μ to be equal. However, $\mu*$ and μ may, in general, be different.

Because $q* = \mu*p*$ and $\mu* = \mu$, Equations (6.215–6.218) lead to a system of four equations (Nosonovsky and Adams 2002):

$$-A_1 \sin\theta_1 + A_2 \cos\theta_2 - A_1 \sin\theta_1 - A_2 \cos\theta_2 = -A_0 \sin\theta_1 \qquad (6.219)$$

$$-A_1 \sin 2\theta_1 + A_2\beta \cos 2\theta_2 - A_1'\left(\frac{G}{G}\right)\left(\frac{\beta}{\kappa\beta'}\right)\sin 2\theta_1' - A_2'\left(\frac{G}{G}\right)\left(\frac{\beta}{\kappa}\right)\cos 2\theta_2'$$

$$= -A_0 \sin 2\theta_1 \qquad (6.220)$$

$$A_1(\beta^2 - 1 - \cos 2\theta_1) - A_2\beta \sin 2\theta_2 - A_1'\left(\frac{G}{G}\right)\left(\frac{\beta}{\kappa\beta'}\right)(\beta'^2 - 1 - \cos 2\theta_1')$$

$$- A_2'\left(\frac{G'}{G}\right)\left(\frac{\beta}{\kappa}\right)\sin 2\theta_2' = -A_0(\beta^2 - 1 - \cos 2\theta_1) \qquad (6.221)$$

$$MA_1(\beta^2 - 1 - \cos 2\theta_1) - A_1 \sin 2\theta_1 + A_2\beta \cos 2\theta_2 - MA_2\beta \sin 2\theta_2$$

$$= -A_0(\sin 2\theta_1 + M(\beta^2 - 1 - \cos 2\theta_1)) \qquad (6.222)$$

Note that Equation (6.222) defines the parameter M, which is a ratio between the contributions of the wave solutions to the tangential and normal stresses, also known as the wave coefficient of friction (Graff 1975). At this point of the analysis, $M = \mu* = \mu$, although it will be shown in the next section that these three quantities can differ. A summary of the definitions of these coefficients is provided in Table 6.1.

For given μ, θ_0 and material properties, Equations (6.219–6.222) can be solved for A_1, A_2, A_1', and A_2', which are each proportional to A_0. Note that the wave friction coefficient M is independent of wavelength, spatial position, and time. In a similar manner, a parameter α can be obtained that characterizes the ratio of shear stress to the change in relative slip velocity (i.e., the wave contribution to the relative slip velocity) and is defined according to Nosonovsky and Adams (2002):

$$\alpha = [\tau_{xy}(x,0,t)/G]/(\dot{u}(x,0,t) - \dot{u}'(x,0,t))/c_2] \tag{6.223}$$

where

$$\dot{u}(x,0,t) - ikc((A_0 + A_1)\cos\theta_1 + A_2\sin\theta_2)\exp[ik(x-ct)]$$
$$\dot{u}'(x,0,t) = -ikc(A_1'\cos\theta_1' + A_2'\sin\theta_2')\exp[ik(x-ct)] \tag{6.224}$$

Thus,

$$\alpha = \frac{2(A_1 - A_0)\sin\theta_1\cos\theta_2 - A_2\cos2\theta_2}{(-A_1'\cos\theta_1' + A_2'\sin\theta_2' + A_1\cos\theta_1 + A_2\sin\theta_2)} \tag{6.225}$$

and α depends only on M, θ_0 and the material parameters of the two bodies; it is independent of wavelength, x and t.

If a more complicated frictional law than the one described by Equation (6.218) is considered (e.g., the one that includes a dependence on sliding velocity), Equations (6.219–6.221) still remain valid. However, Equation (6.222) must be replaced with a different relation.

6.6.2 Incoming Shear Wave

Reflection of an incoming shear wave, of amplitude A_0 and wave number k_0 is described by the following displacement components in the lower body (Nosonovsky and Adams 2002):

$$u(x,y,t) = -A_0\sin\theta_0\exp[ik_0(x\cos\theta_0 + y\sin\theta_0 - c_2t)] +$$
$$A_1\cos\theta_1\exp[ik_1(x\cos\theta_1 - y\sin\theta_1 - c_1t)] +$$
$$A_2\sin\theta_2\exp[ik_2(x\cos\theta_2 - y\sin\theta_2 - c_2t)]$$
$$v(x,y,t) = A_0\cos\theta_0\exp[ik_{01}(x\cos\theta_0 + y\sin\theta_0 - c_1t)] - \tag{6.226}$$
$$A_1\sin\theta_1\exp[ik_1x\cos\theta_1 - y\sin\theta_1 - c_1t)] +$$
$$A_2\cos\theta_2\exp[ik_2(x\cos\theta_2 - y\sin\theta_2 - c_2t)]$$

where

$$\theta_0 = \theta_2 \qquad k_0 = k_2 \tag{6.227}$$

and Equations (6.210–6.213) are still valid. For the upper body, the displacements of the refracted waves are given by Equation (6.208). The normal and shearing stresses in the lower body are given by

$$
\begin{aligned}
\tau_{yy}(x,y,t) &= A_0 G i k_0 \sin 2\theta_0 \, \exp[ik(x + y \tan \theta_0 - ct)] + \\
&\quad A_1 G i k_1 [(c_1 / c_2)^2 - 1 - \cos 2\theta_1] \exp[ik(x - y \tan \theta_1 \\
&\quad ct)] - A_2 G i k_1 \sin 2\theta_2 \, \exp[ik(x - y \tan \theta_2 - ct)] \\
\tau_{xy}(x,y,t) &= A_0 G i k_0 \cos 2\theta_0 \, \exp[ik(x + y \tan \theta_0 - ct)] - \\
&\quad A_1 G i k_1 \sin 2\theta_1 \, \exp[ik(x - y \tan \theta_1 - \\
&\quad ct)] + A_2 G i k_2 \cos 2\theta_2 \, \exp[ik(x - y \tan \theta_2 - ct)]
\end{aligned}
\tag{6.228}
$$

For the upper body, the two last parts of Equation (6.214) remain valid.

The boundary conditions (Equations 6.215–6.218) remain unchanged and yield the following system of equations:

$$-A_1 \sin \theta_1 + A_2 \cos \theta_2 - A_1' \sin \theta_1 - A_2' \cos \theta_2' = -A_0 \cos \theta_2 \tag{6.229}$$

$$
A_1 \sin 2\theta_1 + A_2 \beta \cos 2\theta_2 - A_1'\left(\frac{G'}{G}\right)\left(\frac{\beta}{\kappa \beta'}\right) \sin 2\theta_1' - A_2'\left(\frac{G'}{G}\right)\left(\frac{\beta}{\kappa}\right) \cos 2\theta_2 =
$$
$$
- A\beta \cos 2\theta_2
\tag{6.230}
$$

$$
A_1(\beta^2 - 1 - \cos 2\theta_1) - A_2 \beta \sin 2\theta_2 - A_1'\left(\frac{G'}{G}\right)\left(\frac{\beta}{\kappa \beta'}\right)(\beta^2 - 1 - \cos 2\theta_1') -
$$
$$
A_2\left(\frac{G}{G}\right)\left(\frac{\beta}{\kappa}\right) \sin 2\theta_2 = -A_0 \beta \sin 2\theta_1
\tag{6.231}
$$

$$
MA_1(\beta^2 - 1 - \cos 2\theta_1) - A_1 \sin 2\theta_1 + A_2 \beta \cos 2\theta_2 - MA_2 \beta \sin 2\theta_2 =
$$
$$
- A_0(\beta \cos 2\theta_1 + M\beta \sin 2\theta_1)
\tag{6.232}
$$

For given μ, θ_0 and material properties, Equations (6.229–6.231) can be solved for A_1, A_2, A_1', and A_2', which are each proportional to A_0. Finally, the parameter α is now given by

$$\alpha = \frac{2A_1 \sin\theta_1 \cos\theta_2 - (A_2 + A_0)\cos 2\theta_2}{(-A_1' \cos\theta_1' + A_2' \sin\theta_2' + A_1 \cos\theta_1 + (A_2 - A_0)\sin\theta_2)} \tag{6.233}$$

6.6.3 Discussion

Nosonovsky and Adams (2002) investigated the effect of incident harmonic plane dilatational and shear waves on an elastic half-space that is in sliding frictional contact with another elastic half-space. In their study, the two half-spaces had a contact interface that was governed by the Coulomb friction law. Interacting with the interface, the waves formed two reflected waves (shear and dilatational) in the first half-space and two refracted waves in the second half-space. They also investigated the dependencies of the reflected and refracted wave amplitudes on the incoming wave angles. They showed that the amplitudes of contact normal stress, shear stress, and relative sliding velocity are proportional to each other.

Nosonovsky and Adams (2002) also investigated the interaction of a rectangular interfacial wave pulse, formed by a superposition of harmonic waves, or of an infinite set of pulses, with the frictional interface. They showed that an interfacial stick–slip motion of the two bodies can be a result of this interaction and that this motion can occur either in the same or opposite direction of the remotely applied shear traction or without a shear traction. The velocity of relative motion (for a series of pulses) and the slip distance (for a single pulse) as a function of material parameters, interface friction coefficient, remotely applied shear and normal tractions, and the parameters of the pulse have been given by Nosonovsky and Adams (2002). This phenomenon provides a mechanism of transformation of wave motion into steady sliding motion with the possibility of position control or sliding velocity control.

6.7 Friction Reduction and Self-Organized Patterns due to Friction-Induced Vibrations

An important property of elastic interfacial waves and vibrations is that they can lead to the reduction of the observed or apparent coefficient of friction (Adams 1998) because a train of propagating slip pulses at the interface results in what is observed as overall motion of the elastic bodies in contact (Figure 6.10). The studies within linear elasticity do not provide any quantitative criterion of how significantly the coefficient of friction can be decreased and how much energy can be saved as a result. The phenomenon is important because it fits the general tendency that, after a self-organized regime is formed, friction reduces (the Prigogine principle).

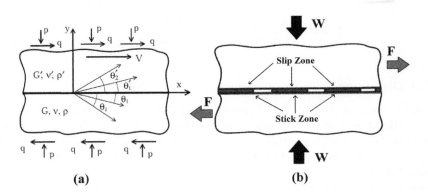

(a)   **(b)**

FIGURE 6.10

A superposition of dilatational and shear waves radiated from the interface can form a train of propagating slip pulses at the interface, thus resulting in apparent friction reduction. (The ratio of the remotely applied stresses q/p is less than m, but the relative motion occurs due to slip zone propagation.) Friction reduction due to self-organized structure is a common phenomenon not limited to mechanical waves.

6.8 Summary

In this chapter we studied stability of the frictional sliding of ideal elastic bodies. First, we discussed the concepts of the contact mechanics related to this effect. We formulated the entropic variational stability criterion of stationary sliding, $\delta^2 \dot{S} > 0$. We discussed in detail several cases of frictional instabilities: the velocity dependency of the coefficient of friction (friction coupled with sliding velocity), the thermoelastic instabilities (friction coupled with thermal expansion of the material), the Adams–Martins instabilities (friction coupled with elasticity), the interaction of frictional interface with incoming elastic waves, and the possibility of generating outgoing waves and propagating slip waves at the frictional interface. In the following chapters we will discuss practical cases of friction-induced vibrations and self-organized structures to which these vibrations can lead, which is useful for the reduction of friction.

References

Achenbach, J. D., and Epstein, H. I. 1967. Dynamic interaction of a layer and a half-space. *Journal of Engineering Mechanics* 5:7–42.

Anh, L.-X. 2003. *Dynamics of mechanical systems with Coulomb friction*. Berlin: Springer.

Anh, V. V., Angulo, J. M., and Ruiz-Medina, M. D. 1999. Possible long-range dependence in fractional random fields. *Journal of Statistical Planning and Inference* 80 (1): 95–110.

Adams, G. G. 1995. Self-excited oscillations of two elastic half-spaces sliding with a constant coefficient of friction. *ASME Journal of Applied Mechanics* 62:867–872

———. 1996. Self-excited oscillations in sliding with a constant friction coefficient: A simple model. *Journal of Tribology* 118 (4): 819–823.

———. 1998. Steady sliding of two elastic half-spaces with friction reduction due to interface stick–slip. *ASME Journal of Applied Mechanics* 65:470–475.

———. 1999. Dynamic motion of two elastic half-spaces in relative sliding without slipping. *Journal of Tribology* 121 (3): 455–461.

Adams, G. G., Barber, J. R., Ciavarella, M., and Rice, J. R. 2005. A paradox in sliding contact problems with friction. *Journal of Applied Mechanics* 72:450.

Adams, G. G., and Nosonovsky, M. 2000. Contact modeling—Forces. *Tribology International* 33 (5): 431–442.

Adams. G. G. 2001. An intersonic slip pulse at a frictional interface between dissimilar materials. *Journal of Applied Mechanics* 68 (1): 81–86.

Afferrante, L., Ciavarella, M., and Barber, J. R. 2006. Sliding thermoelastodynamic instability. *Proceedings of Royal Society A* 462:2161–2176.

Anderson, A. E., and Knapp, R. A. 1990. Hot spotting in automotive friction systems. *Wear* 135:319–337.

Andrews, D. J., and Ben-Zion, Y. 1997. Wrinkle-like slip pulse on a fault between different materials. *Journal of Geophysical Research*—All Series 102:553–571.

Banerjee, B. N., and Burton, R. A. 1979. Experimental studies on thermoelastic effect in hydrodynamically lubricated face seals. *ASME Journal of Lubrication Technology* 101:275–282.

Banerjee, B. N., and Burton, R. A. 1976. Thermo-elastic instability in lubricated sliding between solid surfaces.

Barber, J. R. 1969. Thermoelastic instabilities in the sliding of conforming solids. *Proceedings of Royal Society London A* 312:381–394.

Barber, J. R., Beamond, T. W., Waring, J. R., and Pritchard, C. 1985. Implications of thermoelastic instability for the design of brakes. *AMSE Journal of Tribology*, 107:206–210.

Ben-David, O., Rubinstein, S. M., and Fineberg, J. 2010. Slip–stick and the evolution of frictional strength. *Nature* 463:76–79.

Ben-Zion, Y., and Andrews, D. J. 1998. Properties and implications of dynamic rupture along a material interface. *Bulletin of the Seismological Society of America* 88 (4): 1085–1094.

Berry, G. A., and Barber, J. R. 1984. The division of frictional heat—A guide to the nature of sliding contact. *ASME Journal of Tribology* 106:405–415.

Bhushan, B., and Nosonovsky, M. 2004. Scale effects in dry and wet friction, wear, and interface temperature. *Nanotechnology* 15(7):749.

Bowden, F. P., and Tabor, D. 1950. *The friction and lubrication of solids*, 374. Oxford, England: Clarendon Press.

Bowden, F. P., and Thomas, P. H. 1954. The surface temperature of sliding solids. *Proceedings of the Royal Society of London, Series A. Mathematical and Physical Sciences* 223 (1152): 29–40.

Breitenfeld, M. S., and Beubelle, P. H. 1998. Numerical analysis of dynamic debonding under 2D in-plane and 3D loading. *International Journal of Fracture* 93 (1–4): 13–38.

Brock, L. M. 1996. Some analytical results for heating due to irregular sliding contact of thermoelastic solids. *Indian Journal of Pure Applied Mathematics* 27:1257–1278.

Burton, R. A. 1980. Thermal deformation in frictionally heated contact. *Wear* 59 (1): 1–20.

Burton, R. A., Nerlikar V., and Kilaparti, S. R. 1973. Thermoelastic instability in a seal-like configuration. *Wear* 24:177–188.

Cattaneo, C. 1938. Sul contatto di due corpi elasti: Distribuzione locale degli sforzi. *Accademia Nazionale dei Lincei, Rendiconti* 27 (6): 342–348.

Cochard, A., and Rice, J. R. 2000. Fault repture between dissimilar materials: Ill-posedness, regularization, and slip-pulse response. *Journal of Geophysical Research* 105 (B11): 25891–25.

Christie, I., Griffiths, D. F., Mitchell, A. R. and Zienkiewicz, O. C. 1976. Finite element methods for second order differential equations with significant first derivatives. *International Journal for Numerical Methods in Engineering* 10:1389–1396.

Comninou, M., and Dundurs, J. 1977. Elastic interface waves involving separation. *ASME Journal of Applied Mechanics* 44:222–226.

———. 1978. Elastic interface waves and sliding between two solids. *Journal of Applied Mechanics* 45:325.

Derjaguin, B. V., Muller, V. M., and Toporov, Y. P. 1975. Effect of contact deformations on the adhesion of particles. *Journal of Colloid and Interface Science* 53 (2): 314–326.

Dow, T. A., and Stockwell, R. D. 1977. Experimental verification of thermoelastic instabilities in sliding contact. ASME *Journal of Lubrication Technology* 99:359–364.

Du, S., Zagrodzki, P., Barber, J. R., and Hulbert, G. M. 1997. Finite element analysis of frictionally-excited thermoelastic instability. *Journal of Thermal Stresses* 20:185–201.

Dundurs, J., Tsai, K. C., and Keer, L. M. 1973. Contact between elastic bodies with wavy surfaces. *Journal of Elasticity* 3 (2): 109–115.

Erdogan, F., and Gupta, G. D. A. 1972. On the numerical solution of singular integral equations (singular integral equations numerical solution from Gauss–Chebyshev formulas for mixed boundary value problems). *Quarterly of Applied Mathematics* 29:525–534.

Freund, L. B. 1978. Discussion: Elastic interface waves involving separation. (Comninou, M., and Dundurs, J. 1977, ASME *Journal of Applied Mechanics* 44: 222–226). *Journal of Applied Mechanics* 45:226.

Gelfand, I. M., and Shilov, G. E. 1964. *Generalized functions*, 1. New York: Academic Press.

Genot, F., and Brogliato, B. 1999. New results on Painleve paradoxes. *European Journal of Mechanics A* 18:653–677.

Graf, K. F. 1975. *Wave motion in elastic solids*. Columbus: Ohio State University Press.

Greenwood, J. A., and Williamson, J. B. P. 1966. Contact of nominally flat surfaces. *Proceedings of the Royal Society of London, Series A. Mathematical and Physical Sciences* 295 (1442): 300–319.

Hill, R. 1950. *The mathematical theory of plasticity*. Oxford: Oxford University Press.

Hundy, B. B. 1957. *Railway Steel Topics* 4:19–35.

Hutchinson, J. W. 2000. Plasticity at the micron scale. *International Journal of Solids and Structures* 37 (1): 225–238.

Jang, J. Y., and Khonsari, M. M. 2003. A generalized thermoelastic instability analysis. *Proceedings of Royal Society of London A* 459:309–329, doi: 10.1098/rspa.2002.1030.

Johnson, K. L., Greenwood, J. A., and Higginson, J. G. 1985. The contact of elastic regular wavy surfaces. *International Journal of Mechanical Sciences* 27:383–396.

Johnson, K. L., Kendall, K., and Roberts, A. D. 1971. Surface energy and the contact of elastic solids. *Proceedings of the Royal Society of London, A. Mathematical and Physical Sciences* 324 (1558): 301–313.

Kilaparti, R., and Burton, R. A. 1977. A moving hot-spot configuration for a seal-like geometry, with frictional heating, expansion and wear. *ASLE Transactions* 20 (1): 64–70.

Kuznetsov, E. A. 1976. Periodic contact problem for half-plane allowing for forces of friction. *Soviet Applied Mechanics* 12 (10): 37–44.

———. 1985. Effect of fluid lubricant on the contact characteristics of rough elastic bodies in compression. *Wear* 157:177–194.

Lebeck, A. O. 1980. The effect of ring deflection and heat transfer on the thermoelastic instability of rotating face seals. *Wear* 59 (1): 121–133.

Lee, K., and Barber, J. R. 1993. Frictionally excited thermoelastic instability in automotive disk brakes. *ASME Journal of Tribology* 115:607–614.

Leine, R. I., Brogliato, B., and Nijmeijer, H. 2002. Periodic motion and bifurcations induced by the Painlevé paradox. *European Journal of Mechanics A* 21:869–896.

Li, Q., Tullis, T. E., Goldsby, D., and Carpick, R. W. 2011. Frictional ageing from interfacial bonding and the origins of rate-and-state friction. *Nature* 480 (7376): 233–236.

Linker, M. F., and Dieterich, J. H. 1992. Effects of variable normal stress on rock friction: Observations and constitutive equations. *Journal of Geophysical Research: Solid Earth* (1978–2012) 97 (B4): 4923–4940.

Manners, W. 1998. Partial contact between elastic surfaces with periodic profiles. *Proceedings of Royal Society London, Series A* 454:3203–3221.

Martins, J. A C., Oden, J. T., and Simoes, F. M. F. 1990. A study of static and kinetic fruction. *International Journal of Engineering Science* 28 (1): 29–92.

Martins, J. A. C., Guimaraes, J., and Faria, L. O. 1995. Dynamic surface solutions in linear elasticity and viscoelasticity with frictional boundary conditions. *Journal of Vibration and Acoustics* 117:445–451.

Martins, J. A. C., Oden, J. T., and Simoes, F. M. F. 1990. A study of static and kinetic friction. *International Journal of Engineering Science* 28 (1): 29–92.

Meirovitch, L. 1974. A new method of solution of the eigenvalue problem for gyroscopic systems. *AIAA Journal* 12:1337–1342.

Mindlin, R. D. 1949. Compliance of elastic bodies in contact. *ASME Journal of Applied Mechanics* 16:259–268.

Nix, W. D., and Gao, H. 1998. Indentation size effects in crystalline materials: A law for strain gradient plasticity. *Journal of the Mechanics and Physics of Solids* 46 (3): 411–425.

Nosonovsky, M., and Adams, G. G. 2000. Steady-state frictional sliding of two elastic bodies with a wavy contact interface. *Journal of Tribology* 122 (3): 490–495.

Nosonovsky, M., and Adams, G. G. 2002. Interaction of elastic dilatational and shear waves with a frictional sliding interface. *Journal of Vibration and Acoustics* 124 (1); 33–39.

———. 2004. Vibration and stability of frictional sliding of two elastic bodies with a wavy contact interface. *Journal of Applied Mechanics* 71 (2): 154–161.

Nosonovsky, M., and Adams, G. G. 2001. Dilatational and shear waves induced by the frictional sliding of two elastic half-spaces. *International Journal of Engineering Science* 39 (11): 1257–1269.

Nosonovsky, M., and Esche, S. K. 2008. Multiscale effects in crystal grain growth and physical properties of metals. *Physical Chemistry Chemical Physics* 10 (34): 5192–5195.

Panovko, I., and Gubanova, I. 1979. *Stability and vibrations of elastic systems: Contemporary concepts, paradoxes, and errors*, 3rd rev. ed. Moscow: Izdatel'stvo Nauka, 384 pp. (in Russian).

Parker, R. C., and Marshall, P. R. 1948. The measurement of the temperature of sliding surfaces, with particular reference to railway brake blocks. *Proceedings of the Institution of Mechanical Engineers* 158 (1): 209–229.

Prakash, V., and Clifton, R. J. 1993. Time resolved dynamic friction measurements in pressure-shear. *Asme Applied Mechanics Division-Publications-AMD* 165, 33–33.

Prakash, V. 1998. Frictional response of sliding interfaces subjected to time varying normal pressures. *Journal of Tribology*, 120 (1): 97–102.

Ranjith, K., and Rice, J. R. 1999. Stability of quasi-static slip in a single degree of freedom elastic system with rate-and-state dependent friction. *Journal of Mechanics and Physics of Solids* 47:1207–1218.

———. 2001. Slip dynamics at an interface between dissimilar materials. *Journal of the Mechanics and Physics of Solids* 49 (2): 341–361.

Ranjith, K., and Rice, J. R. 2001. Slip dynamics at an interface between dissimilar materials. *Journal of the Mechanics and Physics of Solids* 49 (2): 341–361.

Renardy, M. 1992. Ill-posedness at the boundary for elastic solids sliding under Coulomb friction. *Journal of Elasticity* 27 (3): 281–87.

Rice, J. R. 1997. Slip pulse at low driving stress along a frictional fault between dissimilar media. *Eos Trans. AGU*, 78:46.

Rice, J. R., Lapusta, N., and Ranjith, K. 2001. Rate-and-state dependent friction and the stability of sliding between elastically deformable solids. *Journal of Mechanics and Physics of Solids* 49:1865–1898.

Rottler, J., and Robbins, M. O. 2005. Unified description of aging and rate effects in yield of glassy solids. *Physical Review Letters* 95 (22): 225504.

Sadowski, M. A. 1928. Zwiedimensionale Probleme der Elastizitatshtheorie. *Zeitschrift fur Angewandte Mathematik und Mechanik, B.* 8 (2): 107–121.

Salant, R. F., and Flaherty, A. L. 1995. Elastohydrodynamic analysis of reverse pumping in rotary lip seals with microasperities. *ASME Journal of Tribology* 117:53–59.

Schallamach, A. 1971. How does rubber slide? *Wear* 17 (4): 301–312.

Shtaerman, I. Ya. 1949. *Contact problem in the theory of elasticity*. Moscow: Gostehizdat (in Russian).

Simoes, F. M. F., and Martins, J. A. C. 1998. Instability and ill-posedness in some friction problems. *International Journal of Engineering Science* 36 (11): 1265–1293.

Stewart, D. 2000. Rigid-body dynamics with friction and impact. *SIAM Review* 42:3–39.

Weertman, J. 1963. Dislocations moving uniformity on the interface between isotropic media of different elastic properties. *Journal of the Mechanics and Physics of Solids* 11 (3): 197–204.

Weertman, J. 1980. Unstable slippage across a fault that separates elastic media of different elastic constants. *Journal of Geophysical Research: Solid Earth* (1978–2012), 85 (B3): 1455–1461.

Westergaard, H. M. 1939. Bearing pressures and cracks. *ASME Journal of Applied Mechanics* 6 (2): A49–A53.

Wickert, J. A., and Mote, C. D. 1990. Classical vibration analysis of axially moving cantinua. *Journal of Applied Mechanics* 57 (3): 738–744.

Yeo, T., and Barber, J. R. 1996. Finite element analysis of the stability of static thermo-elastic contact. *Journal of Thermal Stresses* 19:169–184.

Yi, Y.-B., and Barber, J. R. 2001. Hot spotter: A finite element software package for evaluating the susceptibility of axisymmetric multidisk brakes and clutches to thermoelastic instability (TEI) (http://www-personal.engin.umich.edu/wjbar-ber/hotspotter.html).

Yi, Y.-B., Barber, J. R., and Zagrodzki, P. 2000. Eigenvalue solution of thermoelastic instability problems using Fourier reduction. *Proceedings of Royal Society London A* 456:2799–2821.

Zagrodzki, P. 1990. Analysis of thermomechanical phenomena in multidisc clutches and brakes. *Wear* 140:291–308.

Zagrodzki, P., Lam, K. B., Al-Bahkali, E., and Barber, J. R. 1999. Simulation of a slid-ing system with frictionally excited thermoelastic instability. *Thermal Stresses*, Cracow, Poland.

Zharii, O. Y. 1996. Frictional contact between the surface wave and a rigid strip. *Journal of Applied Mechanics* 63 (1): 15–20.

Bibliography

Bhushan, B. 1998. Contact mechanics of rough surfaces in tribology: Multiple asperity contact. *Tribology Letters* 4 (I-35) 1–35.

Bhushan, B., and Nosonovsky, M. 2004. Scale effects in dry and wet friction, wear, and interface temperature. *Nanotechnology* 15 (7): 749.

Bignari, C., Bertetto, A. M., and Mazza, L. 1999. Photoelastic measurments and com-putation of the stress field and contact pressure in a pneumatic lip seal. *Tribology International* 32:1–13.

Etsion, I., Kligerman, Y., and Halperin, G. 1999. Analytical and experimental investi-gation of laser-textured mechanical seal faces. *Tribology Transactions* 42:511–516.

Hertz, H. R. 1882. Ueber die Beruehrung elastischer Koerper (On contact between elas-tic bodies). In *Gesammelte Werke (Collected works)*, vol. 1, Leipzig, Germany, 1895.

Johansson, L. 1993. Model and numerical algorithm for sliding contact between two elastic half-planes with frictional heat generation and wear. *Wear* 160:77–93.

Kapoor, A., Williams, J. A., and Johnson, K. L. 1994. The steady-state sliding of rough surfaces. *Wear* 175:81–92.

Larsson, J., Biwa, S., and Storakers, B. 1999. Inelastic flattening of rough surfaces. *Mechanics of Materials* 31:29–41.

Ruina, A. 1983. Slip instability and state variable friction laws. *Journal of Geophysical Research* 88 (B12): 10359–10370.

Scholz, C. H. 1998. Earthquakes and friction laws. *Nature* 391:37–42.

Tabor, D. 1981. Friction—The present state of our understanding. *ASME Journal of Lubrication Technology* 103:169–179.

Whitehouse, D. J., and Archard, J. F. 1970. The properties of random surfaces of sig-nificance in their contact. *Proceedings of Royal Society London, Ser. A* 316:97–121.

Wolfram, S. 1991. *Mathematica, a system for doing mathematics by computer*, 2nd ed. Reading, MA: Addison–Wesley.

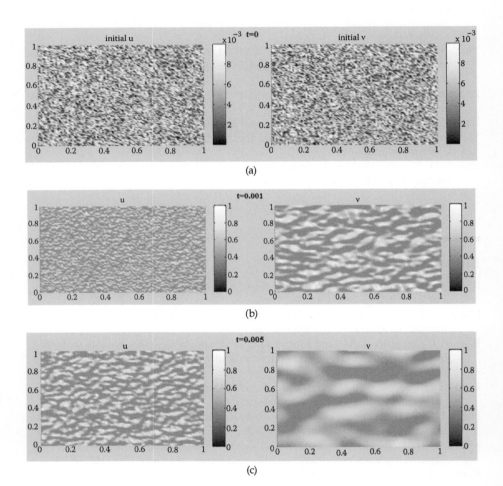

FIGURE 8.16
Time evolution of Turing system components (u and v) in different time steps, at $t = 0$, 0.001, 0.005, 0.01, and 0.1 (first case). (Reprinted with permission from Mortazavi, V., and Nosonovsky, M. 2011a. *Langmuir* 27 (8): 4772–4779. Copyright American Chemical Society.)

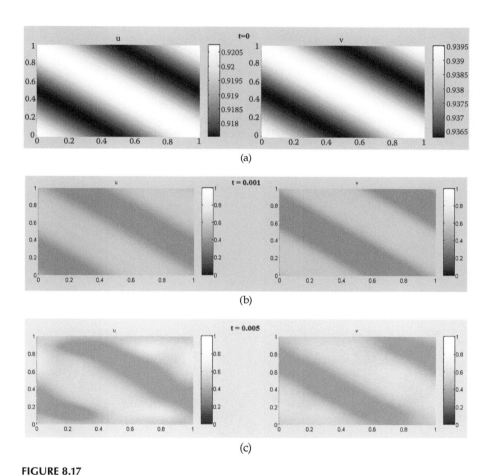

FIGURE 8.17
Time evolution of Turing system components (u and v) in different time steps, $t = 0$, 0.001, 0.005, 0.01, 0.05, and 0.1 (second case). (Reprinted with permission from Mortazavi, V., and Nosonovsky, M. 2011a. *Langmuir* 27 (8): 4772–4779. Copyright American Chemical Society.)

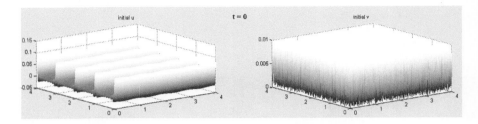

FIGURE 8.18
Time evolution of Turing system components (u and v) in two different time steps (third case, using Equations 8.18 and 8.19). (Reprinted with permission from Mortazavi, V., and Nosonovsky, M. 2011a. *Langmuir* 27 (8): 4772–4779. Copyright American Chemical Society.)

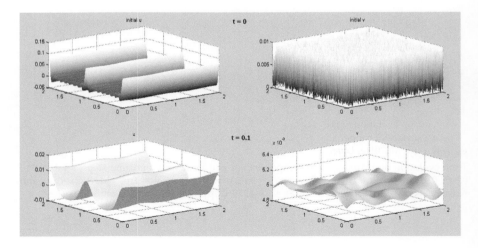

FIGURE 8.19
Time evolution of Turing system components (u and v) in different time steps (third case, using Equations 8.18 and 8.20). (Reprinted with permission from Mortazavi, V., and Nosonovsky, M. 2011a. *Langmuir* 27 (8): 4772–4779. Copyright American Chemical Society.)

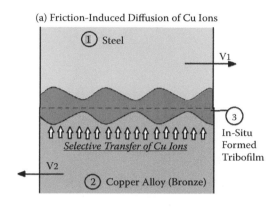

(a) Friction-Induced Diffusion of Cu Ions

① Steel

V_1

③
In-Situ
Formed
Tribofilm

Selective Transfer of Cu Ions

V_2

② Copper Alloy (Bronze)

(b) Top View of In-Situ Formed Tribofilm

Island of Cu
Ions

Bronze Substrate

FIGURE 8.20
Schematic presentation of the selective layer (b) obtained by friction between a surface of steel
and (a) a surface of copper alloy (bronze). (Reprinted with permission from Mortazavi, V., and
Nosonovsky, M. 2011a. *Langmuir* 27 (8): 4772–4779. Copyright American Chemical Society.)

FIGURE 9.26

7

Friction-Induced Vibrations and Their Applications

In the preceding chapter we studied the mathematical foundations of friction-induced vibrations. Here we will review friction-induced vibrations as they occur in engineered applications and in nature. We include four large classes of systems: technical systems (brakes, belts, components, engines, tires); micro- and nano-devices; sound generation, including musical instruments; and systems in nature. The most common manifestation of friction-induced vibrations is noise, such as car brake squeal (Nosonovsky and Adams 2001, 2004; Persson 2001; Akay 2002; Nguyen 2003; Baumberger, Caroli, and Ronsin 2003; Kirillov 2008; Sheng 2008; Brunel et al. 2010). Usually undesirable and unsteady, friction-induced sound in certain situations can form a harmonic self-organized pattern. This phenomenon is used in musical instruments to produce musical sounds.

7.1 Brakes and Vehicles

There are several scenarios how friction can lead to vibrations. According to Akay (2002), there are four main mechanisms involved: (1) geometric instabilities, (2) material non-linearities, (3) thermoelastic instabilities, and (4) instabilities due to decreasing friction with increasing velocity. In geometric and material instabilities, friction has a passive role. Thus, in geometric instabilities, such as brush and commutator contact in electric motors, friction plays a necessary but passive role. Similarly, friction also has a passive role in instabilities caused by system non-linearities (e.g., in cases where material properties exhibit non-linear contact stress–strain behavior. On the other hand, friction plays an active role in instabilities caused, for example, by a decreasing friction force with increasing velocity. In all of these mechanisms, friction is coupled with another process, resulting in a positive feedback loop, as described in the preceding chapters.

An example of geometric instabilities is a rod placed in the grooves of two pulleys, set apart a distance of $2L$ and rotating toward each other (Figure 7.1). If the position of the rod is not perfectly symmetric, the load is not equally distributed between the pulleys, and the longer end of the rod will produce

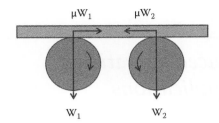

FIGURE 7.1
Vibration of a rod on two pulleys rotating in the opposite directions.

a larger load and friction force; thus, the rod will tend to oscillate about the equilibrium position with a frequency that depends on the coefficient of friction between the pulleys and the rod.

Two other illustrative examples of friction-induced vibrations suggested by Akay (2002) are the sound generation in a wine glass rubbed by a finger and during the extraction of a nail from wood. There are many other interesting examples; however, the most common and best studied occurrences of friction-induced vibrations are in automotive applications, including car brakes, engines, and tire–pavement friction.

Frictional instabilities in a coupled system can lead to the condition called mode lock-in, when the system responds at one of its fundamental frequencies and its harmonics selecting a mode in which it locks depending on the normal load, sliding velocity, and the contact geometry (Akay 2002). This happens, for example, when the wine glass responds, at or near a particular natural frequency of the glass to the exclusion of its other modes. Other typical effects of friction-induced vibrations include mode coupling and the occurrence of complex modes of vibration.

The standard manifestation of friction-induced vibrations is the noise that is produced by car brakes during abrupt braking if the brakes are not properly tuned. The sound is generated by vibrations produced by friction between the brake disk and the pad. During braking, the pads are pushed against the disk and the mechanical energy of motion is converted mostly to heat. However, a small part of energy (usually less than 0.1%) can result in vibrations of the disk, the pad, and other components of the car and the vehicle, generating the ugly noise.

According to historical evidence, as early as 1930, brake noise emerged as one of the top 10 noise problems in New York City. Even today it continues to be a source of trouble, producing warranty costs in North America for brake noise, vibration, and harshness of one billion dollars annually (Akay 2002; Sheng 2008).

A similar phenomenon is known for aircraft and railway brakes. Aircraft brakes are subjected to especially severe conditions because they have to absorb the energy of a moving airplane within several seconds after it touches the runway, leading to very high temperature (about 1000°C) and high-temperature gradients. Aircraft brake disks are typically made of carbon–carbon composites due to their high thermal resistance, while brake disks for regular cars are usually made of cast iron. Recently, other materials, such as SiC-reinforced aluminum, carbon–SiC composites, and sintered carbon, have started to emerge. Most pad materials are organic materials based on a metal fiber-reinforced organic matrix. Other types of friction materials include metallic and carbon based. The additives are included to control the coefficient of friction or wear and lubricants such as graphite are used to decrease the coefficient of friction. Abrasive particles are added to increase the coefficient of friction or disk wear to remove iron oxides from the disk surface, thus providing more consistent braking properties (Sheng 2007).

Brake vibration mechanisms are complex and include several sources, such as in-plane vibrations of the disk, time-dependent variations of the friction force and contact area, geometric instabilities due to non-uniform contact between the disk and the pad, thermoelastic instabilities, and other factors contributing to stick-slip. About 25 different words are used to describe brake noise, depending on the frequency, duration, and mechanism (Akay 2002).

The high-frequency (1–10 kHz) noise includes squelch, squeal, squeak, chirp, and wire brush. Squeak is a short-lived squeal, wire brush describes a randomly modulated squeal, and squelch describes an amplitude-modulated version of squeak noise. A high-frequency squeal involves higher disk modes with 5 to 10 nodal diameters and nodal spacing comparable to or less than pad size. Mode shapes remain stationary with respect to the ground, indicating some constraining effect by the pads, although not excessive enough to alter the mode shapes.

Low-frequency noise involves groan, judder, moan, hum, grunt, crunch, fog horn, and other types. These sounds involve modes with one to four modal diameters and occur at frequencies other than the natural frequencies of the disk.

Various analytical (including complex modal analysis), computational, and experimental methods are used to investigate disk brake squeal (Kinkaid, O'Reilly, and Papaclopoulos 2003; Cao et al. 2004; Massi et al. 2007; Behrendt, Weiss, and Hoffmann 2011) and how it depends on various conditions, such as temperature, pressure, humidity, velocity, and break-in. (Thus, during the break-in for a new brake disk, the coefficient of friction increases slowly, and sometimes squeal can occur when the coefficient of friction reaches a threshold.) Many recommendations and techniques have been suggested to reduce and eliminate brake noise (Sheng 2008).

In addition to brakes, friction-induced vibrations can occur in other parts of a vehicle, including clutch, gears, engine, and tires. Low-frequency torsional vibrations referred to as shudder (judder) occasionally occur due to clutch friction during acceleration at low speed. Shudder exists in dry clutches in a manual transmissions or in wet clutches in an automatic transmissions. Sliding friction has been recognized as a significant source of noise and vibration in gear meshing. A noise called diesel sounding knock is associated with the stick–slip motion of pistons caused by the abrupt change of the coefficient of friction from static to dynamic when the engine is in idling speed. The vibration of the piston in turn excites the crankshaft at its resonant frequency via the connecting rods. Tire–road interaction in road vehicles and wheel–rail interaction in rail vehicles also cause noise. Part of the road vehicle rubber tire noise is attributed to the tire vibrations associated with the adhesion and friction created in the contact patch between the tire and the road surface (Sheng 2008).

As automobiles become increasingly quieter, wiper operational noise that occurs in rubber–glass interface of wiper blades and the interior noise become a quality concern for drivers. Wiper noise is classified into three groups: squeal noise, chattering, and reversal noise. Squeal (squeaky) noise is a high-frequency vibration (about 1000 Hz) that is easily generated before and after the wiper reverses direction. Chattering (or beep) noise is a low-frequency vibration of 100 Hz or less. Reversal noise is an impulsive sound with a frequency of 500 Hz when the wiper reverses direction. The interior noise of a passenger car consists of many elements, such as the irritating noises called "squeak and rattle" comprising buzzes, squeaks, ticks, rattles, etc. Squeaks are friction-induced noises caused by relative motion resulting from a stick–slip phenomenon between interfacing surfaces, usually due to suspension inputs. Rattles are impact-induced noises that occur when there is a relative motion between components with a short-duration loss of contact. They are generally caused by loose or overly flexible elements under forced excitation (Sheng 2008).

In addition to automotive applications, friction-induced vibrations occur in belts, hinges (e.g., door hinges), and other mechanisms. Note that, mathematically, the equation for belt deflection is similar to that for a beam on an elastic foundation, which has been discussed as a "simple model" of elastic medium in the preceding chapter.

$$EI\frac{d^4w}{dx^4} - T\frac{d^2w}{dx^2} + kw = \rho A\frac{d^2w}{dt^2} \tag{7.1}$$

where EI is the bending stiffness of the belt (modeled as a Bernoulli–Euler beam), T is the tension force, k is the elastic force (possibly, $k = 0$), ρ is the density, and A is the cross-section area.

Friction-induced vibrations are also found in micro-devices, including hard-disk drives, atomic-force microscope cantilevers, micro-electromechanical systems (MEMS), and others.

Friction can be used to facilitate motion and to reduce vibration (Akay 2002). A vibratory conveyor consists of a planar vibrating platform at a certain angle with the horizontal surface. A load (a particle or body) is applied to the conveyor. The angle is insufficient to facilitate sliding when the platform does not vibrate and the load remains stuck. However, vibration results in the effective reduction of mass due to the action of the inertia force during the phase of the motion downward. This may result in slip of the load. A somewhat similar principle is used by ultrasonic motors (Nosonovsky and Adams 2002), which use an elastic surface wave-induced piezo-substrate to facilitate the motion of the load. The underlying mechanism of frictional vibration reduction is that contact forces are manipulated to reduce the average friction force during the oscillation cycle.

7.2 Music and Sound Generation

While brake squeal is usually an unpleasant noise, friction can also generate pleasant sounds. Acay (2002) noted that friction sounds are rarely stationary and are rather unsteady or transient. This is related to the very "dual" nature of friction, which simultaneously dissipates energy and transfers it from one body to another.

Most musical instruments are designed to produce harmonic sounds, and many of them use vibrations, excited by friction for this purpose. The most common example is the bowed string instrument, such as violin and cello. The player usually controls bow velocity, position, and pressure. When the bow is drawn across the string with a constant velocity and pressure, the string vibrates with its own natural frequency. During the half-period, when the string moves in the same direction as the bow, the relative velocity is lower than that during the half-period when the string moves in the opposite direction. The coefficient of friction is dependent on the relative sliding velocity and tends to decrease with increasing velocity. As a result, the friction between the bow and the string is higher when they move in the same direction than when they move in the opposite directions, and some energy is pumped into the string, keeping it vibrating. The stick–slip motion is also possible (Woodhouse and Galluzzo 2004; Sheng 2008).

The bowing action divides a string into two approximately straight-line sections. The string moves with the bow until the tension in the string exceeds the friction force, at which time the string separates from the bow until they reconnect. The "corner" formed by the bow travels along the string at the transverse speed of string waves. At each instant, the string maintains the two approximately straight-line segments between the corner and the fixed ends. Thus, the complete motion of the string during bowing consists of the circulating corner, which is the homogeneous solution of the equations of motion, superposed on the particular integral that describes a stationary corner at the bow contact (Akay 2002).

Various models for violin string vibration have been suggested, and these models emphasize the importance of the torsional degree of freedom and the effect of temperature (Askenfelt 1989; Woodhouse, Schumacher, and Garoff 2000; Sheng 2008). The effects are coupled with the actual vibration of the string. It is also noted that musical sound generation in a violin is a complex process. The vibrations of the string are transferred through the "bridge" into the body of the instrument, which acts as a resonator amplifying specific frequencies.

Friction-induced vibrations play a role also in certain musical instruments without strings. Akay (2002) mentioned that the cuica, the Brazilian friction drum, presents an example of an instrument in which friction excitation occurs. The operation of the cuica is similar to that of a bowed string instrument. A bamboo stick vibrates the membrane as a bow does with a string, and the player's finger on the membrane selects the mode of vibration as does placing a finger on the string. Another example of an instrument is the Tibetan singing bowl, which is in a sense similar to a wineglass vibration (Inacio, Henrique, and Antunes 2006; Terwagne and Bush 2011).

In addition to musical instruments, sound synthesis is used to produce the realistic sound effects mimicking such friction-induced noises as the squeal of parquet floors (Ekimov and Sabatier 2006), door hinges, chalk on a blackboard, and the squeak of snow during walking (Avanzini, Serafin, and Rocchesso 2005).

7.3 Nature

Friction induced-vibrations are found in nature in many situations. The most interesting examples in non-living nature include earthquakes, landslides, avalanches, and similar events. Examples in living nature include stridulating insects and other creatures.

Earthquakes are caused by frictional instabilities of various kinds (Zheng and Rice 1998; Ben-Zion 2001; Ben-Zion and Sammis 2003). Most earthquakes occur as a result of sudden slippage of a tectonic plate along a pre-existing fault, rather than due to the formation of a new crack or fault. The slip is governed by the friction law, and the state-and-rate friction law (Equations 4.30 and 4.31 in Chapter 4) is one of the most popular to study the onset of instabilities associated with earthquakes (Nosonovsky 2007). One of the popular state-and-rate laws is

$$\mu = \mu_0 + a \ln \frac{V}{V_0} + b \ln \frac{V_0 \theta}{L} \tag{7.2}$$

where a, b, L, V_0, and μ_0 are parameters; V is the sliding velocity; and θ is the state parameter (the typical age of contact) evolving in accordance with the differential equation

$$\frac{d\theta}{dt} = 1 - \frac{V\theta}{L} \tag{7.3}$$

The coefficient of friction is equal to μ_0 in the steady state ($\theta = L/V_0$, $V = V_0$) case. If the sliding velocity is changed, the coefficient of friction evolves with time to the new value of $\mu = \mu_0 + (a-b)\ln(V/V_0)$. The stability depends on the difference of $a - b$, which is a material property (Sheng 2008).

The sound of sand in the desert is apparently another example of friction-generated sound. There are two types of sand sounds: short (<0.25 s) high-frequency (500–2500 Hz) squeaking or whistling and the much less frequent low-frequency (50–300 Hz) booming sound associated with avalanching and found in large, dry, isolated dunes in deserts (Sheng 2008).

Living organisms such as fish, insects, spiders, snakes, and lobsters also use friction to generate sound for warning and for calling mates. Many of these organisms have special body structures used for stridulation or sound production by rubbing. The anatomical organs used for stridulating vary. The stridulatory apparatus has two parts: the pars stridens with a ridged surface and the plectrum (the scraper). The location of these parts can differ. Many insects, such as grasshoppers, have a hind leg scraper rubbed against the adjacent forewing. Beetles use their head's motion for stridulation, while other insects can use their wings, back, or abdomen. The friction sounds produced by stridulating are quite diverse and cover frequency ranges from 1 to 100 kHz (Sheng 2008). Invertebrates such as venomous snakes stridulate to display threat by rubbing coils of their body and producing a sizzling sound. Larger creatures are known to stridulate as well. These include the Caribbean spiny lobster, which creates a loud buzzing sound to deter predators (Patek 2000), certain birds (e.g., the club-winged manakin living in the Andes), fish, and apparently even some mammals (the lowland streaked tenrec in Madagascar).

Friction sounds can also develop in humans and include pleural and pericardial friction sounds resulting from inflammation of tissues surrounding the lungs and heard correspondingly (Akay 2002), as well as in hips (Walter et al. 2008). Human skin sound was investigated by Zahouani et al. (2009). Another interesting example of relevance to biology is the "turkey friction calls" used by hunters to imitate turkeys. These sounds are generated by friction of wooden components.

7.4 Summary

In this chapter we reviewed several examples of friction-induced vibrations and noises that occur in technical systems (with the emphasis on automotive applications), in sound generation (musical instruments), and in nature.

References

Akay, A. 2002. Acoustics of friction. *Journal of Acoustical Society America* 111:1525.

Askenfelt, A. 1989. Measurement of the bowing parameters in violin playing. Part II: Bow-bridge distance, dynamic range, and limits of bow force. *Journal of Acoustical Society America* 86 (2): 503.

Avanzini, F., Serafin, S., and Rocchesso, D. 2005. Interactive simulation of rigid body interaction with friction-induced sound generation. *IEEE Transactions Speech Audio Processing* 13:1073–1081.

Baumberger,T., Caroli, C., and Ronsin, O. 2003. Self-healing slip pulses and the friction of gelatin gels. *European Physics Journal E* 11:85–93.

Behrendt, J., Weiss, C., and Hoffmann, N. P. 2011. A numerical study on stick–slip motion of a brake pad in steady sliding. *Journal of Sound Vibration* 330:636–651.

Ben-Zion, Y. 2001. Dynamic ruptures in recent models of earthquake faults. *Journal of Mechanics Physics Solids* 49:2209–2244.

Ben-Zion, Y., and Sammis, C. G. 2003. Characterization of fault zones. *Pure Applied Geophysics* 160:677–715.

Brunel, J. F., Dufrenoy, P., Charley J., et al. 2010. Analysis of the attenuation of railway squeal noise by preloaded rings inserted in wheels. *Journal of Acoustical Society America* 127:1300–1306.

Cao, Q., Ouyang, H., Friswell, M. I., et al. 2004. Linear eigenvalue analysis of the disc-brake squeal problem. *International Journal of Numerical Methods Engineering* 61:1546–1563.

Ekimov, A., and Sabatier, J. M. 2006. Vibration and sound signatures of human footsteps in buildings. *Journal of Acoustical Society America* 120:762–768.

Inacio, O., Henrique, L. L., and Antunes, J. 2006. The dynamics of Tibetan singing bowls. *Acta Acustica* 92:637–653.

Kinkaid, N. M., O'Reilly, O. M., and Papaclopoulos, P. 2003. Automotive disc brake squeal. *Journal of Sound Vibration* 267:105–166.

Kirillov, O. N. 2008. Subcritical flutter in the acoustics of friction. *Proceedings Royal Society A* 464:2321–2339.

Massi, F., Baillet, L., Giannini, O., et al. 2007. Brake squeal: Linear and nonlinear numerical approaches. *Mechanical Systems and Signal Processing* 21:2374–2393.

Nguyen, Q. S. 2003. Instability and friction. *Comptes Rendus Macanique* 331:99–112.

Nosonovsky, M. 2007. Modeling size, load, and velocity-dependence on friction at micro/nanoscale. *International Journal of Surface Science Engineering* 1:22–37.

Nosonovsky, M., and Adams, G. G. 2001. Dilatational and shear waves induced by the frictional sliding of two elastic half-spaces. *International Journal of Engineering Science* 39:1257–1269.

———. 2002 Interaction of elastic dilatational and shear waves with a frictional sliding interface. *ASME Journal of Vibrations and Acoustics* 124:33–39.

———. 2004. Vibration and stability of frictional sliding of two elastic bodies with a wavy contact interface. *ASME Journal of Applied Mechanics* 71:154–300.

Patek, S. N. 2000. Spiny lobsters stick and slip to make sound. *Nature* 411:153.

Persson, B. N. J. 2001. Elastic instabilities at a sliding interface. *Physical Review B* 63:104–101.

Sheng, G. 2008. *Friction-induced vibrations and sound: Principles and applications.* Boca Raton, FL: CRC/Taylor & Francis.

Terwagne, D., and Bush, J. W. M. 2011. Tibetan singing bowls. *Nonlinearity* 24:R51–R66.

Walter, W. L., Waters, T. S., Gillies, M., et al. 2008. Squeaking hips. *Journal of Bone Joint Surgery—America* 90A:102–111.

Woodhouse, J., and Galluzzo, P. M. 2004. The bowed string as we know it today. *Acta Acoustica* 90:579–589.

Woodhouse, J., Schumacher, R. T., and Garoff, S. 2000. Reconstruction of a bowing point friction force in a bowed string. *Journal of Acoustical Society of America* 108:1.

Zahouani, H., Vargiolu, R., Boyer G., et al. 2009. Friction noise of human skin in vivo. *Wear* 267:1274–1280.

Zheng, G., and Rice, J. R. 1998. Conditions under which velocity-weakening friction allows a self-healing versus a cracklike mode of rupture. *Bulletin of Seismological Society America* 88:1466–1483.

8

From Frictional Instabilities to Friction-Induced Self-Organization

In the preceding chapters we formulated thermodynamic stability criterion for the stationary non-equilibrium regime (Equation 3.37). Destabilization can lead to self-organization (i.e., formation of new structures at the frictional interface, such as in situ formed tribofilms or patterns). Furthermore, in Chapter 3, we discussed the minimum entropy production principle. In a simplified form, this principle states that a dynamical system tends to evolve to the path of least resistance—in other words, minimum friction and wear. In Chapters 6 and 7 we studied various types of frictional instabilities and vibrations. Here we will review several mechanisms of frictional instabilities and self-organized patterns that can arise from such instabilities.

8.1 Self-Organization, Instabilities, and Friction

Friction and wear are usually viewed as irreversible processes, which lead to energy dissipation (friction) and material deterioration (wear). However, it is known that under certain circumstances frictional sliding can result in the formation (self-organization) of spatial and temporal patterns. Study of the concept of self-organization in tribology is not an old research area, nor is it old in other scientific fields. The development of this concept is one of the achievements of modern physics, which deals with complex processes of nature (Gershman and Bushe 2006). The concept is based on the ideas of irreversible thermodynamics (Nosonovsky and Bhushan 2009; Prigogine 1980). Major progress in this area is associated with Prigogine (1980).

Thermodynamically, friction is a process that occurs in the open and non-equilibrium systems. This is an irreversible process—a process that cannot reverse itself and restore the system to its initial state. A reversible process is defined as a process that can be reversed without leaving any trace on the surroundings.

Clausius in the 1850s introduced entropy, S, as a measure of irreversibility of the thermodynamic system:

$$dS = \frac{dQ}{T} \tag{8.1}$$

where T is the temperature and Q is the heat. When heat dQ is transferred from a body with temperature T_1 to a body with temperature T_2, the entropy grows by the amount $dS = dQ/T_1 - dQ/T_2$. Thus, if heat is transferred from a hotter body to a colder one ($T_1 > T_2$), the net entropy grows ($dS > 0$). This provides a convenient formal basis for the second law of thermodynamics, stating that the net entropy of a closed system either remains constant (for a reversible process) or grows (for an irreversible process).

During friction, an irreversible, dissipative process, heat is generated. For example, when the friction force F is applied to a body that passes the distance dx, the energy $dQ = Fdx$ is dissipated into the surroundings, and the entropy of the surroundings increases for the amount of $dS = (F/T)dx$.

In 1877, L. Boltzmann suggested another definition of entropy using the statistical thermodynamics approach and the concept of micro-states:

$$S = k \ln \Omega \tag{8.2}$$

where k is Boltzmann's constant and Ω is the number of micro-states corresponding to a given macro-state. Micro-states are arrangements of energy and matter in the system that are distinguishable at the atomic or molecular level, but are indistinguishable at the macroscopic level (Craig 1992). A system tends to evolve into a less ordered (more random) macro-state that has a greater number of corresponding micro-states, and thus the entropy given by Equation (8.2) increases.

From the preceding definition, it can be found that entropy depends on the energy of the system. Hence, irreversible processes are accompanied not only by the growth of entropy but also by the change of energy within the system (internal energy). In this case, we have to consider the reduction of internal energy because, in an equilibrium state, entropy has the possible maximum values, whereas energy has the possible minimum values (Ebeling 1999). Thus, irreversible processes should be accompanied by the dissipation of energy.

So, the change of entropy dS consists of two parts (Kostetsky 1970). It includes a flow of entropy d_eS, caused by an interaction with the environment, and d_iS, a part of entropy change due to the processes that are taking place within the system (entropy production):

$$dS = d_eS + d_iS \tag{8.3}$$

and always,

$$d_iS \geq 0 \tag{8.4}$$

The entropy production according to Equation (8.4) is a general definition of irreversibility, which states entropy of an irreversible system always increases. It is worth noting that the division of entropy change on two components (Equation 8.3) allows the establishment of a distinction between closed (isolated) and open systems. Distinction is shown with d_eS, which takes into account the change of entropy due to an exchange of matter in open systems.

The concept of self-organization is based on the ideas of irreversible thermodynamics (Prigogine 1961). Prigogine's theory states that while the net entropy grows in most thermodynamic systems, some of them may lead to a process to decrease entropy production. To do that, these systems should be thermodynamically open and operate far from thermodynamic equilibrium and can exchange energy, matter, and entropy with the environment. The physical meaning of Prigogine's theorem is as follows: A system that cannot come to equilibrium attempts to enter a condition with the lowest energy dissipation. That is why we can expect that the wear rate under stable conditions will be minimal with an extremely low contribution to entropy production (Gershman and Bushe 2006).

The mechanism leading into this low-entropy production is forming of some new structures that are called dissipative or secondary structures, and the process of dissipative structure formation is called self-organization. These dissipative structures can be spatial, temporal, or functional (Kondepudi and Prigogine 2000). Spontaneous formation of dissipative structures is a result of perturbations that can be realized only in open systems that exchange energy, matter, and entropy with their environments (Kondepudi and Prigogine 2000).

The concept of self-organization of systems far from equilibrium is currently widely used for fundamental research in many areas of science—for example:

- Benard cells in boiling liquid (Koschmieder 1993)
- Reaction and diffusion systems (Mortazavi and Nosonovsky 2011a)
- The instability that arises at fusion
- The solidification caused by pulse laser irradiation of crystals (Ebeling et al. 1990)
- Space–time ordering at the liquid–phase interface during mechanical deformation (Prigogine 1997)
- Structures growing at the liquid–solid interface in the course of the growth of a crystal (Prigogine and Kondepudi 1990)
- Dissipative structures that are formed during the unidirectional growth of binary alloys from liquid (Prigogine 1997)
- Spatial heterogeneity as domains with various chemical concentration in rocks (Ebeling et al. 1990)
- Chemical reactions in biochemical systems (Prigogine and Kondepudi 1990)

According to the theorem by Prigogine (1961), only after passing through instability can the process of self-organizing begin. During regular deviation from equilibrium (i.e., when an increase occurs in a parameter that describes a system's condition or its external influence), the steady (and close to equilibrium) state is stable. However, if a certain critical value of one parameter is exceeded, the state of the system could become unstable. The new order that has been established within the system could correspond to an ordered state—that is, a new state with smaller entropy in comparison to a chaotic condition (Gershman and Bushe 2006).

The irreversible processes of structure formations are associated with nonequilibrium phase transformations. These transformations are connected with specific bifurcation or instability points where the macroscopic behavior of the system changes qualitatively and may leap either into chaos or into greater complexity and stability (Capra 1996). In the latter case, as soon as the system passes these specific points, its properties change spontaneously because of self-organization and formation of dissipative structures (Fox-Rabinovich 2007).

As it was described, friction and wear could be characterized by exchange of matter and energy with the environment. The flow of heat, entropy, and material away from the interface during the dry friction and wear can lead to formation of secondary structures. The secondary structures are either patterns that form at the interface (e.g., stick and slip zones) or those formed as a result of mutual adjustment of the bodies in contact. Formation of these structures and the transition to the self-organized state with low friction and wear occur through the destabilization of the steady-state (stationary) sliding. The entropy production rate reaches its minimum at the self-organized state. Therefore, the self-organization is usually beneficial for the tribolological system, as it leads to the reduction of friction and wear (Nosonovsky and Bhushan 2009).

As was mentioned, concepts of self-organization and the destabilization of the steady state are related together. The next sections of this chapter will concentrate on mechanisms of self-organization and instabilities at the frictional interface. Then, the practical problems that can be categorized as evidences of these concepts will be introduced. Each of these problems will be the main focus in one of the next sections.

Nosonovsky (2010a, 2010b), Nosonovsky and Bhushan (2009), and Nosonovsky et al. (2009) suggested entropic criteria for friction-induced self-organization on the basis of the multi-scale structure of the material (when self-organization at the macro-scale occurs at the expense of the deterioration at the micro-scale) and coupling of the healing and degradation thermodynamic forces. Table 8.1 summarizes their interpretation of various tribological phenomena, which can be interpreted as self-organization. In addition, self-organization is often a consequence of coupling of friction and wear with other processes, which creates a feedback in the tribosystem.

TABLE 8.1

Self-Organization Effects in Tribosystems

Effect	Mechanism/ Driving Force	Condition to Initiate	Final Configuration
Stationary micro-topography distribution after running in	Feedback due to coupling of friction and wear	Wear affects micro-topography until it reaches the stationary value	Minimum friction and wear at the stationary micro-topography
In situ tribofilm formation	Chemical reaction leads to the film growth	Wear decreases with increasing film thickness	Minimum friction and wear at the stationary film thickness
Slip waves	Dynamic instability	Unstable sliding	Reduced friction
Self-lubrication	Embedded self-lubrication mechanism	Thermodynamic criteria	Reduced friction and wear
Surface healing	Embedded self-healing mechanism	Proper coupling of degradation and healing	Reduced wear

Source: Nosonovsky, M. 2010b.

8.2 Stability of Frictional Sliding with Coefficient of Friction Dependent on Temperature

8.2.1 Introduction

Different types of instabilities were discussed in Chapter 6. This section suggests and investigates another type of instability due to the temperature dependency of the coefficient of friction. Mortazavi, Wang, and Nosonovsky (2012) formulated a stability criterion and performed a case study of a brake disk. They showed that the mechanism of instability can contribute to poor reproducibility of aircraft disk brake tests reported in the literature. They also proposed a method to increase the reproducibility by dividing the disk into several sectors with decreased thermal conductivity between the sectors.

One area where frictional instabilities are particularly important is the design of disk brakes. Frictional instabilities in car disk brakes have been an object of investigation mostly because of the disk brake squeal (Lee and Barber 1993; Kincaid et al. 2003; Paliwal et al. 2003). Aircraft disk brakes often have a similar construction to car disk brakes; however, the instability constitutes an even more crucial problem for the aircraft brakes.

Aircraft disk brakes are designed to dissipate very large amounts of energy in order to stop the plane within a short time after an aircraft touches the runway. Large amounts of heat are generated in aircraft disk brakes within seconds, resulting in high temperatures as well as very high-temperature gradients. The disks should be made of a material that is light, wear resistant, and able to absorb huge amounts of heat without melting or breaking. In brief, the brake materials should have good heat sinking ability in order to minimize the interface temperature at the braking surface arising out of frictional heat (Venkataraman and Sundararajan 2002). The materials that provide a compromise between weight, strength, and heat transfer are the carbon–carbon (CC) composites (Stimson and Fisher 1980; Byrne 2004).

A number of studies have been conducted in the past decade to investigate the tribological performance of the CC composite disk brake material at various sliding velocities, temperatures, and levels of humidity (Gomes et al. 2001; Luo et al. 2004; Yuan et al. 2005; Kasem et al. 2007). Venkataraman and Sundararajan (2002) showed that CC composites exhibit a transition from a low coefficient of friction during the "normal" wear regime to a high coefficient of friction during the "dusting" wear regime, when the normal pressure times the sliding velocity exceeds a critical value. The transition is associated with the attainment of a critical temperature at the interface between the two CC composite bodies sliding against each other. Yen and Ishihara (1994) showed that two types of surface morphology can be distinguished on the sample surface and argued that the thermoelastic instabilities (TEIs) are responsible for this effect.

Most theoretical studies of the instabilities have concentrated on investigating the onset of the instability and stability criteria. However, the quantitative study of the unstable motion is also of great practical importance. Due to various safety requirements, the aircraft brake disks should demonstrate highly reproducible performance. However, the instability of frictional sliding between a disk brake and a pad may result in the high sensitivity of the tribological system to initial random perturbations. As a result, the time and distance required to stop a plane may vary significantly even under the same conditions (such as the mass and initial velocity of the plane).

8.2.2 Mathematical Modeling

Most of current models of friction-induced instabilities ignore the temperature dependency of the coefficient of friction. These models usually assume that either elastodynamic or thermoelastic effects can give rise to friction-induced instabilities and vibrations. The timescales of these effects differ considerably, so it is usual to neglect the coupling between them (i.e., to neglect the thermal effects in elastodynamic analyses) and to use the quasi-static approximation in thermoelastic analyses. In addition, these models assume that the coefficient of friction is constant (i.e., not varying with temperature).

On the other hand, there is experimental evidence that CC composites undergo a transition from a low to a high value of the coefficient of friction depending on the temperature change (Venkataraman and Sundararajan 2002; Filip 2002; Roubicek et al. 2008).

To investigate the possibility that the temperature dependency of the coefficient of friction leads to instability, Mortazavi et al. (2012) considered the one-dimensional (1D) heat conduction equation in rectangular coordinates with a heat generation source resulting from friction in a slab of length L. They assumed that the 1D region represents a two-dimensional (2D) slab with a small thickness w, so that the heat propagates instantly throughout the thickness and a 1D approximation is valid:

$$\frac{\partial^2 T(x,t)}{\partial x^2} + \frac{1}{wk} g(x,t) = \frac{1}{\alpha} \frac{\partial T(x,t)}{\partial t} \quad \text{in } 0 < x < L, \, t > 0$$

$$-k\frac{\partial T}{\partial x} + h(T - T_0) = 0 \qquad\qquad \text{at } x = 0, \, t > 0$$

$$k\frac{\partial T}{\partial x} + h(T - T_0) = 0 \qquad\qquad \text{at } x = L, \, t > 0$$

$$T(x,0) = F(x) \qquad\qquad\qquad \text{for } 0 \leq x \leq L, \, t = 0$$

(8.1)

where
  $T(x,t)$ is the temperature
  x and t are the spatial coordinate and time
  k and α are the thermal conductivity and diffusivity, respectively
  h is the coefficient of convective heat transfer

Mortazavi et al. (2012) assumed that the slab is initially at a temperature $F(x)$ and, for times $t > 0$, it dissipates heat into the environment with constant temperature T_0.

The heat generation term due to friction is given by

$$g(x,t) = PV\mu$$

(8.2)

where P is pressure, V is sliding velocity, and μ is the coefficient of friction, which was assumed to be linearly temperature dependent (Mortazavi et al. 2012):

$$\mu = \mu_0[1 + \lambda(T - T_0)]$$

(8.3)

where λ is the constant of proportionality.

It is convenient to consider Equation (8.1) in the non-dimensional form by defining dimensionless parameters:

$$\theta(X,\tau) = \frac{T - T_0}{T_0}, \, \tau = \frac{\tau\alpha}{L^2}, \, X = \frac{x}{L}, \, G = \left(\frac{PVL^2}{wk}\right)\left(\frac{\mu_0}{T_0}\right), \, H = \frac{Lh}{k}, \, \varepsilon = \lambda T_0 \quad (8.4)$$

Equation (8.1) becomes (Mortazavi and Nosonovsky 2012b)

$$\frac{\partial^2 \theta}{\partial X^2} + G(1 + \varepsilon\theta) = \frac{\partial \theta}{\partial \tau} \quad \text{in } 0 < X < 1, \tau > 0$$

$$-\frac{\partial \theta}{\partial X} + H\theta = 0 \qquad \text{at } X = 0, \tau > 0$$

$$\frac{\partial \theta}{\partial X} + H\theta = 0 \qquad \text{at } X = 1, \tau > 0$$

$$\theta(X, 0) = f(X) \qquad \text{for } 0 \le X \le 1, \tau = 0$$

(8.5)

The non-homogeneous term $G \in \theta$ in Equation (8.5) is due to the non-dimensional parameter ε, which defines the temperature dependency of the coefficient of friction. Equation (8.5) has a steady solution $\theta_s(x)$. Mortazavi et al. (2012) investigated the stability of this steady state by considering the possibility that a small perturbation in the temperature can grow with time. Thus,

$$\theta(X, \tau) = \theta_s(X) + \tilde{\theta}(X, \tau) \tag{8.6}$$

where $\tilde{\theta}(X, \tau)$ is the perturbation. Now, substituting Equation (8.6) into Equation (8.5; Mortazavi and Nosonovsky 2012b),

$$\frac{\partial^2 \tilde{\theta}}{\partial X^2} + G \in \tilde{\theta} = \frac{\partial \tilde{\theta}}{\partial \tau} \quad \text{in } 0 < X < 1, \tau > 0$$

$$-\frac{\partial \tilde{\theta}}{\partial X} + H\tilde{\theta} = 0 \qquad \text{at } X = 0, \tau > 0$$

$$\frac{\partial \tilde{\theta}}{\partial X} + H\tilde{\theta} = 0 \qquad \text{at } X = 1, \tau > 0$$

$$\tilde{\theta}(X, \tau) = \tilde{f}(X) \qquad \text{for } 0 \le X \le 1, \tau = 0$$

(8.7)

Solution of Equation (8.7) can be expressed as a sum of normal modes, each of which has the general form of

$$\tilde{\theta}(X, \tau) = \psi(X)e^{S\tau} \tag{8.8}$$

where S is a constant. Substituting Equation (8.8) into Equation (8.7):

$$\frac{d^2\psi}{dX^2} + (G\varepsilon - S)\psi = 0 \quad \text{at } 0 < X < 1$$

$$-\frac{d\psi}{dX} + H\psi = 0 \qquad \text{at } X = 0 \tag{8.9}$$

$$\frac{d\psi}{dX} + H\psi = 0 \qquad \text{at } X = 1$$

Solution of Equation (8.9) is in the form of

$$\psi(X) = C_1 \sin(m\pi X) + C_2 \cos(m\pi X) \tag{8.10}$$

where C_1 and C_2 are two constants that can be obtained from the boundary conditions and m is a natural number. Substituting Equation (8.10) into Equation (8.9), we get

$$S = G\varepsilon - (m\pi)^2 \tag{8.11}$$

Mortazavi et al. (2012) concluded from Equation (8.8) that the stability of a small perturbation depends on whether the exponential term grows or decays with time τ, which, in turn, depends upon the sign of S for any m. Thus, from Equation (8.11), decaying in any mth mode was observed for $G\varepsilon < (\mu m)^2$. Since m is increasing, the stability condition should be satisfied for the first term $G\varepsilon < \pi^2$. In the case of $\varepsilon = 0$ (i.e., when there is no temperature dependency of the coefficient of friction), with increasing time, $\theta(X, \tau)$ will approach the steady temperature distribution, $\theta_s(X)$. However, in the case of $\varepsilon \neq 0$, the stability condition requires (Mortazavi et al. 2012)

$$\varepsilon < \pi^2 \left(\frac{wk}{PVL^2}\right)\left(\frac{T_0}{\mu_0}\right) \tag{8.12}$$

Defining

$$\varepsilon_{critical} = \pi^2 \left(\frac{wk}{PVL^2}\right)\left(\frac{T_0}{\mu_0}\right)$$

Mortazavi et al. (2012) concluded that if $\varepsilon > \varepsilon_{critical}$, the solution is unstable. Note that $\varepsilon_{critical}$ depends on the geometry (L), pressure (P), and sliding velocity (V) in our system. This is in agreement with the experimental report by Venkataraman and Sundararajan (2002), who found that there were different regimes of the temperature dependency of the coefficient of friction and that these three parameters had considerable influence on the regimes.

Note also that for the negative $\varepsilon < 0$, the stability condition is always satisfied. That means that if the coefficient of friction decreases with temperature, there is no unstable behavior. This is because the instability is caused by the positive feedback between the coefficient of friction and temperature (i.e, a small positive local fluctuation of temperature would cause a local increase of the coefficient of friction, which, in turn, would cause further growth of temperature). When the feedback is negative ($\varepsilon < 0$ and, therefore, $\lambda < 0$), this type of unstable behavior does not occur (Mortazavi et al. 2012).

To find the exact solution of Equation (8.5), one can use the transformation

$$\theta(X,\tau) = \psi(X,\tau)e^{G\varepsilon\tau} \tag{8.13}$$

which yields, on substituting,

$$\frac{\partial^2 \psi}{\partial X^2} + Ge^{G\varepsilon\tau} = \frac{\partial \psi}{\partial \tau} \qquad \text{in } 0 < X < 1, \ \tau > 0$$

$$-\frac{\partial \psi}{\partial X} + H\psi = 0 \qquad \text{at } X = 0, \ \tau > 0$$

$$\frac{\partial \psi}{\partial X} + H\psi = 0 \qquad \text{at } X = 1, \ \tau > 0 \tag{8.14}$$

$$\psi(X,0) = f(X) \qquad \text{for } 0 \leq X \leq 1, \ \tau = 0$$

Equation (8.14) can be solved analytically, using first the method of the separation of variables and then the Green's function (Özisik 1993). First, Mortazavi et al. (2012) found the solution of the homogeneous equation without the term $G\exp(G \in \tau)$:

$$\psi'(X,\tau) = 2\sum_{m=1}^{\infty} e^{-(m\pi)^2\tau} sin(m\pi X)\left(\int_{X'=0}^{1} sin(m\pi X')f(X')dX'\right) \tag{8.15}$$

Here, Mortazavi et al. (2012) assumed that both sides of the slab were kept at the constant temperature T_0 (i.e., $\psi(0,\tau) = \psi(1,\tau) = 0$ and $H \to \infty$). This assumption simplifies the solution without affecting the stability analysis, since the stability is governed by the exponential term of the solution, which is not affected by the boundary conditions.

The solution of the homogeneous equation can be written in terms of Green's function (Özişik 1993):

$$\psi'(X,\tau) = \int_{X'=0}^{1} \gamma(X,\tau;X',\tau'=0)f(X')dX' \tag{8.16}$$

Comparing Equation (8.16) with Equation (8.15), one can construct the Green's function by replacing τ by $\tau - \tau'$:

$$\gamma(X,\tau;X',\tau'0 = 2\sum_{m=1}^{\infty} e^{-(m\pi)^2(\tau-\tau')}\sin(m\pi X)\sin(m\pi X') \tag{8.17}$$

where X' and τ' are integration parameters. Then, the solution of the non-homogeneous problem is given in terms of Green's function (Mortazavi et al. 2012):

$$\psi(X,\tau) = \psi'(X,\tau) + \int_{\tau'=0}^{\tau} d\tau' \int_{X'=0}^{1} G\exp^{G\in\tau}\gamma(X,\tau-\tau',X')dX' \tag{8.18}$$

Substituting Equation (8.17) into Equation (8.18) yields the solution of Equation (8.14):

$$\psi(X,\tau) = 2\sum_{m=1}^{\infty} e^{-(m\pi)^2\tau}\sin(m\pi X)\left(\int_{X'=0}^{1} \sin(m\pi X')f(X')dX'\right)$$

$$+2\sum_{m=1}^{\infty} \frac{G}{(m\pi)^2-G\int} e^{-G\int\tau}\sin(m\pi X)X(1-\cos(m\pi)) \quad \text{for } 0 < X < 1 \tag{8.19}$$

Finally, substituting Equation (8.18) into Equation (8.13), Mortazavi et al. (2012) found the solution of Equation (8.5):

$$\theta(X,\tau) = 2\sum_{m=1}^{\infty} e^{(G\int-(m\pi)^2)\tau}\sin(m\pi X)\left(\int_{X'=0}^{1} \sin(m\pi X')f(X')dX'\right)$$

$$+2\sum_{m=1}^{\infty} \frac{G}{(m\pi)^2-G\int}\sin(m\pi X)(1-\cos(m\pi)) \quad \text{for } 0 < X < 1 \tag{8.20}$$

The stability of a small perturbation of the solution given by Equation (8.19) is governed by Equation (8.12). This can be observed directly from Equation (8.19), noting that the stability depends on whether the exponential term grows or decays with time τ, which, in turn, depends upon whether $G\varepsilon - (m\pi)^2 < 0$ for any m and should be satisfied for the first term $G\varepsilon < \pi^2$ as it was explained previously.

To examine stability condition (Equation 8.12), numerical results for solution of Equation (8.7) is presented in Figure 8.1 (Mortazavi et al. 2012). The simulation was performed using the MATLAB® software package. In order to investigate the evolution of a localized hot/cold spot, Mortazavi et al. (2012) introduced a small random perturbation imposed over a constant temperature field and confined between $0.49 < X < 0.5$ (i.e., at the center of the domain). The perturbation was a random function built by assigning random numbers with the amplitude between zero and two to the values in $0.49 < X < 0.5$ and zero otherwise. Spatial and temporal step size for the numerical simulation was 0.01 and 0.005, respectively.

Figure 8.1 shows the response of the system to small perturbation for different values of ε (0.0001, 0.00005, and 0.00001). Transient temperature is presented for four different values of the dimensionless time (0, 0.00125, 0.0025, 0.05). The parameters of Equation (8.12) were chosen according to the experimental values reported in the literature on CC composites in disk brakes (for example, Zhao et al. 2007): $P = 1\ MPa$, $k = 50\ W/mK$, $L = 0.5\ m$, $V = 1000\ m/s$. These values correspond to $\varepsilon_{critical} = 3.88 \times 10^{-5}$. It is observed from Figure 8.1 that the solution grew unboundedly in the first two cases corresponding to $\varepsilon > \varepsilon_{critical}$ and decayed in the third case, $< \varepsilon_{critical}$, in agreement with the stability criterion of Equation (8.12; Mortazavi et al. 2012).

Thus, Mortazavi et al. (2012) found that when the temperature dependency of the coefficient of friction is introduced, the solution for the temperature field is unstable for $\varepsilon > \varepsilon_{critical}$. The comparison with experimental data shows that in practical cases the value of ε is comparable with $\varepsilon_{critical}$ (Zhao et al. 2007). Therefore, the effect of the temperature dependency of the coefficient of friction should be taken into account when the stability of the frictional sliding with heat generation is analyzed. Although a rectangular slab was studied in this section, a similar effect is expected with a circular disk, as will be discussed next.

8.2.3 Reproducibility of Disk Brake Test Results

In this section we present a model for an aircraft or car disk brake in contact with a pad with a temperature-dependent coefficient of friction between them and show that frictional instabilities affect the reproducibility of brake test results (e.g., the time needed to stop the car or aircraft).

8.2.3.1 Numerical Model

Let us consider a rigid brake disk with the outer and inner radii of R_{out} and R_{in} in contact with a rigid pad pressed together by the pressure P (Figure 8.2). The torque created by the disk is (Mortazavi and Nosonovsky 2012b)

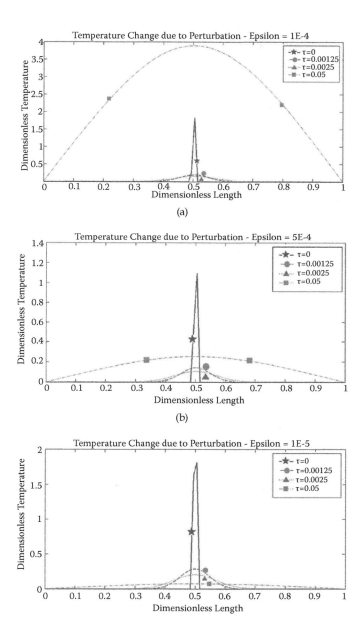

FIGURE 8.1
Response of the system to different values of ε: (a) 0.0001; (b) 0.00005; and (c) 0.00001. τ is the dimensionless time. The solution grew unboundedly in the first two cases corresponding to ε > ε$_{critical}$ and decayed in the third case ε < ε$_{critical}$. (Reprinted with permission from Mortazavi, V., Wang, C., and Nosonvsky, M. 2012. Stability of frictional sliding with the coefficient of friction depended on the temperature. *Journal of Tribology* 134:041601. Copyright ASME.)

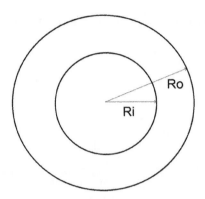

FIGURE 8.2
Schematic of a brake disk. (Reprinted with permission from Mortazavi, V., Wang, C., and Nosonvsky, M. 2012. Stability of frictional sliding with the coefficient of friction depended on the temperature. *Journal of Tribology* 134:041601. Copyright ASME.)

$$M = \int_{r=R_{in}}^{R_{out}} 2\pi r^2 \mu_m P \, dr = \frac{2}{3}\pi\mu_m P(R_{out}^3 - R_{in}^3) \qquad (8.21)$$

where μ_m is the mean coefficient of friction throughout the entire disk surface. The aircraft or car brake is usually equipped with n disks, so the force that decelerates the aircraft or car is given by

$$F = nC\frac{M}{R_{out}} = \frac{2}{3R_{out}}nC\pi\mu_m P(R_{out}^3 - R_{in}^3) \qquad (8.22)$$

where C is a non-dimensional coefficient dependent on the radii of the disks and pads. In Equation (8.22), it is assumed that the pad radius and brake radius are identical. For an aircraft or vehicle of mass m and initial velocity V, the time required to stop is given by

$$t = \frac{Vm}{F} = \frac{3VmR_{out}}{2nC\pi\mu_m P(R_{out}^3 - R_{in}^3)} \qquad (8.23)$$

Let us assume now that the coefficient of friction depends on temperature as described by Equation (8.3) and that T_m is mean temperature at a certain time during the transient stage so that $\mu_m = \mu(T_m)$. Mortazavi et al. (2012) assumed a simple linear dependency given by Equation (8.23) within a certain domain between the minimum and the maximum temperatures, so that the range of the coefficient of friction is $\mu_{min} < \mu(T) < \mu_{max}$ (Figure 8.3).

The time required to stop the aircraft is now in the range $t_{min} < t < t_{max}$, with corresponding values calculated from Equation (8.23):

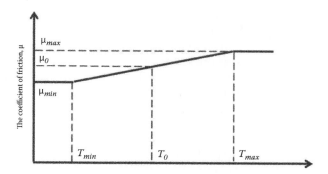

FIGURE 8.3

Temperature dependence of the coefficient of friction. (Reprinted with permission from Mortazavi, V., Wang, C., and Nosonvsky, M. 2012. Stability of frictional sliding with the coefficient of friction depended on the temperature. *Journal of Tribology* 134:041601. Copyright ASME.

$$t_{min} = \frac{3VmR_{out}}{2nC\pi\mu_{max}P(R_{out}^3 - R_{in}^3)}$$

$$t_{max} = \frac{3VmR_{out}}{3nC\pi\mu_{min}P(R_{out}^3 - R_{in}^3)}$$ (8.24)

Most of the mechanical energy is converted into heat. The temperature field at the interface should satisfy the heat conduction equation, which is written here in polar coordinates:

$$\frac{1}{r}\frac{\partial}{\partial r}\left(r\frac{\partial T}{\partial r}\right) + \frac{1}{r^2}\frac{\partial^2 T}{\partial \phi^2} + \frac{1}{wk}g(T) = \frac{1}{\alpha}\frac{\partial T}{\partial t}$$ (8.25)

Stability of the solution of Equation (8.17) should be analyzed now (Mortazavi et al. 2012).

8.2.3.2 Stability Analysis

As discussed in the preceding sections, a small perturbation in the temperature field distribution (an elevated or reduced temperature) can grow unboundedly if the solution is unstable. The stability can depend on the sign of λ in Equation (8.3). For $\lambda > 0$, the coefficient of friction will grow with temperature, and additional heat will be generated leading to a further increase of μ. The heat will also be conducted to neighboring points, so the coefficient of friction at those points will grow as well, and the unstable behavior with a growing size of the "hot spot" will be observed. For $\lambda < 0$, quite oppositely, friction will decrease and the temperature will eventually drop to the steady-state level. In practical conditions, the coefficient of friction can either grow or decrease with temperature.

Thus, Roubicek et al. (2008) observed that the coefficient of friction decreased with increasing temperature during the SAE J2430 friction test, which is frequently used in the United States for evaluating brake performance. This is attributed to physical and chemical changes in the friction layer that forms on the friction surface (Filip 2002). As it was mentioned earlier, Venkataraman and Sundararajan (2002) found that the dependency of the coefficient of friction varied within different working ranges of the sliding velocity and load.

Then, Mortazavi et al. (2012) concentrated on the potentially unstable case of $\lambda > 0$. In the case of unstable behavior, any small perturbation grows and the maximum value of the coefficient of friction, μ_{max} (in the case of a positive perturbation), or its minimum value, μ_{min} (in the case of a negative perturbation), will be reached within a short time. Furthermore, the hot spot will spread to neighboring regions of the surface. The area of the hot spot can be estimated as the thermal diffusivity × time $t\alpha$. Since the perturbation can be either positive or negative, the disk area will be divided into N domains of either the maximum or minimum coefficient of friction. The number of domains can be estimated by dividing the total area by the size of the region of perturbation:

$$N = \frac{n\pi(R_{out}^2 - R_{in}^2)}{t\alpha} \tag{8.26}$$

The probability distribution function for a large number of trials of equal probability (e.g., coin flips) is given by the normal distribution

$$p(x) = \frac{1}{\sqrt{2\pi}\sigma} \exp\left(\frac{(x - \mu_{mean})^2}{2\sigma^2}\right) \tag{8.27}$$

where σ is the standard deviation and $\mu_{mean} = (\mu_{max} + \mu_{min})/2$ is the mean value. The value of σ is given by (Mortazavi and Nosonovsky 2012b)

$$\sigma = \frac{(t_{max} - t_{min})}{N}\sqrt{\frac{N}{4}} = (t_{max} - t_{min})\sqrt{\frac{t\alpha}{4n\pi(R_{out}^2 - R_{in}^2)}} \tag{8.28}$$

in which t is defined by Equation (8.23):

$$\sigma = \frac{(t_{max} - t_{min})}{2n\pi}\sqrt{\frac{3Vm\alpha R_{out}}{2C\mu_m P(R_{out}^2 - R_{in}^2)(R_{out}^3 - R_{in}^3)}} \tag{8.29}$$

For the reproducibility of the results, it is desirable that the standard deviation be as small as possible (Figure 8.4). One possible way to decrease σ is to

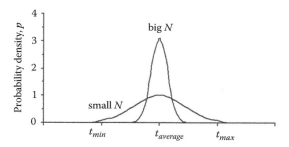

FIGURE 8.4
Effect of the number of domains, N, on the reproducibility of the brake test. (Reprinted with permission from Mortazavi, V., Wang, C., and Nosonvsky, M. 2012. Stability of frictional sliding with the coefficient of friction depended on the temperature. *Journal of Tribology* 134:041601. Copyright ASME.)

decrease the thermal diffusivity α or to increase the total working disk area $n\pi(R_{out}^2 - R_{in}^2)$ and the frictional traction $\mu_m P$. Another approach to increase the reproducibility may be texturing the disk surface so that it is artificially divided into a significant number N of domains (Mortazavi et al. 2012).

8.2.4 Results and Discussion

The heat conduction equation in polar coordinates (Equation 8.25) is written in the dimensionless form (Mortazavi et al. 2012) as

$$\frac{1}{r^*}\frac{\partial}{\partial r^*}\left(r\partial\frac{\partial\theta}{\partial r^*}\right)+\frac{1}{r^2}\frac{\partial^2\theta}{\partial\phi^{*2}}+G(1+\varepsilon\theta)=\frac{\partial\theta}{\partial\tau} \tag{8.30}$$

where r^* and ϕ^* are the dimensionless radius and angle, respectively. Moreover, Mortazavi et al. (2012) assumed that the coefficient of friction depends linearly on temperature, and that the values of temperature are limited by $T_{\min} < T < T_{\max}$. A small perturbation of the steady solution $\theta_s(r^*, \phi^*)$ of Equation (8.30) is given by

$$\theta(r^*,\phi^*,\tau) = \theta_s(r^*,\ \phi^*)+\tilde{\theta}(r^*,\phi^*,\tau) \tag{8.31}$$

where $\tilde{\theta}(r^*,\phi^*,\tau)$ is the small perturbation of the dimensionless temperature. Substituting Equation (8.31) in Equation (8.30) yields

$$\frac{1}{r^*}\frac{\partial}{\partial r^*}\left(r^*\frac{\partial\tilde{\theta}}{\partial r^*}\right)+\frac{1}{r^2}\frac{\partial^2\tilde{\theta}}{\partial\phi^{*2}}+G\varepsilon\tilde{\theta}=\frac{\partial\tilde{\theta}}{\partial\tau} \tag{8.32}$$

The feedback loop is created due to the coupling of the temperature and the coefficient of friction, and response of the system to onset of perturbation

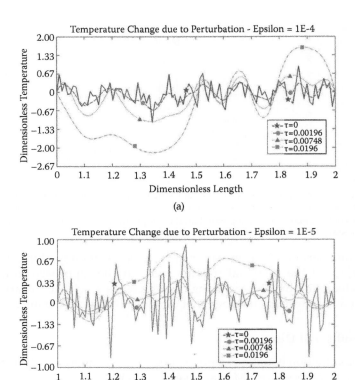

FIGURE 8.5
Response of system to perturbation in the whole domain for different values of ε: (a) 0.0001; (b) 0.00001. τ is the dimensionless time. (Reprinted with permission from Mortazavi, V., Wang, C., and Nosonvsky, M. 2012. Stability of frictional sliding with the coefficient of friction depended on the temperature. *Journal of Tribology* 134:041601. Copyright ASME.)

in the whole domain is shown in Figure 8.5. The results are presented for following parameters: $P = 1$ MP, $k = 50$ W/mK, $R_{in} = 1$ m, $R_{out} = 2$ m, $n = 2000$ rpm. Figure 8.5 clearly shows how the value of ε affects growing or decaying instabilities caused by perturbation in the whole domain.

Then Mortazavi et al. (2012) focused on the reproducibility of test results. Figure 8.6 and Figure 8.7 compare the results of system response in two different cases: first for rigid brake, and then for a brake divided to different sectors. For simplicity, data presented in these two figures assume $\theta = \theta(\phi^*, \tau)$. A random initial perturbation at every point of the disk results in initially positive perturbations that tend to propagate and grow into the positive area, while initially negative perturbation tend to propagate and grow into the negative area (Figure 8.6a). As a result, the disk surface after 1000 time steps of simulation was divided into several domains with the maximum and minimum values of

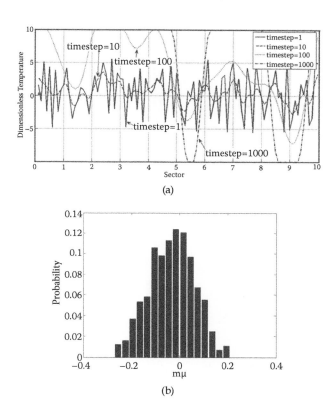

(a)

(b)

FIGURE 8.6
Simulation results for temperature (T) at 1, 10, 100, and 1000 time steps: (a) random fluctuation; (b) distribution of the average value of the coefficient of friction after 100 runs. (Reprinted with permission from Mortazavi, V., Wang, C., and Nosonvsky, M. 2012. Stability of frictional sliding with the coefficient of friction depended on the temperature. *Journal of Tribology* 134:041601. Copyright ASME.)

temperature. The average value of the friction force was calculated by averaging the coefficient of friction at the entire disk and then by all time steps. The simulation was run 100 times and a histogram showing a probability distribution of the average μ was produced (Figure 8.6b; Mortazavi et al. 2012).

After that, it was assumed that the disk was divided into 10 sectors in ϕ direction with zero thermal conductivity between the sectors. The same simulations were run and the results are shown in Figure 8.7. It is observed that a random initial perturbation at any point results in the formation of a number of domains (identical with the sectors) with maximum or minimum temperature (Figure 8.7a). Again, the histogram showing the average value of μ was produced on the basis of 100 simulations (Figure 8.7b). The results show that the deviation of the average μ is much lower in this case (namely, the variance $\sigma^2 = 0.0027$) than in the first case ($\sigma^2 = 0.0086$). This is understandable since, due to the insulation of the sectors, the instability cannot

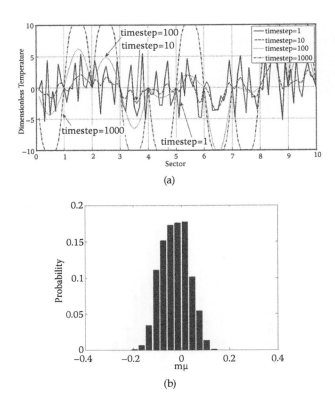

FIGURE 8.7
Simulation results for temperature (*T*) at 1, 10, 100, and 1000 time steps: (a) random fluctuation; (b) distribution of the average value of the coefficient of friction after 100 runs. (Reprinted with permission from Mortazavi, V., Wang, C., and Nosonvsky, M. 2012. Stability of frictional sliding with the coefficient of friction depended on the temperature. *Journal of Tribology* 134:041601. Copyright ASME.)

propagate throughout the entire area of the disk. It is therefore, suggested that texturing the disk or dividing it into sectors can increase the reproducibility of the results (Mortazavi et al. 2012).

To summarize, we studied the stability of frictional sliding with the temperature-dependent coefficient of friction. We presented a mathematical model without including effects of transforming layer and chemical/physical property changes with temperature and formulated the stability condition governing whether the perturbations imposed on the surface temperature in the frictional sliding can grow or decay depending upon working conditions such as pressure, sliding velocity, and geometry. Although it is usually ignored in most disk-pad contact models, this temperature dependency can have a significant effect on stability. The temperature dependency of the coefficient of friction leads to the formation of hot and cold spots on the brake disks. The number of these spots or domains depends upon the thermal diffusivity of the disk material and affects the reproducibility of the brake test results. A larger number of spots is desirable for better reproducibility. It can be achieved by decreasing the thermal diffusivity (although this approach comes in conflict with the need of high dissipation rates), by increasing the disk area, or alternatively, by texturing the surface and dividing it artificially into domains.

8.3 Running-In as a Self-Organized Process

8.3.1 Introduction

In this section we study running-in as a friction-induced self-organized process in the sense that sliding surfaces adjust to each other during this transient process. Furthermore, we suggest using Shannon entropy as a surface roughness parameter that provides a simple test for self-organization. We introduce a theoretical model of "running-in" and compare its results with experimental data on surface roughness evolution and change of the coefficient of friction during the running-in ultra-high vacuum friction tests for WC pin versus Cu substrate (Mortazavi and Nosonovsky 2011b).

When frictional sliding of two solid bodies is initiated, friction and wear usually remain high during a certain initial transient period, referred to as the running-in period (Blau 1989). After the running-in, friction and wear tend to decrease to their stationary values. This decrease of kinetic friction after the initiation of the motion is a general trend. The opposite trend—increasing friction during the running-in—is rarely reported in the literature. Furthermore, the coefficient of kinetic friction (i.e., after sliding has been initiated) is normally higher than the coefficient of static friction (i.e., before sliding has been initiated). In addition, in the models of dynamic friction, such as the state-and-rate models, it is usually assumed, on the basis of experimental observations, that with any change of the sliding velocity, the friction force tends to rise abruptly and then to decrease to its stationary value during a transient process (Nosonovsky 2007). Therefore, there is a general tendency for friction to decrease during the transient, or running-in, process.

There are numerous phenomenological explanations of the running-in phenomenon dealing with various mechanisms, such as viscoplastic deformation, creep, and deformation of asperities. Researchers have paid most of their attention to the empirical change laws and the characterization of surface topography in running-in process (Blau 1989; Begtsson and Ronnenberg 1986; Chowdhury et al. 1979). However, a general thermodynamic description of the running-in, which would explain its underlying physical principles, is absent from the literature.

Entropy is the measure of irreversibility and dissipation. Friction and wear are irreversible dissipative processes. Therefore, it is natural to consider them from the point of view of entropy production. In accordance with the second law of thermodynamics, the thermodynamic entropy tends to increase during dissipative processes. However, in systems that operate far from the thermodynamic equilibrium, the orderliness at the interface can actually increase, leading to friction-induced self-organization (Nosonovsky and Bhushan 2009; Nosonovsky 2010a, 2010b).

During the running-in, surface topography evolves until it reaches a certain stationary state referred to as the equilibrium roughness distribution. The surfaces adjust to each other, leading to a more ordered state with lower dissipation rates. In the present chapter, we consider the running-in as a self-organized process leading to a more ordered state of the surface corresponding to lower dissipation rates. We introduce a surface roughness parameter (the Shannon entropy), which characterizes quantitatively the degree of surface orderliness, and investigate how the evolution of surface roughness during friction affects the rates of dissipation (i.e., friction) and surface deterioration (i.e., wear).

8.3.2 Shannon Entropy as a Characteristic of a Rough Surface

Friction and wear are two processes that occur during the sliding of solid surfaces and are intimately related to each other. They both reflect the tendency of materials to deteriorate irreversibly and energy to dissipate. During friction, heat is generated and dissipated, so the thermodynamic entropy grows. During wear, the surface is deteriorated so that parts of the material are removed as wear particles. The integrity of the material is measured by the so-called configurational entropy. There is evidence that wear rate and therefore rates of configurational entropy are proportional to the rates of thermodynamic entropy, thus making friction and wear two sides of the same process.

The configurational entropy is related to the number of ways of arranging all the particles of the system while maintaining some overall set of specified system properties, such as energy. The concept is closely related to the idea of bits of information needed to describe the interface. If the position of the surface profile $y(x_n)$ (where $n = 1$ to N) at any point x_n is independent, then $I = N\log_2(B)$ bits of information are needed to describe the state of the interface in general, where B is the range of y (number of "bins" of the vertical resolution), Figure 8.8. However, if the interface is not completely random, less information will be needed.

Fleurquin et al. (2010) proposed the Shannon entropy of a rough profile as a measure of profile's randomness

$$S = -\sum_{j=1}^{B} p_j \ln(p_j)$$

(8.33)

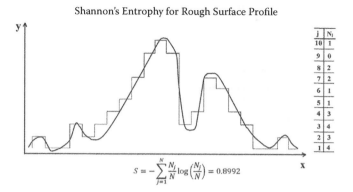

FIGURE 8.8
Example of how Shannon entropy could be calculated for rough surface profile.

where N_j is the number of appearances in the bin j, N is the total number of data points, $p_j = N_j/N$ is the probability of appearance of a height in the bin j, and B is the total number of bins. The Shannon entropy is a generalization of the thermodynamic entropy suggested by L. Boltzmann for the information theory. Fleurquin et al. (2010) used the so-called "Robin-Hood model" to investigate the evolution of the surface roughness due to wear, assuming diffusion-like kinetic law (height is "redistributed" between neighboring points, similarly to the wealth redistribution by Robin Hood).

A surface profile with lower Shannon entropy is "more ordered" (or "less random") than a profile with a higher Shannon entropy, and therefore decreasing Shannon entropy during the transient process is an indication of self-organization. Thus, a smooth surface ($B = 1$, $p_1 = 1$) has the lowest possible entropy $S = 0$, and for a periodic rectangular profile (Figure 8.9), we have $0 < S < 0.3$.

Note that the Shannon entropy, as defined by Equation (8.33), takes into consideration only the height distribution of the profile and does not depend upon the spatial distribution. In this regard, the Shannon entropy is similar to the root mean square (RMS) or the standard deviation (σ)—the standard roughness parameters popular among the tribologists—measured over the range L, where m is defined as the mean value:

$$\sigma^2 = \frac{1}{L}\int_0^L [y(x) - m]^2\, dx, \qquad m = \frac{1}{L}\int_0^L v(x)\, dx \qquad (8.34)$$

The RMS, similarly to the Shannon entropy, does not take into consideration the spatial distribution and correlation of heights between neighboring points. However, such spatial correlation is an important characteristic of the

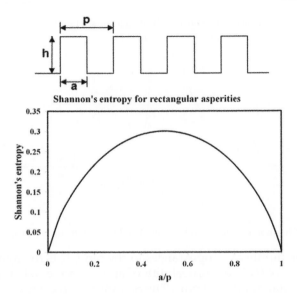

FIGURE 8.9
For a periodic rectangular profile, the Shannon entropy is between 0 and 0.3.

profile. The parameter known as the correlation length, β, is used sometimes and it is defined as the length scale at which correlation between the heights of the neighboring points decays.

Note that the profile entropy as a measure of randomness is not the same as other surface roughness parameters in a traditional sense, such as the RMS. The profile can have the same degree of randomness at different values of the RMS, as well as different RMS for the same level of randomness.

8.3.3 Thermodynamic Model for Surface Roughness Kinetics

We will now investigate the rough profile evolution, assuming that its kinetics is driven by two types of processes or events:

1. The smoothening event is typical for the deformation-driven friction and wear. High asperities fracture due to the deformation (abrasive wear), resulting in smoothening of the profile.
2. The roughening event is typical for the adhesive-driven friction and wear, which has a higher rate for smoother surfaces.

When a particle is removed from the interface due to abrasive or adhesive wear, the data point $y(x)$ is driven from a high bin j into a lower bin i, so the Shannon entropy would change by

$$\Delta S = \frac{N_j - 1}{N} \ln \frac{N_j - 1}{N} - \frac{N_j}{N} \ln \frac{N_j}{N} + \frac{N_{j-1}}{N} \ln \frac{N_{j-1}}{N} - \frac{N_{j-1}}{N} \ln \frac{N_{j-1}}{N} \qquad (8.35)$$

Using $\ln(x + \Delta x) - \ln(x) \approx \Delta x / x$ yields

$$\Delta S = \frac{1}{N} \ln \frac{N_i}{N_j} \qquad (8.36)$$

Smoothening occurs when a particle moves from a less populated bin j into a more populated bin i, and the change of entropy is negative. In the opposite case of roughening, the change of entropy is positive.

As discussed in the previous section, two types of processes occur at the interface: roughening and smoothening. Furthermore, for a very rough surface, smoothening is significant; for a very smooth surface, roughening is significant, so that a certain value of equilibrium roughness can exist. While a comprehensive model of these processes would involve many complex factors, we suggest here a phenomenological model. We assume that the entropy rate due to smoothening is proportional to the value of Shannon entropy, whereas the entropy rate due to roughening is inversely proportional to the value of entropy:

$$\dot{S} = -AS + B / S \qquad (8.37)$$

where A and B are phenomenological coefficients.

The first term in Equation (8.37) represents the tendency of a rough surface to smoothen, which is typical for the deformational/abrasive wear (high asperities are destroyed due to deformation). The second term represents the tendency of a smooth surface to roughen, which is typical for the adhesive wear (smoother surface has higher adhesion). The solution of Equation (8.37) has a stationary point that corresponds to $\dot{S} = 0$, and it is given by $S = \sqrt{B/A}$.

Sliding friction is caused by a number of mechanisms. The intensity of some of these mechanisms (such as the deformation of asperities) grows with roughness, whereas the intensity of others (such as adhesion) decreases with roughness. Again, we assume a simple phenomenological model, in which the coefficient of friction is related to roughness as

$$\mu = C_{def} S + C_{adh} / S \qquad (8.38)$$

where C_{def} and C_{adh} are phenomenological coefficients related to the deformation and adhesion. The coefficients of friction and wear that correspond to the stationary state are given by

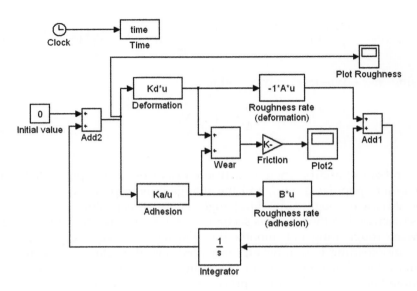

FIGURE 8.10
A feedback loop model (top) and its presentation in Simulink (bottom). Two simultaneous processes (adhesion and deformation) affect the surface roughness parameter in different manners. Consequently, an equilibrium value of roughness exists that corresponds to minimum
friction.

$$\mu = C_{def}\sqrt{B/A} + C_{adh}/\sqrt{B/A} \qquad (8.39)$$

Note that the minimum friction corresponds to $S = \sqrt{C_{adh}/C_{def}}$. The stationary point of Equation (8.37) corresponds to the minimum friction if B/A
$= C_{adh}/C_{def}$. The assumption can be justified if the coefficients A and B are
proportional to wear rates due to deformation and adhesion, which in turn
are proportional to friction components due to deformation and adhesion as
designated by C_{def} and C_{adh}.

A control model, using the concept specified previously, was developed
with Matlab/Simulink software (Figure 8.10). The model shows two feedback loops due to the deformation and adhesion mechanisms. The first feedback tends to decrease roughness when it is high, whereas the second tends
to increase roughness when it is low.

Results of simulations are displayed in Figure 8.11. The time dependence of
the coefficient of friction and roughness during the running-in is simulated
with Simulink for $C_{def} = C_{adh} = 1$ and for two cases: $A = B$ and $A \neq B$. Several
initial values of roughness were chosen and the evolution of roughness and
the coefficient of friction was determined. In both cases, following the transient running-in process, roughness reaches its equilibrium value. For $A = B$,
the coefficient of friction always decreases and reaches its minimum value as
predicted by Equation (8.36). However, for $A \neq B$ the coefficient of friction can

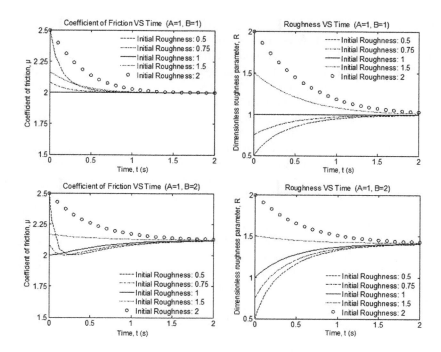

FIGURE 8.11

The time dependence of the coefficient of friction and roughness parameter during the running-in simulated with Simulink for $A = B$ and $A \neq B$. For $A = B$, while roughness reaches its equilibrium value, the coefficient of friction always decreases. Therefore, self-organization of the rough interface results in the decrease of friction and wear. For $A \neq B$ the coefficient of can decrease or increase depending on the initial value of roughness.

decrease or increase depending on the initial value of roughness. For most practical situations, the initial roughness state of surface is rougher than the equilibrium steady-state roughness, so both the roughness and the coefficient of friction decrease during the running-in transient period and the self-organization takes place (Mortazavi and Nosonovsky 2011a).

8.3.4 Experimental Study

To investigate the evolution of roughness parameters in a transient running-in process, an experiment was conducted for the sliding friction of a copper (Cu) pin versus tungsten carbide (WC) substrate in collaboration with Prof. W. Tysoe from the University of Wisconsin-Milwaukee. In order to eliminate the effects of oxidation and contamination, the experiment was carried out in an ultra-high vacuum (UHV) chamber tribometer, operating at pressure $\sim 3.5 \times 10^{-10}$ atm.

The testing sample was made of 99.99% pure copper, while the material of pin was tungsten carbide (WC) selected due to its hardness and stiffness.

The Cu sample (about $20 \times 10 \times 1$ mm thickness) was mounted to a sample manipulator, which was oriented horizontally on the vacuum chamber and on the opposite of the tribometer. This allowed the sample to be moved to the center of the chamber for tribological experiments to be carried out. Moreover, the sample could be placed in front of an ion bombardment source for sample cleaning (Mortazavi and Nosonovsky 2011b).

The vacuum tribometer consisted of a tribopin mounted to the end of an arm that could be moved horizontally (in the x- and y-directions) and also vertically (in the z-direction), either toward or away from the sample, which was oriented horizontally during experiment. These motions were controlled by external servomotors, which allowed the pin position to be precisely determined, and normal (z-direction) and lateral or friction (x-direction) forces were simultaneously measured by strain gauges mounted to thinned sections of the triboarm. The apparatus was under computer control, so the loads, scan speed, scan area, and scan pattern could be selected. More details of the apparatus are presented in Wua et al. (2002).

Measurements were made using a single pass in the same direction with the sliding speed of 4 mm/s with a normal load of 0.9 N, and the experiment was conducted at room temperature. The aim was to produce scratches with a different number of scans (rubbings), each one representing a different period of transient running-in process. To do so, the number of scans of tribopin on the surface of the sample respectively increased for scratch numbers 1 to 5—namely, 1, 4, 8, 16, and 32 scans. Two series of these scratches, each one with the length of 4 mm, were produced.

After conducting the scratching experiments, atomic force microscopy (AFM) images of different scratches were obtained using Pacific Nanotechnology Nano-R™ AFM. The AFM included a motorized zoom/focus video microscope, an AFM scanner, three motors for moving the probe toward the sample, a sample holder, and a motorized X-Y positioning stage. The AFM cantilever was made of silicon with Si_3N_4 coating, in conical shape and with the tip radius of about 10 nm. The topography images were produced in the contact mode. Scans (512×215 pixels) were collected at a minimum of three different locations on each scratch. Then, image processing was performed using a SPIP software package (by Image Metrology) to obtain figures of different roughness parameters such as RMS, roughness average (R_a), extreme-value height descriptor (R_t), skewness (SK), and kurtosis (K; Mortazavi and Nosonovsky 2011b).

8.3.5 Results

Using direct measuring of both the normal and lateral forces (the normal force, however, was kept constant at 0.9 N, as described), the coefficient of friction during the running-in process was calculated and is displayed in Figure 8.12. As expected, the coefficient of friction was reduced during running-in until it reached the steady-state value of 0.43.

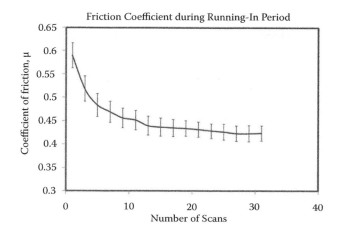

FIGURE 8.12
Friction coefficient change during running-in process.

As described earlier, five roughness parameters—R_a, RMS, R_t, SK, and K—were obtained. The variations of these roughness parameters are displayed in Figure 8.13(a–e).

The root mean square, roughness average, extreme-value height, and surface skewness decreased during the transient process, showing the progressive change in the surface topography. The dramatic decrease of these parameters, which occurred in the very early scans, can be attributed to removing the predominant peaks of initial roughness. As discussed theoretically, the surfaces were observed to be smoothened steadily until equilibrium roughness was achieved. However, the trend with the kurtosis was different. Figure 8.13(e) displays the increase of surface kurtosis during first scans from K = 4.349 to K = 6.198, and, reaching average of K = 6.522 in last scans. Increasing the value of kurtosis means increasing the degree of pointedness of the surface profile. With obtaining probability density function, the Shannon entropy values during the running-in stage could be calculated (Mortazavi and Nosonovsky 2011b).

To compute the Shannon entropy during running-in stage, we proceeded as follows. We obtained the height probability distribution for each location using previously mentioned software. We took a fixed range of heights in this profile and divided this interval in B bins. We numerically found that for a number of bins more than 1,000 the amount of error in calculation of Shannon entropy was less than 1%; thus, we subsequently chose B = 1,000. Then, using Equation (8.33), the results were plotted as shown in Figure 8.14. It is observed that the entropy tended to decrease, indicating that self-organization had occurred during the run-in period. This indicates that self-organization occurred during the transient run-in process due to mutual adjustment of surface roughness topography.

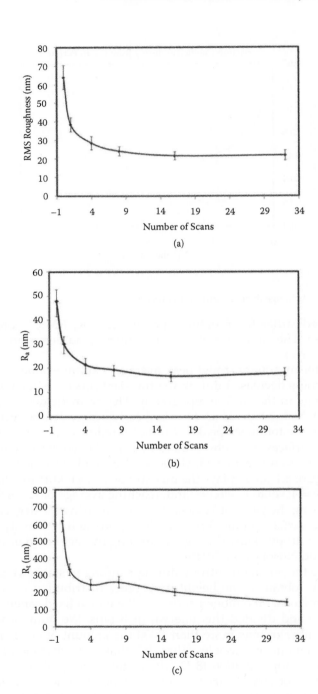

FIGURE 8.13
Variations of different roughness parameters during running-in period: (a) Root mean square,
(b) roughness average, (c) extreme-value height, (d) skewness, and (e) kurtosis.   *Continued*

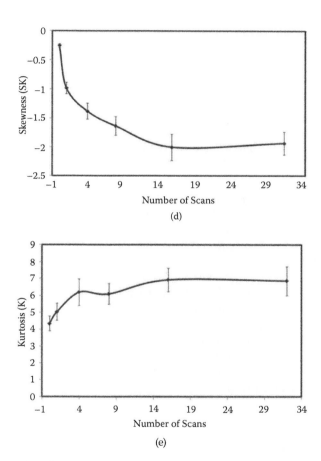

FIGURE 8.13 (*Continued*)
Variations of different roughness parameters during running-in period: (a) Root mean square, (b) roughness average, (c) extreme-value height, (d) skewness, and (e) kurtosis.

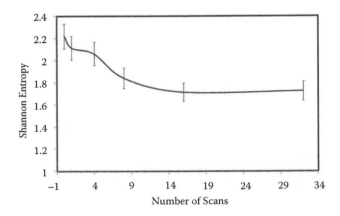

FIGURE 8.14
Shannon entropy evolution during running-in period.

8.3.6 Discussion

In this section, we developed a theoretical approach to study a running-in period as an example of friction-induced self-organization. We concluded that friction, due to non-linear effects, can lead to a less random surface profile, and we proposed Shannon entropy as a measure of profile self-organization. The adjustment of surface roughness to an equilibrium value during the running-in transient process leads to the minimization of friction and wear, and it was investigated by considering a feedback loop due to the coupling of two processes or mechanisms (in our example, adhesion and deformation). We found experimental evidence for our theoretical considerations. We observed that Shannon entropy as a characteristic of a rough surface profile that quantifies the degree of orderliness of the self-organized system decreased during the running-in period, indicating to reach a smoother and more ordered surface profile.

8.4 Frictional Turing Systems

8.4.1 Introduction

In this section, we investigate the possibility of Turing-type pattern formation during friction. Turing or reaction-diffusion systems describe variations of spatial concentrations of chemical components with time due to local chemical reactions coupled with diffusion. During friction, the patterns can form at the sliding interface due to mass transfer (diffusion), heat transfer, various tribochemical reactions, and wear. We present simulation results showing the possibility of such pattern formation. We also discuss existing experimental data that suggest in situ tribofilms can form at the frictional interface due to a variety of friction-induced chemical reactions (oxidation, the selective transfer of Cu ions, etc.; Mortazavi and Nosonovsky 2011a).

8.4.2 Turing Systems and Self-Organization

The reaction-diffusion (RD) systems and their important class called the "Turing systems" constitute a different type of self-organization mechanism. Kagan (2010) suggests that RD systems can describe certain types of friction-induced pattern formation involving heat transfer and diffusion-like mass transfer due to wear. Adams-Matrins instabilities (AMI) and TEI involve wave propagation (hyperbolic) partial differential equations (PDEs), which describes dynamic behavior of elastic media; the RD system involves parabolic PDEs, typical for diffusion and heat propagation problems. The RD systems describe evolution of concentrations of reagents in space and time due to local chemical reactions and the diffusion of the product of reactions (Leppanen 2004). The RD system of PDEs is given by

$$\frac{\partial w_j}{\partial t} = f_i(w_j) + d_{ij} \Delta w_j \tag{8.40}$$

where w_j is the vector of concentrations, f_i represent the reaction kinetics, and d_{ij} and Δ are a diagonal matrix of diffusion coefficients and Laplace operator, respectively. Alan Turing (1952) showed that a state that is stable in the local system can become unstable in the presence of diffusion, which is counter-intuitive, since diffusion is commonly associated with a stabilizing effect. The solutions of RD equations demonstrate a wide range of behaviors including the traveling waves and self-organized patterns, such as stripes, hexagons, and spirals.

While parabolic RD equations cannot describe elastic deformation, they may be appropriate for other processes, such as viscoplastic deformation and interface film growth. For a system of two components, u and v, Equation (8.41) has the following form:

$$\frac{\partial u}{\partial t} = f(u, v) + d_u \left(\frac{\partial^2 u}{\partial x^2} + \frac{\partial^2 u}{\partial y^2} \right) \tag{8.41}$$

$$\frac{\partial v}{\partial t} = g(u, v) + d_v \left(\frac{\partial^2 v}{\partial x^2} + \frac{\partial^2 v}{\partial y^2} \right) \tag{8.42}$$

where f and g are the reaction kinetics functions and d_{ij} is a diagonal matrix ($d_{11} = d_u$ and $d_{22} = d_v$).

Suppose that u represents the non-dimensional temperature at the sliding interface and v is the local slip velocity, also non-dimensional. The non-dimensional values of u, v, x, t, and other parameters are obtained from the dimensional values by division of the latter by corresponding scale parameters. Then Equation (8.41) is interpreted as the description of heat transfer along the interface, and Equation (8.42) describes the flow of viscous material along the interface. In the present chapter, we discuss several types of kinetic functions that can lead to the formation of periodic patterns (Mortazavi and Nosonovsky 2011a).

8.4.3 Numerical Simulation of Frictional Turing Systems

In this section, we define Turing systems, which can describe pattern formation at the frictional interface and a numerical scheme to simulate such systems. We suggest two interpretations of Equations (8.41) and (8.42), which can describe the frictional contact of non-elastic materials and tribofilm formation. First, during friction, heat is generated and mass transfer occurs. The function $f(u, v)$ characterizes the heat generation due to friction and can be taken, for example, in the form of (Mortazavi and Nosonovsky 2011a)

$$f(u,v) = w_0(\mu_0 + \alpha_1 u + \beta_1 v)u \tag{8.43}$$

where $\mu_0 + \alpha_1 u + \beta_1 v$ is the coefficient of friction dependent on the temperature u and local slip velocity v, and the non-dimensional coefficients α_1, β_1, and w_0 are constant. The function $g(u, v)$ characterizes rheological properties of the material and depends on its viscous and plastic properties.

The second interpretation of Equations (8.41) and (8.42) can be used if the growth of a tribofilm (a thin interfacial layer activated by friction) is considered. Whereas u still represents the interfacial temperature, v can be interpreted as the non-dimensional thickness of the tribofilm formed at the interface. The tribofilm can grow due to the material transfer to the interface via diffusion activated by friction, due to precipitation of a certain component (e.g., a softer one) in an alloy or composite material, or due to a chemical reaction, temperature gradient, etc. For example, during the contact of bronze versus steel, a protective Cu tribofilm can form at the interface, which significantly reduces the wear. Such in situ tribofilm has protective properties for the interface since it is formed dynamically and compensates the effect of wear. Furthermore, if wear is a decreasing function of the tribofilm thickness, it is energetically profitable for the film to grow, since growing film reduces wear and further stimulates its growth, forming a feedback loop until a certain equilibrium thickness is attained. Therefore, such tribofilms can be used for machine tool protection and other applications, as discussed in the literature (Aizawa et al. 2005). The growth of the film is governed by interfacial diffusion and by a local kinetic function $g(u, v)$ dependent upon the temperature and local film thickness.

One of the standard forms for functions $f(u, v)$ and $g(u, v)$ based on transferring original reaction-diffusion equations and proper scaling was proposed to be (Dufiet and Boissonade 1991, 1992):

$$f(u, v) = \gamma(a - u + u^2 v) \tag{8.44}$$

$$g(u, v) = \gamma(b - u^2 v) \tag{8.45}$$

In order to investigate the possibility of pattern formation in a Turing system, a stability analysis should be performed. In the absence of diffusion, u and v can approach a linearly stable, uniform steady state. However, under certain conditions, spatially inhomogeneous patterns can evolve by the diffusion-driven instability (Murray 1989). A reaction-diffusion system exhibits diffusion-driven instability or Turing instability if the homogeneous steady state is stable to small perturbations in the absence of diffusion; however, it is unstable to small spatial perturbations when diffusion is introduced. Diffusion results in the spatially inhomogeneous instability and determines that spatial patterns that evolve.

From the linear stability analysis, one can show that, when $d_v > 1$ and $d_u = 1$, a solution with pattern formation (Turing pattern) can exist, if the parameters satisfy the following conditions simultaneously (Dufiet and Boissonade 1991; Murray 1989):

$$0 < b - a < (a + b)^3 \tag{8.46}$$

$$[d_v(b - a) - (a + b)^3]^2 > 4d_v(a + b)^4 \tag{8.47}$$

$$d_v(b - a) > (a + b)^3 \tag{8.48}$$

The results of the stability analysis are summarized in Figure 8.15 for $d_v = 20$ (Dufiet and Boissonade 1992; Murray 1989). The plot shows two curves corresponding to the inequalities introduced by Equations (8.46) and (8.47), whereas the third inequality (Equation 8.48) corresponds to a curve that, for the given values, lies outside the domain of the plot. Therefore, the inequalities divide the parameter space (a, b) into three regions. In region 1, the steady state is stable to any perturbation. In region 2, the steady state exhibits an oscillating instability. In region 3, the steady state is destabilized by inhomogeneous perturbations. Along the border (1, 3) transitions from the uniform state to stationary spatial patterns occur through a so-called Turing bifurcation. Region 3 is sometimes called the "Turing space" and it is of interest for us because it corresponds to pattern formation.

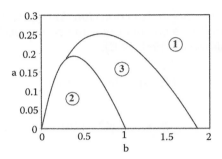

FIGURE 8.15
Linear stability analysis: region 1: uniform steady state; region 2: oscillating instability; region 3: Turing patterns. (Reprinted with permission from Mortazavi, V., and Nosonovsky, M. 2011a. *Langmuir* 27 (8): 4772–4779. Copyright American Chemical Society.)

After determining the region of parameters appropriate for pattern formation from the stability analysis, we need to perform a numerical simulation. To solve the system of Equations (8.41) and (8.42), a number of numerical schemes have been proposed. Approximating the diffusive term by second order backward difference, we used the second order semi-implicit backward difference formula (BDF) scheme proposed by Ruth (1995):

$$\frac{1}{2\Delta t}(3u^{n+1} - 4u^n + u^{n-1}) = 2f^n - f^{n-1} + d_u \Delta_h u^{n+1} \qquad (8.49)$$

$$\frac{1}{2\Delta t}(3v^{n+1} - 4v^n + v^{n-1}) = 2g^n - g^{n-1} + d_v \Delta_h v^{n+1} \qquad (8.50)$$

where n corresponds to the current time step, and $n-1$ and $n+1$ show the previous and next time steps.

Numerical simulations with different initial conditions are performed with the numerical scheme of Equations (8.49) and (8.50) with spatial step size, $\Delta x = 0.001$, and time step of $\Delta t = 10^{-6}$. The length and width of system are considered to be equal by 1 ($L = 1$).

8.4.4 Results

In this section we discussed several exemplary cases of parameter values and initial conditions that can lead to patterns. Cases 1 and 2 are the classical examples of the Turing systems, whereas case 3 is justified by the frictional mechanism of heat generation and mass transfer.

For the first case, we selected the following values of the parameters in Equations (8.44) and (8.45), based on discussion of the stability presented in the preceding section (Mortazavi and Nosonovsky 2011a):

$$\gamma = 10000, d_v = 20, a = 0.02, \text{ and } b = 1.77 \tag{8.51}$$

For the initial conditions we set a random distribution function (using Matlab 7.0 standard package) for both values of u and v. Time evolution of u and v at different time steps is shown in Figure 8.16.

The results in Figure 8.16 present five consecutive snapshots of system presented by Equations (8.41) and (8.42) corresponding to different time steps (t = 0, 0.001, 0.005, 0.01, and 0.1). It is observed that the initially random pattern (t = 0) evolves with time into a pattern with spatial details of steadily growing characteristic length (t = 0.001, 0.005, 0.05) and, finally, into a so-called hexagonal-like (or honeycomb) pattern (t = 0.1), which indicates that the pattern formation occurs. It is observed from Figure 8.16 that the concentrations of u and v at each point have opposite phases (when u has a high value, v has a low one, and vice versa). Moreover, we find that u and v evolve in different rates: While small size grains could be observed in initial time steps for u, large-scale spots formed for v in the initial time steps. However in the next time steps the patterns show similar trends. Such a difference in initial time steps could occur because of different values of the diffusion coefficient in Equations (8.41) and (8.42) ($d_u = 1, d_v > 1$).

For the second case, we used the following values for the parameters in Equations (8.44) and (8.45) based on previously mentioned points (Mortazavi and Nosonovsky 2011a)

$$\gamma = 10000 \text{ and } d_v = 20 \text{ and } a = 0.07 \text{ and } b = 1.61 \tag{8.52}$$

For initial conditions, we considered the following harmonic functions:

$$u(x, y) = 0.919145 + 0.0016\cos(2\pi(x + y)) \tag{8.53}$$

$$v(x, y) = 0.937903 + 0.0016\cos(2\pi(x + y)) \tag{8.54}$$

The time evolution of u and v at different time steps ($t = 0, 0.001, 0.005, 0.01, 0.05,$ and 0.1) is shown in Figure 8.17.

The results in Figure 8.17 present five consecutive snapshots of the system described by Equations (8.41) and (8.42) corresponding to different time steps. The results here show how gradually, from initially striped patterns, the reaction-diffusion mechanism leads to new periodic patterns. For example, a comparison of the concentrations of u and v demonstrates how an almost perfectly stripped patterns ($t = 0.001$) starts to break down at $t = 0.005$. Similarly to what has been observed in Figure 8.16, the phase difference between u and v values in each point is found: When u has a high value, v has a low one, and vice versa (Mortazavi and Nosonovsky 2011a).

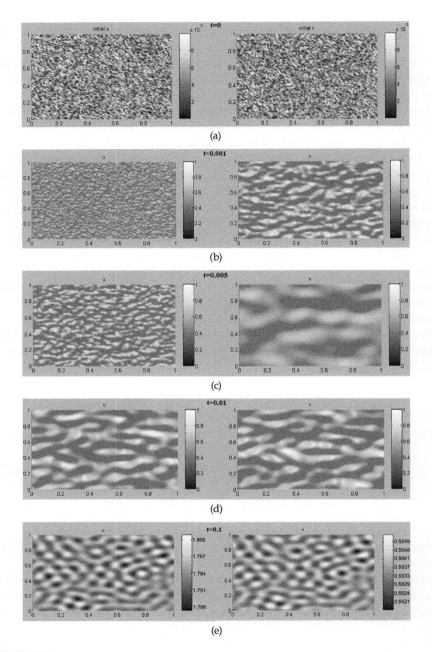

(a)

(b)

(c)

(d)

(e)

FIGURE 8.16
(See color insert.) Time evolution of Turing system components (u and v) in different time steps, at t = (a) 0, (b) 0.001, (c) 0.005, (d) 0.01, and (e) 0.1 (first case). (Reprinted with permission from Mortazavi, V., and Nosonovsky, M. 2011a. *Langmuir* 27 (8): 4772–4779. Copyright American Chemical Society.)

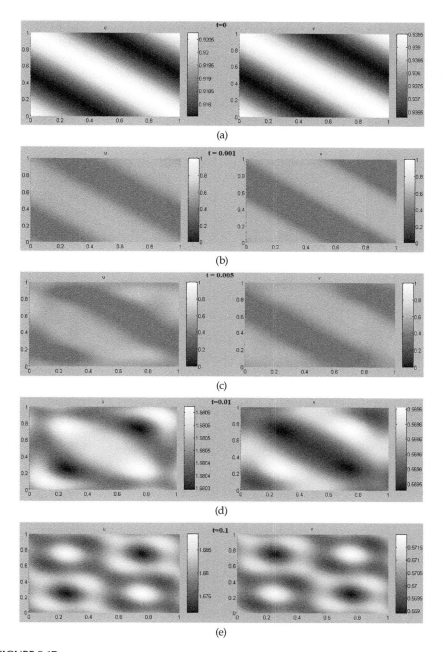

FIGURE 8.17
(See color insert.) Time evolution of Turing system components (u and v) in different time steps, $t =$ (a) 0, (b) 0.001, (c) 0.005, (d) 0.01, 0.05, and (e) 0.1 (second case). (Reprinted with permission from Mortazavi, V., and Nosonovsky, M. 2011a. *Langmuir* 27 (8): 4772–4779. Copyright American Chemical Society.)

Whereas the two first cases showed that our model is capable of capturing the self-organized patterns in Turing systems, in the third case we tried to use more specific functions of reaction kinetics, which are expected to characterize friction-induced reaction mechanisms. Let us consider u as a non-dimensionalized temperature and v as a non-dimensionalized tribofilm thickness (Mortazavi and Nosonovsky 2011a):

$$\frac{\partial u}{\partial t} = f(u,v) + d_u \left(\frac{\partial^2 u}{\partial x^2} + \frac{\partial^2 u}{\partial y^2} \right) \tag{8.55}$$

$$\frac{\partial v}{\partial t} = g(u,v) + d_v \left(\frac{\partial^2 v}{\partial x^2} + \frac{\partial^2 v}{\partial y^2} \right) \tag{8.56}$$

And let us assume functions $f(u,v)$ and $g(u,v)$ to be in the following forms:

$$f(u,v) = w_0(\mu_0 + \alpha_1 u + \beta_1 v)u \tag{8.57}$$

$$g(u,v) = (\alpha_2 u + \beta_2 v) \tag{8.58}$$

or

$$g(u,v) = (\alpha_2 u + \beta_2 v)u \tag{8.59}$$

As initial conditions, we used the following equation as a temperature distribution:

$$u(x,y) = 0.002\cos(2\pi(x+y)) + 0.01\sum_{j=1}^{8}\cos(2\pi jx) \tag{8.60}$$

and random distribution function as an initial roughness for $v(x,y)$. Moreover, we considered the following values for parameters of Equations (8.57–8.59):

$$w_0 = 10^2,\ \alpha_1 = 10^{-4},\ \beta_1 = 10^{-4},\ \mu_0 = 5*10^{-1},$$
$$\alpha_2 = 10^{-4},\ \beta_2 = 10^{-4} \tag{8.61}$$

The results for different time steps, using Equations (8.57) and (8.58) are shown in Figure 8.18. Similar results considering Equation (8.59) as $g(u,v)$, a non-linear function, are shown in Figure 8.19.

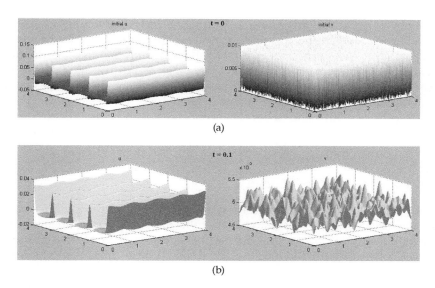

FIGURE 8.18
(See color insert.) Time evolution of Turing system components (u and v) in two different time steps (third case, using Equations 8.18 and 8.19). (Reprinted with permission from Mortazavi, V., and Nosonovsky, M. 2011a. *Langmuir* 27 (8): 4772–4779. Copyright American Chemical Society.)

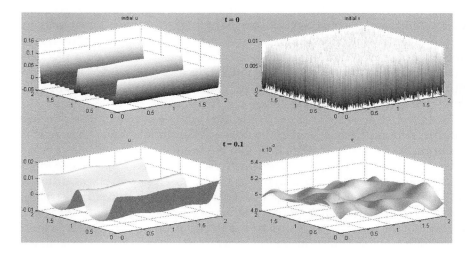

FIGURE 8.19
(See color insert.) Time evolution of Turing system components (u and v) in different time steps (third case, using Equations 8.18 and 8.20). (Reprinted with permission from Mortazavi, V., and Nosonovsky, M. 2011a. *Langmuir* 27 (8): 4772–4779. Copyright American Chemical Society.)

It is obvious that patterns found in Figure 8.18 and 8.19 are not the same as the patterns found in the two previous cases. We observe similar trends in the evolution of tribofilm thickness (v) from an initially random distribution of roughness to a more organized and more patterned distribution. However, the investigation of pattern formation based on solving complete three-dimensional heat and mass transfer equations in tribofilm is needed to show a more realistic picture of how and under what conditions such patterns could occur. In the next section we will discuss possible experimental evidence of friction-induced pattern formation (Mortazavi and Nosonovsky 2011a).

8.4.5 Discussion

The modeling analysis in the preceding sections shows that properly selected functions $f(u,v)$ and $g(u,v)$ can lead to pattern formation. The question remains on whether any experimental data can be interpreted as friction-induced patterns formed by the RD mechanism. It would be appropriate to look for this type of self-organization in processes involving viscoplastic contact or diffusion-dominated effects (e.g., in situ tribofilms). There are several effects that can be interpreted in this way. The so-called "secondary structures" can form at the frictional interface due to self-organization (Fox-Rabinovich and Totten 2006), and some of these structures can have a spatial pattern.

One example is the so-called "selective transfer" suggested by Garkunov (2000). This term refers to a dynamically formed protective tribofilm, which is formed due to a chemical reaction induced by friction. The film protects against wear, since the processes of wear and formation of the film are in dynamic equilibrium (Nosonovsky 2010a). While it is generally assumed that the film is homogeneous with constant thickness at any given moment of time, some data suggest that spatial structures can form. The non-homogeneity of the tribofilm may be caused for various reasons; a similar phenomenon is well known when destabilization of a thin liquid layer occurs during de-wetting (Herminghaus et al. 1998) and it was found also in solid films due to the interplay of adhesion and friction (Shenoy and Sharma 2001).

The selective layer (an in situ tribofilm due to the selective transfer) formed during friction between steel and a copper alloy (bronze) was investigated experimentally by Ilie and Tita (2007). In the presence of glycerin or a similar lubricant, the ions of copper were selectively transferred from bronze to the frictional interface forming the copper tribofilm. This copper was different in its structure from the copper that falls out through normal electrolytic procedures. Ilia and Tita investigated the selective layer using the AFM and found that the layer formed a micro-island pattern with size on the order of 1 μm, rather than a uniform film of constant thickness. A schematic figure of such layer can be seen in Figure 8.20 (Mortazavi and Nosonovsky 2011a).

Another important example of pattern formation reported in the literature is related to the self-adaptive mechanisms improving the frictional properties of hard coatings (e.g., during dry cutting), by tailoring their oxidation

(a) Friction-Induced Diffusion of Cu Ions

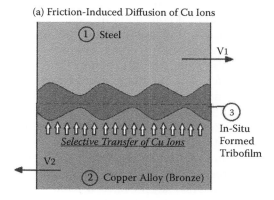

(b) Top View of In-Situ Formed Tribofilm

Bronze Substrate

FIGURE 8.20
(See color insert.) Schematic presentation of the selective layer (b) obtained by friction between a surface of steel and (a) a surface of copper alloy (bronze). (Reprinted with permission from Mortazavi, V., and Nosonovsky, M. 2011a. *Langmuir* 27 (8): 4772–4779. Copyright American Chemical Society.)

behavior (Erdemir et al. 1996; Erdemir, Erck, and Robles 1991; Lovell et al. 2010). Thus, boric acid formation on boron carbide is a potential mechanism for reaching ultra-low friction. Such a mechanism uses the reaction of the boric oxide (B_2O_3) with ambient humidity (H_2O) to form a thin boric acid (H_3BO_3) film. The low friction coefficient of boric acid is associated with its layered triclinic crystal structure (Erdemir 2001; Singer et al. 2003). The layers consist of closely packed and strongly bonded boron, oxygen, and hydrogen atoms, but the layers are widely separated and attracted by van der Waals forces only. During sliding, these atomic layers can align themselves parallel to the direction of relative motion and slide easily over one another (Erdemir et al. 1996).

The tribological behavior of protective coatings formed by both ex situ and in situ transfer films was studied by Singer et al. (2003). Coatings that exhibit long life in sliding contact often do so because the so-called "third body" forms and resides in the sliding interface. The concept of the "third body" as a separate entity, different from the two contacting bodies, is very similar to the concept of the tribofilm. Ex situ surface analytical studies identified the composition and structure of third bodies and provided possible scenarios for their role in accommodating sliding and controlling friction. In situ Raman spectroscopy clearly identified the third bodies controlling frictional behavior during sliding contact between a transparent hemisphere and three solid lubricants: The amorphous Pb–Mo–S coating was lubricated by an MoS_2 transfer film, the diamond-like carbon/nano-composite (DLC/DLN) coating by a graphite-like transfer film, and the annealed boron carbide by H_3BO_3 and/or carbon couples. In situ optical investigations identified third body processes with certain patterns responsible for the frictional behavior (Singer et al. 2003).

TiB_2 thin films are well known for their high hardness, which makes them useful for wear-resistant applications. Mayrhofer et al. (2005) showed that over-stoichiometric $TiB_{2.4}$ layers have a complex self-organized columnar nano-structure precursor. A selected area electron diffraction (SAED) pattern from a $TiB_{2.4}$ layer showed a texture near the film/substrate interface with increased preferred orientation near the film surface (Mayrhofer et al. 2005). The film has a dense columnar structure with an average column diameter of ~20 nm and a smooth surface with an average RMS roughness essentially equal to that of the polished substrate surface, ~15 nm.

Aizawa et al. (2005) investigated in situ TiN and TiC ceramic coating films utilized as a protective coating for dies and cutting tools. They found that chlorine ion implantation assists these lubricious oxide (TiO and Ti_nO_{2n-1}) film to be formed in situ during wearing. They also performed the microscopic analysis, observed worn surfaces and wear debris, and found microscale patterns.

Lin and Chu (2009) described Benard cell-like surface structures found from the observation of transmission electron microscopy (TEM) images of the scuffed worn surface as a result of lubricated steel versus steel contact. They attributed the cells to high temperatures (800°C) and very strong fluid convection or even evaporation occurring inside the scuffed surface. However, the possibility of a diffusion-driven-based pattern formation should not be ruled out.

The experimental evidence of pattern formation, which can, at least theoretically, be attributed to the reaction-diffusion mechanism, is summarized in Table 8.2. Further evidence is needed to rule out alternative explanations.

We conclude that the evolution of a reaction-diffusion system can describe the formation of certain types of friction-induced interfacial patterns. These patterns can form at the sliding interface due to the processes of mass transfer (diffusion and wear), heat transfer, and various tribochemical reactions. On the other hand, existing experimental data suggest that in situ formation of tribofilms due to a variety of friction-induced chemical reactions (such as boric acid formation, oxidation, and selective transfer of Cu ions from the bulk to the interface) or wear can result in the formation of interfacial patterns (islands or honeycomb type domains). This pattern formation can be attributed to the Turing systems.

TABLE 8.2

Summary of Different Experimental Pattern Formation Evidence Discussed in Literature

	Materials in Contact	Tribofilm Material	Mechanism	Pattern	Ref.
1.	Steel versus copper alloy (bronze)	Copper	The selective transfer	Cu islands (~1 μm)	Ilie and Tita (2007)
2.	Boric oxide (B_2O_3)	Thin boric acid (H_3BO_4) film	Oxidation	Layered pattern	Erdemir et al. (1991, 1996); Lovell et al. (2010)
3.	Glass versus Pb–Mo–S	MoS_2 films	Transfer film	MoS_2 islands (nano-scale)	Singer et al. (2003)
4.	Sapphire against a DLC/DLN coating	Silicone-formed DLC/DLN coating	Transfer film	Silicone islands (nano-scale)	Singer et al. (2003)
5.	TiN versus AlN	TiB_2 thin films	Oxidation	Columnar structure (~20 nm)	Mayrhofer et al. (2005)
6.	Steel versus steel	TiN or TiC films of 1 μm	Oxidation	Micro-scale patterns	Aizawa et al. (2005)
7.	Steel versus steel	α-Fe inside the substrate of the unscuffed surface	Wear	Benard cell like patterns	Lin and Chu (2009)

Source: Mortazavi, V., and Nosonovsky, M. 2011a. *Langmuir* 27 (8): 4772–4779.

8.5 Modeling of the Formation of Tribofilms

8.5.1 Introduction

In this section, we will discuss tribofilm formation during friction as an example of friction-induced self-organization. Experimental data suggest that tribofilms can form at the frictional interface due to a variety of friction-induced chemical reactions (oxidation, the selective transfer of Cu ions, etc.). These tribofilms as well as other frictional "secondary structures" can form various patterns such as islands or honeycomb domains. In this study, we propose the cellular Potts model (CPM) to simulate formation of the tribofilm through the self-organization of a two-phase alloy with a hard matrix. Such structure is typical for a bearing Al–Sn-based alloy. Self-organization of this alloy consists in soft-phase accumulation on the friction surface and forming of a new layer. We will discuss dependency of tribofilm formation on the generated heat due to the friction and other different parameters.

8.5.2 Criterion for Self-Organization

The stability condition for the thermodynamic system is given in the variational form by

$$\frac{1}{2}\delta^2\dot{S}\sum_k \delta X_k \delta J_k \geq 0 \tag{8.62}$$

where $\delta^2\dot{S}$ is the second variation of entropy production rate (Nosonovsky and Bhushan 2009) and k is the number of the generalized forces and flows. Equation (8.62) states that the energy dissipation per unit time at the steady state should be at its minimum, or the variations of the flow and the force should be of the same sign. When Equation (8.62) is not satisfied, the system is driven away from equilibrium, which creates the possibility for self-organization.

Another form of Equation (8.62) is given by (Fox-Rabinovich et al. 2007):

$$\frac{\partial}{2\partial t}(\delta^2 S) = \frac{1}{2}\delta^2\left(\frac{(\mu WV)^2}{\lambda T^2}\right) = \delta X \delta J = \delta(\mu WV)\delta\left(\frac{\mu WV}{\lambda T^2}\right) \geq 0 \tag{8.63}$$

Equation (8.63) is a powerful tool to study frictional contact. It involves the coefficient of friction, thermal conductivity, and the sliding velocity. In the case of any inter-dependence between these values, the stability of the system should be analyzed. It can be assumed (Nosonovsky and Bhushan 2009) that the coefficient of friction depends on V and T, while the thermal conductivity depends on V:

$$\mu = \mu(V, T) \qquad \lambda = \lambda(V) \tag{8.64}$$

The stability condition given by Equation (8.62) takes the form of (Nosonovsky and Bhushan 2009)

$$\frac{1}{2}\delta^2\dot{S} = \frac{W^2}{\lambda T^2}\left(\frac{\partial\mu}{\partial V}V+\mu\right)\left(\frac{\partial\mu}{\partial V}V+\mu-\frac{\mu V}{\lambda}\frac{\partial\lambda}{\partial V}\right)(\delta V)^2$$
$$-\left(\frac{2V^2 W^2\mu}{\lambda T^3}\right)\frac{\partial\mu}{\partial T}(\delta T)^2 \geq 0 \tag{8.65}$$

The stability condition is violated if either the coefficient of friction grows with temperature ($\partial\mu/\partial T > 0$) or, in the first term, the parentheses have different signs. The latter is possible if $\partial\mu/V < 0$, whereas $\partial\mu/\partial V > 0$, or if $\partial\mu/\partial V > 0$, whereas $\partial\lambda/\partial V < 0$. When the stability condition is violated, the tribological system is likely to enter the self-organizing regime, with reduced friction and wear.

When the coefficient of friction and the thermal conductivity depend upon a material's micro-structure, ϕ, it may be convenient to introduce a parameter, ψ, that characterizes the micro-structure of the surface (for example, the density of a micro-pattern) so that

$$\mu = \mu\,(\psi, \phi) \tag{8.66}$$

$$\lambda = \lambda\,(\psi, \phi)$$

The stability condition given by Equation (8.62) takes the form of

$$\frac{1}{2}\delta^2\dot{S} = \frac{V^2 W^2}{T^2}\frac{\partial\mu}{\partial\phi}\left(\frac{1}{\lambda}\frac{\partial\mu}{\partial\phi} - \frac{\mu}{\lambda^2}\frac{\partial\lambda}{\partial\phi}\right)(\delta\phi)^2 \geq 0 \tag{8.67}$$

The stability condition can be violated if

$$\frac{\partial\mu}{\partial\phi}\frac{\partial\lambda}{\partial\phi} < 0 \tag{8.68}$$

It is known from non-equilibrium thermodynamics that when the secondary structure is formed, the rate of entropy production reduces (Fox-Rabinovich et al. 2007). Therefore, if Equation (8.68) is satisfied, the frictional force and wear can reduce. By selecting appropriate values of ψ (e.g., the density of a micro-pattern), the condition of Equation (8.67) can be satisfied. Note, that the wear rate is related to the rate of surface entropy production:

$$\frac{dw}{dt} = B\frac{dS}{dt} = YJ \tag{8.69}$$

It is suggested to use the theory presented in this section to optimize the micro-structure of a composite material in order to ensure that the self-organized regime occurs. For that end, the dependencies of Equation (8.67) should be investigated experimentally and their derivatives obtained. Following that, the value of ψ, which provides the best chances for the transition to the self-organized regime should be selected (Nosonovsky 2010a).

8.5.3 Formation of a Tribofilm

As was discussed in previous chapters, sliding friction and wear lead to irreversible energy dissipation and material degradation. However, under certain circumstances, friction can also result in self-organization or formation of patterns and structures at the frictional interface (Nosonovsky and Bhushan 2009, 2012; Bershadski 1992; Gershman and Bushe 2006; Nosonovsky 2010a). These patterns and structures are necessary for energy dispersion during its transformation from a friction zone into a frictional body (Gershman and Bushe 2006).

The self-organized patterns or "secondary structures" (Fox-Rabinovich and Totten 2006) can include a broad range of phenomena, such as tribofilms formed in situ (Aizawa et al. 2005), patterns of surface topography (Maegawa and Nakano 2010; Nosonovsky and Adams 2001), and other interfacial patterns including propagating trains of stick and slip zones formed due to dynamic sliding instabilities (Nosonovsky and Adams 2001).

Among various scenarios leading to self-organization that were discussed in the literature are the destabilization of the stationary or steady state sliding due to tribo–chemical reactions; heat generation and transfer induced by friction; or coupling of wear with friction, as well as the synergetic action of several friction mechanisms, such as the adhesion and deformation, and their effect on surface topography (Nosonovsky 2010a). Recent studies have concentrated on identifying criteria for self-organization and relating the microstructure of composite materials to their ability for self-organization and formation of protective tribofilms in situ (Gershman 2006; Fox-Rabinovich et al. 2007; Nosonovsky and Bhushan 2009). In addition, new mechanisms of self-organization have been suggested, including self-organized criticality (Zypman et al. 2003), which is a special type of self-organization related to stick–slip effects, and reaction-diffusion systems (Kagan 2010; Mortazavi and Nosonovsky 2011a).

Frictional sliding generates heat, and heat tends to lead to the thermal expansion of materials. This can result in so-called thermoelastic instabilities, which have been studied extensively in the past 30 years (Barber 1969). The effect of TEI on sliding is similar to that of AMI. In addition to heat generation, friction can be coupled with other effects, such as, for example, wear, which can lead to similar instabilities. Also, the coefficient of friction can depend upon the sliding velocity, load, and interface temperature, which can also lead to complex dynamic behavior (Mortazavi et al. 2012).

Tribological compatibility is the concept associated with self-organization and can be defined as the ability of a tribosystem to provide optimum conditions within a given range of operating parameters by the chosen criteria (Bushe and Gershman 2006). If the elements of the tribosystem are compatible, they quickly adapt to each other, depending on the specific operational and functional requirements.

Dry (lubricant-free) friction is one of the situations in which tribological compatibility can be investigated (Bushe and Gershman 2006). This mode of operation is typical for tribosystems working in vacuum, brake engineering, current collection, dry machining, etc. Tribological compatibility ensures satisfactory work under this situation once surface layers participate in the process of friction.

For instance, stable friction of the bearing alloys containing a soft phase occurs because of the formation of a film of soft material (Gershman and Bushe 2004). This film serves as a solid lubricant, enhancing the operation of the friction surfaces. Once seizure and corresponding metal transfer occur, some severe types of surface damage develop, for example, galling and jamming. These damages can result in the termination of the relative movement of frictional components and in their failure.

It is important to consider the physical mechanism of seizure for the severe friction conditions. The physical substance of the seizure process can be evaluated based on the energy hypothesis (Semenov 1976), the film hypothesis (Ainbider 1957), the diffusion hypothesis (Kazakov 1968), and the dislocation movements during these processes (Astrov 1965).

These processes can be non-equilibrium chemical reactions in which the material of a frictional body cannot participate as a reagent (Gershman and Bushe 2006). It can be partition of phase components, such as allocation of a soft phase in a surface layer. Such effects are observed in bronze–plastic bearings and also in bearings from aluminum or copper alloys (Konchits 1984). In these materials, the soft phase had a uniform distribution in the structure of the alloy. Owing to the occurrence of the aforementioned relations, the soft phase concentrates within a surface layer (Konchits). After that, the distribution ceases to be equal, so an ordering of structure occurs. It should be accompanied by a decrease in entropy.

To get more insight into such experimental observations, more rigorous simulation techniques are required. In this research, using the cellular Potts model, we propose an approach to model the self-organization of a two-phase alloy with a hard matrix. The CPM (Graner 1992; Glazier and Graner and Glazier 1992) is a lattice-based computational modeling method to simulate the behavior of cellular structures, and it has been applied to problems where the dynamics is driven by energy minimization arising from interfacial tensions between different media. In recent years, the model has been used in different studies to simulate biological tissues (Glazier and Graner 1993), grain growth (Glazier 1990), foam structure (Jiang et al. 1990), coarsening (Glazier et al. 1990), drainage (Jiang and Glazier 1996), fluid flow, and reaction-advection diffusion systems (Dan et al. 2005). More recently, we used the model for simulation of super-hydrophobic behavior of liquid droplets and bubbles and for study of the contact angle hysteresis (Mortazavi et al. 2012).

8.5.4 A Cellular Potts Model for Formation of a Tribofilm

8.5.4.1 Cellular Potts Modeling

The cellular Potts model is a method to model meso-scale phenomenon using statistical techniques. The simulation domain is discretized using a lattice, as shown in Figure 8.21. Each lattice site has a spin. Contiguous

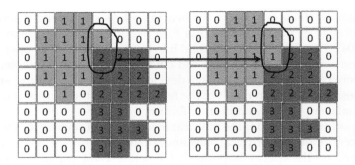

FIGURE 8.21
Schematic of a 2D lattice and a spin flipping attempt in CPM.

simply connected lattice sites with the same spin constitute a generalized cell. A generalized cell can be a bubble, a biological cell, or a metal grain. The configuration of the simulation domain evolves one lattice site at a time based on a set of probabilistic rules. Configuration changes are based on a Hamiltonian function that is used for calculating the probability of accepting configuration changes (Graner and Glazer 1992). The CPM Hamiltonian is given by

$$H = H_J + H_v + H_o \tag{8.70}$$

where H_J is the adhesive term, H_v is the volume restriction term, and H_o is the potential energy term. The adhesive term is given by

$$H_J = \sum_{i,j...neighbors} J(\tau(\sigma(i)),\ \tau(\sigma(j)))(1 - \delta(\sigma(i),\sigma(j))) \tag{8.71}$$

Here, $J(\tau(\sigma(i)),\tau(\sigma(j)))$ is the interfacial energy term between cells of type τ encompassing neighboring lattice sites i, j, and $\sigma(i)$ is the spin of lattice site i. The summation is over a pre-specified radius centered on the lattice site i. The volume restriction term is given by

$$H_v = \sum_{\sigma} \lambda_{volume}[V(\sigma) - V_{target}(\sigma)]^2 \tag{8.72}$$

where $V(\sigma(i))$ is the current volume of the cell encompassing the lattice site i, $V_{target}(\sigma(i))$, is the target volume of the same cell, and λ_{volume} is the Lagrange multiplier for the volume term. The potential energy term is given by

$$H_o = \sum_i \vec{F} \bullet \vec{R}(i)$$

(8.73)

where $\vec{F}$ is a uniform potential field and $\vec{R}(i)$ is the position of the lattice with respect to the reference frame.

Monte Carlo simulations of Potts models have traditionally used local algorithms such as that of Metropolis (Graner and Glazer 1992). A lattice site (i,j) is chosen at random and a new trial spin is also chosen at random from one of its surroundings to test local boundary energy minimization (Figure 8.21). The probability P of accepting such a reassignment is (Graner and Glazer 1992)

$$P = \begin{cases} 1 & \Delta H \leq 0 \\ e^{-\Delta H/kT} & \Delta H \leq 0 \end{cases}$$

(8.74)

where ΔH is the difference between total energy before and after the spin reassignment, T is the magnitude of fluctuations, and k is Boltzman constant.

If $\Delta H > 0$, a second random number, r, is generated and the change of state occurs if $e^{-\Delta H/kT} > r$; otherwise, the label is not changed. Simulation time is measured by Monte Carlo steps (MCS), where one MCS corresponds to as many spin flip attempts as the total number of lattice sites N. The updating rules are exactly the same for each lattice site and the evolution is continuous.

8.5.4.2 Simulation and Results

Our model consists of a 2D lattice with dimensions of $L^2 = 400 \times 400$ sites and two different media: $\sigma = 0$ for matrix and $\sigma = 1$ for soft phase (Figure 8.22). The adhesive term of Hamiltonian, H_J, as defined in Equation (8.71), is based on the interaction between neighboring cells and involves interfacial energies. When neighboring cells belong to the same medium, we have $J(\tau, \tau')$, which means the Potts energy imposes that there is an interfacial energy only between cells of different media (i.e., between neighboring hard matrix/soft phase).

The volume restriction term, H_v, in Equation (8.72) is responsible for the compressibility of the medium. When a spin flip is attempted, one cell will increase its volume and another cell will decrease. Thus, the overall energy of the system may change. Volume constraint essentially ensures that cells maintain the volume close to their target volume. In simulations, for both media, the value of $\lambda_{volume} = 1$, and the target volume for each cell is its initial volume. Since we did not require imposing any prior constraint on surfaces, we set the value of $\lambda_{surface} = 0$. Moreover, we assume that there is a linear potential field applying to the whole lattice downward due to the heat source (friction) at top of the lattice; thus Equation (8.73) becomes

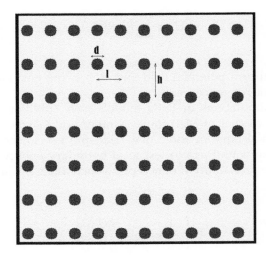

FIGURE 8.22
Initial arrangement of the lattice with hard matrix and inclusion of soft phase.

$$H_o = \sum_i f\vec{R}(i) \tag{8.75}$$

where f is a constant, and $\vec{R}(i)$ is the position of the lattice with respect to the reference frame.

The range of interactions for each site in computing the adhesive term in the Hamiltonian was specified to be the fifth Moore neighbor. We observed that increasing this range to six or seven neighbors did not affect the final results. Since all values in CPM are dimensionless, we were free to set different parameters as a means to fix the length scale and time scale of the simulations in the order of micron and millisecond, respectively.

The initial configuration for the simulation is shown in Figure 8.22. We considered hard matrix with inclusion of soft phase shown with circles of radius d and distance of l and h in two different directions. We changed the structure of the solid surface in the next simulations to investigate the effect of the different parameters on formation of tribofilm. Simulations were performed using CompuCell3D software (Cickovski et al. 2005), which is an open source modeling environment, primarily used to study cellular behavior, and built with the C++ programming language.

For the first simulation, the surface energy between two different media and potential energy's constant (f) were considered to be equal to 10 and 100, respectively. Figure 8.23 shows configuration of the lattice after 70, 230, 400, and 600 MCS of simulation. Grains of soft phase under the influence of temperature gradient due to energy potential field move upward and gradually form the layer on top of the surface. Formation of this newly formed layer was completed in 600 MCS.

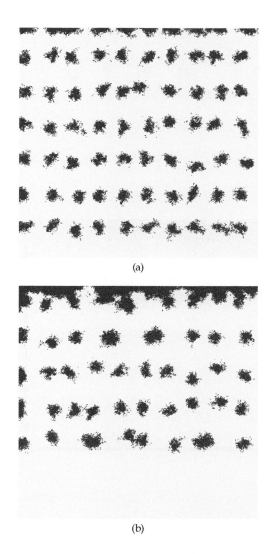

(a)

(b)

FIGURE 8.23
(a) Configuration of the lattice after 70 Monte Carlo time steps (MCS) of simulation; (b) configuration of the lattice after 230 MCS of simulation.   *Continued*

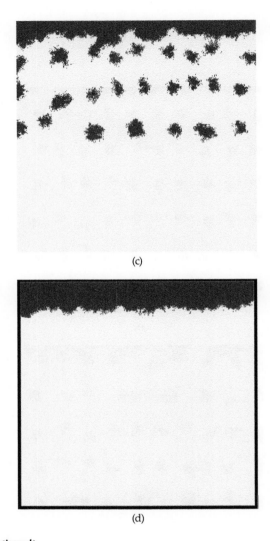

(c)

(d)

FIGURE 8.23 *(Continued)*
(c) Configuration of the lattice after 400 MCS of simulation; (d) configuration of the lattice after 612 MCS of simulation.

To see the effect of temperature gradient on formation of the tribofilm, we repeated the simulation with potential energy's constant (f) considered to be equal to 10. Figure 8.24 shows that, while there is similar movement of soft phase grains upward, it does not lead to formation of a new layer at the top surface of the lattice. Figure 8.25 shows the effect of the potential energy constant (f) at the time of the formation of the layer and its thickness. For $f = 10$, 20, and 30, no layer is formed. The figure clearly shows that, with increasing

FIGURE 8.24
Configuration of the lattice with potential energy's constant (*f*) considered to be equal to 10.

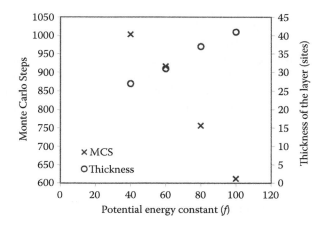

FIGURE 8.25
Effect of potential energy constant (f) on time of the formation of the layer and its thickness. For f = 10, 20, and 30, no layer is formed.

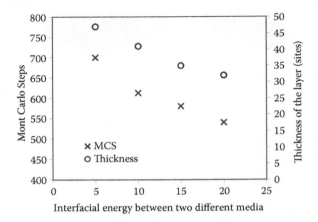

FIGURE 8.26
Effect of interfacial energy on time of the formation of the layer and its thickness.

f, time of formation of the layer decreases; however, the thickness of the layer increases with f.

We also studied the effect of interfacial energy between two media on formation of the layer and its thickness. Figure 8.26 shows that, both the time needed for layer formation and thickness of layer decrease with increasing interfacial energy.

8.5.5 Discussion

We used the CPM to model the self-organization of a two-phase alloy with a hard matrix. Such a structure is typical for a bearing Al–Sn-based alloy. Self-organization of this alloy consists in soft-phase accumulation on the friction surface and forming of a tribofilm. We discussed dependence of tribofilm formation on the interfacial energy between two media and a potential field applied to the lattice. The results show that, for formation of the layer on the top surface of the lattice, the potential field needs to have a higher gradient.

8.6 Stick–Slip Motion and Self-Organization

8.6.1 Introduction

Interfacial patterns including propagating trains of stick and slip zones formed due to dynamic sliding instabilities can be categorized as self-organized patterns or dissipative secondary structures (Fox-Rabinovich and

Totten 2006). In this section, we will discuss how interfacial patterns form, how transition between stick and slip zones occurs, and which parameters affect them.

8.6.2 Self-Organized Criticality and Stick–Slip Motion

Self-organized criticality (SOC) is a concept in the theory of dynamic systems that was introduced in the 1980s (Bak 1996). The best studied example of SOC is the "sandpile model," representing grains of sand randomly placed into a pile until the slope exceeds a threshold value, transferring sand into the adjacent sites, and increasing their slope in turn (Figure 8.27). Placing a random grain at a particular site may have no effect, or it may trigger an avalanche that will affect many sites at the lattice. Thus, the response does not depend on the details of the perturbation (Nosonovsky and Bhushan 2009). It is worth mentioning that the scale of the avalanche is much greater than the scale of the initial perturbation.

There are typical external signs of an SOC system, such as the power-law behavior (the magnitude distribution of the avalanches) and the "one-over-frequency" noise distribution (Bak 1996). The concept has been applied to such diverse fields as physics, cellular automata theory, biology, economics, sociology, linguistics, and others (Nosonovsky and Bhushan 2009).

In the case of dry frictional sliding, it has been suggested that a transition between the stick and slip phases during dry friction may be associated with the SOC, since the slip is triggered in a similar manner to the sandpile avalanches and earthquake slides. Zypman et al. (2003) showed that, in a traditional pin-on-disc experiment, the probability distribution for the size of the slip zones follows the power law. In a later work, the same group

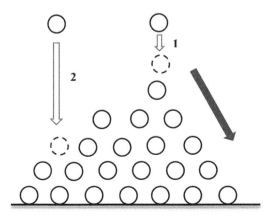

FIGURE 8.27
The sandpile model. Addition of a grain to the sandpile can have (1) the avalanche effect or (2) the local effect.

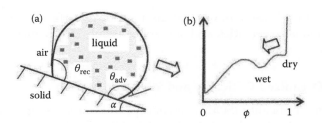

FIGURE 8.28
(a) A water drop on an inclined surface. (b) As the tilt angle *a* grows, the critical state can be reached with a disappearing energy barrier between the wetted and dry states, so that the liquid advances.

found nano-scale SOC-like behavior during atomic force microscopic studies of at least some materials (Zypman et al. 2003; Buldyrev, Ferrante, and Zypman 2006). Thus, "stick" and "slip" are two phases, and the system tends to achieve the critical state between them: In the stick state, elastic energy is accumulated until slip is initiated, whereas energy release during slip leads, again, to the stick state.

During wetting, SOC also apparently plays a role in wetting behaviors (di Meglio 1992; Nosonovsky and Bhushan 2008c). Frictional stick–slip motion and liquid spreading are examples of avalanche-like behavior, when a small input into the system leads to a big change. For example, a droplet on an inclined surface starts its motion when the tilt angle exceeds a certain critical value, so that the gradient of gravity exceeds the energy barriers associated with the pinning of the triple line (Nosonovsky and Bhushan 2009). Wetted and dry states are two stable states with an energy barrier between them (Figure 8.28). When the barrier vanishes, the liquid spreads. For example, when a sessile droplet is placed on an inclined surface, increasing the tilt angle increases the energy gradient due to the gravity, which overcomes the energy barrier at a certain critical value. The liquid starts to spread.

8.6.3 Different Models of Stick–Slip Motion

As was discussed in Chapter 4, the stick–slip phenomenon occurs if the coefficient of static friction is greater than the coefficient of kinetic friction. The stick–slip phenomenon depends not only on the friction force, but also on different system parameters such as inertia, stiffness, surface roughness, and mass of the moving parts. There have been several detailed models for stick–slip friction that include the effects of molecular or asperity size, sliding velocity, various relaxation times, previous history, and other system parameters as mentioned previously. It has been suggested (Berman et al. 1996) that we can categorize three different models of stick–slip: surface topology (roughness) models, distance-dependent models, and velocity-dependent models. However, this classification is not exclusive, and molecular mechanisms of real systems may exhibit aspects of these three different models, simultaneously (Berman et al. 1996).

The surface topology (roughness) models explain stick–slip in terms of the topology or roughness of the sliding surfaces (Berman et al. 1996; Rabinowicz 1965). As the slider climbs an asperity on the substrate, a resisting force is encountered. Once the peak is reached, the slider will slide down rapidly into the valley, resulting in a slip. The measured friction trace with time will show regular or irregular stick–slip "spikes" depending on whether the surface corrugations are themselves regular, as for a lattice plane, or irregular, as for a randomly rough surface. The controlling factors of this type of stick–slip are the topology of the surface (the 2D amplitude and periodicity of protrusions) and the elastic and inertial properties of the sliding surfaces that determine the rate of slip (Berman et al. 1996).

In addition, a stiffer material will have shorter slips because of the shorter recoil to elastic equilibrium. This would allow sticking to more of the smaller asperities, resulting in a richer stick–slip spectrum. With increasing hardness of the materials, plastic deformations during sliding are reduced and the friction pattern approaches a true "contour trace" of the surfaces. Topological stick–slip is observed in the sliding of macroscopically rough surfaces, as well as in atomic force microscopy (AFM) experiments where the intermittent motion of the slider (AFM tip) is a measure of the molecular scale roughness or atomic scale corrugations of the surface lattice (Berman et al. 1996).

Dependence of stick–slip motion to other topological parameters was also suggested (Menezes, Kailas, and Lovell 2010, Tanaka, Kato, and Matsumoto 2003; Gee et al. 1990; Demirel and Granick 1998). Menezes et al. observed that for the case of Al under lubricated conditions, the amplitude of stick–slip oscillations depends on the grinding angle. The amplitude of oscillations starts at 25° and increases through 90°. They argued that the amplitude of stick–slip motion primarily depends on the plowing component (rather than on the adhesion component) of friction. At 90°, the condition at the asperity level is plane strain, leading to higher stresses being generated during sliding. The lubricant present at the interface will experience higher compression and increased viscosity. The increased viscosity will ultimately lead to stick–slip motion. As the grinding angle decreases, the degree of plane strain condition decreases, which reduces the amplitude of stick–slip motion.

Tanaka et al. (2003) reported that under high load conditions, the lubricant film, confined and sheared between two solid walls, is compressed and solidified, which results in stick–slip motion due to molecular deformation. Gee et al. (1990) observed that the occurrence of solid-like characteristics is due to the ordering of the liquid molecule into discrete layers. In general, these properties depend not only on the nature of the liquid but also on the atomic structure of the surfaces, the normal pressure, and the direction and velocity of sliding (Gee et al. 1990). Demirel and Granick (1998) observed liquid-like response at low deformation rates and stick–slip-like response at high deformation rates. For the case of Mg, the stick–stick phenomenon was observed by Menezes et al. (2010) under both dry and lubricated conditions.

When comparing the properties of Al and Mg, the two major differences are the hardness and the number of slip systems. Mg has a higher hardness and lower number of slip systems. A lower number of slip systems could itself promote stick–slip motion as some of the grains in contact at the asperity level would be difficult or easy, depending on the orientation of the slip planes. Such a difference in slip within the contact zone would promote stick–slip, which could explain why stick–slip is observed under both dry and lubricated conditions for Mg (Menezes et al. 2010).

Distance-dependent models of stick–slip suggest that two rough macroscopic surfaces adhere through their microscopic asperities of characteristic length (Berman et al. 1996; Rabinowicz 1958, 1965). During shearing, each surface must first creep a distance of characteristic length—the size of the contacting junctions—after which the surfaces continue to slide, but with a lower (kinetic) friction force than the original (static) value. The reason for the decrease in the friction force is that even though, on average, new asperity junctions should form as rapidly as the old ones break, the time-dependent adhesion and friction of the new ones will be lower than those of the old ones (Berman et al. 1996).

The friction force therefore remains high during the creep stage of the slip, but once the surfaces have moved the characteristic distance, the friction rapidly drops to the kinetic value (Berman et al. 1996). This type of friction has been observed in a variety of mainly dry systems such as paper-on-paper (Heslot et al. 1994; Baumberger, Heslot, and Perrin 1994) and steel-on-steel (Rabinowicz 1951, 1958; Heymann, Rabinowicz, and Rightmire 1954) systems. This model is also used extensively in geology to analyze rock-on-rock sliding (Yoshizawa, Chen, and Israelachvili 1993).

Velocity-dependent models suggest different stick–slip mechanisms. Especially, they can be used for surfaces with thin liquid films (Berman et al. 1996). One of the simplest models is a pure velocity-dependent friction law. In this case there is a high static friction when the surfaces are at rest because the film has solidified. Once the shearing force exceeds this value, the surfaces slide with a lower kinetic friction force, because the film is now in the molten, liquid-like state. Stick–slip sliding proceeds as the film goes through successive freezing–melting cycles. Freezing and melting transitions at surfaces or in thin films may not be the same as the freezing–melting transitions between the bulk solid and liquid phases (Berman et al. 1996). Such films may be considered to alternate between two states, characterized by two friction forces, or they can have a rich friction–velocity spectrum, as proposed by Persson (1994).

The tribological and other dynamic properties of such confined films have been extensively studied by Robbins and Thompson (1991), Landman,

Luedtke, and Ribarsky (1989), and Landman, Luedtke, and Ringer (1992) using computer simulations. Carlson and Batista (1996) have developed a comprehensive rate-and-state dependent friction force law. This model includes an analytic description of the time-dependent freezing and melting transitions of a film, resulting in a friction force that is a function of sliding velocity in a natural way. This model predicts a full range of stick–slip behavior observed experimentally.

It was observed (Berman et al. 1996) that the experimental results of sliding with liquid lubricants in a number of systems involving smooth surfaces are consistent with velocity- and time-dependent phase transition models for stick–slip friction. For these idealized systems, there was no need to invoke a distance-dependent model, nor the classical model based on a negative friction force–velocity profile. In general, more theoretical and mathematical work is needed to give a complete picture of friction mechanisms involved in stick–slip phenomena and to answer how self-organization occurs through these mechanisms.

8.7 Summary

In this chapter we studied several types of friction-induced self-organization, including in situ tribofilm formation, reaction-diffusion systems, stick–slip patterns and self-organized criticality, and mutual adjustment of surfaces during the running-in transient process. Two common features of these diverse processes should be mentioned. First, they all arise from instabilities of stationary states. These instabilities can be studied with the use of the criterion $\delta^2 \dot{S} > 0$ formulated in Chapter 3 (Equation 3.77). Second, these processes tend to lead to friction and wear reduction. This is in line with the minimum entropy production principle as stated in Chapter 3. The property of friction and wear reduction may be beneficial for the development of novel, smart, self-lubricating materials, as will be discussed in the following chapter.

References

Ainbinder, S. B. 1957. *Cold welding of metals*. Science: Moscow.

Aizawa, A., Mitsuo, A., Yamamoto, S., Sumitomo, T., and Muraishic, S. 2005. Self-lubrication mechanism via the in situ formed lubricious oxide tribofilm. *Wear* 259:1–6, 708–718.

Astrov, E. I. 1965. *Plated multilayered metals*. Metallurgy: Moscow.

Bak, P. 1996. *How nature works: The science of self-organized criticality*. New York: Springer.

Barber, J. R. 1969. Thermoelastic instabilities in the sliding of conforming solids. *Proceedings of Royal Society London A* 312:381–394.

Baumberger, T., Heslot, F., and Perrin, B. 1994. *Nature* 367:544.

Begtsson, E. J., and Ronnenberg, A. 1986. The absolute measurement of running-in, *Wear* 109:329–342.

Baumberger, T., Heslot, F., and Perrin, B. 1994. Crossover from creep to inertial motion in friction dynamics. 544–546.

Berman, A. D., Ducker, W. A., and Israelachvili, J. N. 1996. Origin and characterization of different stick–slip friction mechanisms. *Langmuir* 12 (19): 4559–4563.

Bershadski, L. I. 1992. On the self-organization and concepts of wear resistance in tribosystems. *Trenie I Iznos (Friction and Wear)* 13:1077–1094 (in Russian).

Blau, P. J. 1989. *Friction and wear transitions of materials: Break-in, run-in, wear-in.* Park Ridge, NJ: Noyes Publications.

Buldyrev, S. V., Ferrante, J., and Zypman, F. R. 2006. Dry friction avalanches: Experiments and theory. *Physics Reviews E* 74 (066): 110 (doi:10.1103/PhysRevE.74.066110).

Bushe, N., and Gershman, I. S. 2006. Compatibility of tribosystems. In *Self-organization during friction. Advanced surface-engineered materials and systems design,* ed. Fox-Rabinovich, G. S. and Totten, G. E., 59–80. Boca Raton, FL: CRC Taylor & Francis.

Byrne, C. 2004. Modern carbon composite brake materials. *J. Compos. Mater.* 38:1837–1850.

Carlson, J. M., and Batista, A. A. 1996. Constitutive relation for the friction between lubricated surfaces. *Physical Review E* 53 (4): 4153.

Chowdhury, S. K., Kaliszer, H., and Rowe, G.W. 1979. An analysis of changes in surface topography during running-in of plain bearings. *Wear* 16:331–343.

Cickovski, T. M., Huang, C., Chaturvedi, R., Glimm, T., Hentschel, H. G. E., Alber, M. S., Glazier, J. A., Newman, S. A., and Izaguirre, J. A. 2005. A framework for three-dimensional simulation of morphogenesis, *IEEE/ACM Trans. Comput. Biol. Bioinf.* 2:3.

Craig, Norman C. 1992. *Entropy analysis: An introduction to chemical thermodynamics.* New York: VCH publishers.

Dan, D., Mueller, C., Chen, K., and Glazier, J. A. 2005. *Phys. Rev. E: Stat. Phys., Plasmas, Fluids, Relat. Interdiscip. Top.* 72:041909.

Demirel, A. L., and Granick, S. 1998. Transition from static to kinetic friction in a model lubricated system. *Journal of Chemical Physics* 109 (16): 6889–6897.

Dufiet, V., and Boissonade, J. 1991. Conventional and nonconventional Turing patterns. *Journal of Chemical Physics* 96:664–673.

———. 1992. Numerical studies of Turing pattern selection in a two-dimensional system. *Physica A* 188:158–171.

Ebeling, W., Engel, A., Feistel. R. 1990. *Physik der evolutionsprozesse.* Springer-Verlag: Berlin.

Erdemir A. 2001. Solid lubricants and self-lubricating solids. In *Modern tribology handbook,* vol. II, ed. Bhushan, B. Boca Raton, FL: CRC.

Erdemir, A., Bindal, C., Zuiker, C., and Savrun, E. 1996. Tribology of naturally occurring boric acid films on boron carbides. *Surface Coat Technology* 86–87:507–510.

Erdemir, A., Erck, R. A., and Robles J. 1991. Relationship of Hertzian contact pressure to friction behavior of self-lubricating boric acid films. *Surface Coat Technology* 49:435–438.

Filip, P., Weiss, Z., and Rafaja, D. 2002. On friction layer formation in polymer matrix composite materials for brake applications. *Wear* 252:189–198.

Fleurquin, P., Fort, H., Kornbluth, M., Sandler, R., Segall, M., and Zypman, F. 2010. Negentropy generation and fractality in dry friction of polished surfaces. *Entropy* 12:480–489.

Fox-Rabinovich, G. S., and Totten, G. E., eds. 2006. *Self-organization during friction.* Boca Raton, FL: CRC Press.

Fox-Rabinovich, G. S., Veldhuis, S. C., Kovalev, A. I., Wainstein, D. L., Gershman, I. S., Korshunov, S., Shuster, L. S., and Endrino, J. L. 2007. Features of self-organization in ion modified nanocrystalline plasma vapor deposited AlTiN coatings under severe tribological conditions. *Journal of Applied Physics* 102 (074): 305 (doi:10.1063/1.2785947).

Garkunov, D. N. 2000. *Triboengineering (wear and non-deterioration).* Moscow: Agricultural Academy Press 2000 (in Russian).

Gee, M. L., McGuiggan, P. M., Israelachvili, J. N., and Homola, A. M. 1990. Liquid to solid-like transitions of molecularly thin films under shear. In Liu, A. J. and Nagel, S. R. 2001. Jamming and rheology: Constrained dynamics on microscopic and macroscopic scales. Boca Raton, FL: CRC Press.

Gershman, J. S., and Bushe, N. A. 2004. Thin films and self-organization during friction under the current collection conditions. *Surf. Coat. Technol.* 186:405–411.

Gershman, I. S. 2006. Formation of secondary structures and self-organization process of tribosystems during friction with the collection of electric current. In *Self-organization during friction. Advanced surface-engineered materials and systems design,* ed. Fox-Rabinovich, G. S. and Totten, G. E., 197–230. Boca Raton, FL: CRC Taylor & Francis.

Gershman, I. S., and Bushe, N. 2006. Elements of thermodynamics of self-organization during friction. In *Self-organization during friction. Advanced surface-engineered materials and systems design,* ed. Fox-Rabinovich, G. S. and Totten, G. E., 13–58. Boca Raton, FL: CRC Taylor & Francis.

Glazier, J. A., and F. Graner, F. 1993. *Physics Review E* 47:2128–2154.

Glazier, J. A., and Graner, F. 1993. Simulation of the differential adhesion driven rearrangement of biological cells. *Physical Review E* 47(3):2128.

Gomes, J. R., Silva, O. M., Silva, C. M., Pardini, L. C., and Silva, R. S. 2001. The effect of sliding speed and temperature on the tribological behaviour of carbon-carbon composites. *Wear* 249:240–245.

Graner, F., and Glazier, J. A. 1992. Simulation of biological cell sorting using a two-dimensional extended Potts model. *Physical Review Letters* 69(13):2013–2016.

Graner, F., and Glazier, J. A. 1992. *Physics Review Letters* 69:2013–2016.

Herminghaus, S. et al. 1998. Spinodal de-wetting in liquid crystal and liquid metal films. *Science* 282:916.

Heslot, F., Baumberger, T., Perrin, B., Caroli, B., and Caroli, C. 1994. Creep, stick–slip, and dry-friction dynamics: Experiments and a heuristic model. *Physical Review E* 49 (6): 4973.

Heymann, F., Rabinowicz, E., and Rightmire, B. G. 1954. Friction apparatus for a very low-speed sliding studies. *Review of Scientific Instruments* 26:56

Ilie, F., and Tita, C. 2007. Investigation of layers formed through selective transfer with atomic force microscopy. Paper presented in ROTRIB '07, 10th International Conference on Tribology, Bucharest, Romania.

Jiang, Y., Swart, P. J., Saxena, A., Asipauskas, M., and Glazier, J. A. 1999. Hysteresis and avalanches in two-dimensional foam rheology simulations. *Phys. Rev. E: Stat. Phys., Plasmas, Fluids, Relat. Interdiscip. Top.* 59:5819–5832.

Jiang, Y., and Glazier, J. A. 1996. Extended large-Q potts model simulation of foam drainage. *Philos. Mag. Lett.* 74:119–128.

Kagan, E. 2010. Turing systems, entropy, and kinetic models for self-healing surfaces. *Entropy* 12:554–569.

Kasem, H., Bonnamy, S., Rousseau, B., Estrade-Szwarckopf, H., Berthier, Y., and Jacquemard, P. 2007. Interdependence between wear process, size of detached particles and CO_2 production during carbon/carbon composite friction. *Wear* 263:1220–1229.

Kinkaid, N. M., O'Reilly, O. M., and Papadopoulos, P. 2003. Automotive disc brake squeal. *J. Sound Vib.* 267:105–166.

Konchits, V. V. 1984. Friction interaction and current collection in sliding electric contact of composite with metal. *Journal of Friction and Wear* 5(1):59–67.

Kondepudi, D. Prigogine, I. 2000. *Modern thermodynamics.* John Wiley & Sons: Chichester.

Koschmieder, Ernst L. 1993. *Bénard cells and Taylor vortices.* Cambridge University Press.

Landman, U., Luedtke, W. D., and Ribarsky, M. W. 1989. *Journal of Vacuum Science Technology* A7:2829.

Landman, U., Luedtke, W. D., and Ringer, E. M. 1992. Atomistic mechanisms of adhesive contact formation and interfacial processes. *Wear* 153:3.

Lee, K., and Barber, J. L. 1993. Frictionally excited thermoelastic instability in automotive disk brakes. *ASME J. Tribol.* 115:607–614.

Leppanen, T. 2004. Computational studies of pattern formation in Turing systems. Dissertation for the degree of doctor of science in technology, Helsinki University of Technology.

Lin, J. F., and Chu, H. Y. 2009. Analysis of the Benard cell-like worn surface type occurred during oil-lubricated sliding contact. *Proceedings of the ASME/STLE 2009 International Joint Tribology Conference,* IJTC2009, Memphis, Tennessee.

Liu, A. J., and Nagel, S. R. 2001. Jamming and rheology: constrained dynamics on microscopic and macroscopic scales. Boca Raton, FL: CRC Press.

Lovell, M. R., Kabir, M. A., Menezes, P. L., and Higgs, C. F. III. 2010. Influence of boric acid additive size on green lubricant performance. *Philosophical Transactions of Royal Society A* 368 (1929): 4851–4868

Luo, R., Liu, T., Li, J., Zhang, H., Chen, Z., and Tian, G. 2004. Thermophysical properties of carbon/carbon composites and physical mechanism of thermal expansion and thermal conductivity. Carbon, 42:2887–2895.

Maegawa, S., and Nakano, K. 2010. Mechanism of stick–slip associated with Schallamach waves. *Wear* 268:924–930.

Mayrhofer, P. H., Mitterer, C., Wen, J. G., Greene, J. E., and Petrov, I. 2005. Self-organized nanocolumnar structure in superhard TiB_2 thin films. *Applied Physics Letters* 86:131909.

Meglio, J. M 1992. Contact angle hysteresis and interacting surface defects. *Europhys. Lett.* 17:607–612.

Menezes, P. L., Kailas, S. V., and Lovell, M. R. 2010. Response of materials as a function of grinding angle on friction and transfer layer formation. *International Journal of Advanced Manufacturing Technology* 49 (5): 485–495.

Mortazavi, V., D'Souza, R. M, and Nosonovsky, M. 2012. Study of contact angle hysteresis using cellular Potts model. *Physical Chemistry Chemical Physics* 15:2749–2756 (doi:10.1039C2CP44039C).

Mortazavi, V., and Nosonovsky, M. 2011a. Friction-induced pattern formation and Turing systems. *Langmuir* 27 (8): 4772–4779.

———. 2011b. Wear-induced microtopography evolution and wetting properties of self-cleaning, lubricating and healing surfaces. *Journal of Adhesion Science and Technology* 25 (12): 1337–1359.

Mortazavi, V., Wang, C., and Nosonovsky, M. 2012. Stability of frictional sliding with the coefficient of friction depended on the temperature. *Journal of Tribology* 134:041601.

Murray J. D. 1989. *Mathematical biology,* 2nd ed. Berlin: Springer–Verlag.

Nosonovsky, M. 2010a. Entropy in tribology: In search of applications. *Entropy* 12 (6): 1345–1390 (doi:10.3390/e12061345).

———. 2010b. Self-organization at the frictional interface for green tribology. *Philosophical Transactions of Royal Society A* 368 (1929): 4755–4774.

Nosonovsky, M., and Adams, G. G. 2001. Dilatational and shear waves induced by the frictional sliding of two elastic half-spaces. *International Journal of Engineering Science* 39:1257–1269 (doi:10.1016/S0020-7225(00)00085-9).

Nosonovsky, M. 2007. Modelling size, load and velocity effect on friction at micro/nanoscale. *International Journal of Surface Science and Engineering* 1(1):22-37.

Nosonovsky, M., and Bhushan, B. 2008. Do hierarchical mechanisms of superhydrophobicity lead to self-organized criticality? *Scr. Mater.* 59:941–944.

Nosonovsky, M., and Bhushan, B. 2009. Thermodynamics of surface degradation, self-organization and self-healing for biomimetic surfaces. *Philosophical Transactions of the Royal Society A: Mathematical, Physical and Engineering Sciences* 367(1893):1607–1627.

Nosonovsky, M., Amano, R., Lucci, J. M., and Rohatgi, P. K. 2009. Physical chemistry of self-organization and self-healing in metals. *Physical Chemistry Chemical Physics* 11:9530–9536.

Nosonovsky, M., and Bhushan, B. 2009. Thermodynamics of surface degradation, self-organization and self-healing for biomimetic surfaces. *Philosophical Transactions of Royal Society A* 367:1607–1627.

Özışık, M. N. 1993. Heat Conduction, Wiley-IEEE, New York.

Paliwal, M., Mahajan, A., and Filip, P. 2003. Analysis of high frequency squeal in a disc-brake system using a stick-slip friction model. *ASME 2003 International Mechanical Engineering Congress and Exposition, Dynamic Systems and Control.* Vols. 1 and 2.

Persson, B. N. 1994. Theory of friction: The role of elasticity in boundary lubrication. *Journal of Physics Review B* 50:4771.

Prigogine, I. 1980. *From being to becoming.* San Francisco, CA: W. H. Freeman and Company.

Rabinowicz, E. 1951. The nature of the static and kinetic coefficients of friction. *Journal of Applied Physics* 22(11):1373–1379.

Rabinowicz, E. 1958. The intrinsic variables affecting the stick-slip process. *Proceedings of the Physical Society* 71(4):668.

———. 1965. *Friction and wear of materials.* New York: John Wiley & Sons.

Robbins, M. O., and Thompson, P. A. 1991. Origin of stick-slip motion in boundary lubrication. *Science* 253:916.

Roubicek, V., Raclavska, H., Juchelkova, D., and Filip, P. 2008. Wear and environmental aspects of composite materials for automotive braking industry. *Wear* 265:167–175.

Ruuth, S. J. 1995. Implicit-explicit methods for reaction-diffusion problems in pattern formation. *Journal of Mathematical Biology* 34(2):148–176.

Shenoy, V., and Sharma, A. 2001. Pattern formation in a thin solid film with interactions. *Physical Review Letters* 89:119–122.

Singer, I. L., Dvorak, S. D., Wahl, K. J., and Scharf, T. W. 2003. Role of third bodies in friction and wear of protective coatings. *Journal of Vacuum Science Technology A* 21:232–240.

Stimson, I. L., and Fisher, R. 1980. Design and engineering of carbon brakes. *Philos. Trans. R. Soc. London Sect. A* 294:583–590.

Tanaka, K., Kato, T., and Matsumoto, Y. 2003. Molecular dynamics simulation of vibrational friction force due to molecular deformation in confined lubricant film. *Journal of Tribology* 125 (3): 587–591.

Turing, A. M. 1952. The chemical basis of morphogenesis. *Philosophical Transactions of Royal Society B* 237:37–72.

Venkataraman, B., and Sundararajan, G. 2002. The influence of sample geometry on the friction behaviour of carbon-carbon composites. *Acta Mater.* 50:1153–1163.

Wua, G., Gao, F., Kaltchev, M., Gutow, J., Mowlem, J. K., Schramm, W. C., Kotvis, P. V., and Tysoe, W. T. 2002. An investigation of the tribological properties of thin KCl films on iron in ultrahigh vacuum: Modeling the extreme-pressure lubricating interface. *Wear* 252:595–606.

Yen, B. K., and Ishihara, T. 1994. The surface morphology and structure of carbon-carbon composites in high-energy sliding contact. *Wear* 174:111–117.

Yoshizawa, H., Chen, Y. L., and Israelachvili, J. 1993. *Journal of Physical Chemistry* 97:4128.

Yuan, Y., Luo, R., Zhang, F., Li, J., Liu, T., and Zhou, W. 2005. Influence of high temperature heat treatment on the friction properties of carbon/carbon composites under wet conditions. *Mater. Sci. Eng. A* 402:203–207.

Zhaoa, S., Hilmas, G. E., and Dharani, L. R. 2008. Behavior of a composite multi-disk clutch subjected to mechanical and frictionally excited thermal load. *Wear*, 264:1059–1068.

Zypman, F., Ferrante, J., Jansen, M., Scanlon, K., and Abel, P. 2003. Evidence of self-organized criticality in dry sliding friction. *Journal of Physics: Condensed Matter* 15:L191–L196 (doi:10.1088/0953-8984/15/12/101).

9

Self-Lubrication

In the preceding chapters we discussed several self-organizing mechanisms induced by friction. It is important to understand structure–property relationships involved in these mechanisms in order to be able to embed such mechanisms into novel smart materials. Such materials with the capacity for self-organization have begun to emerge. They are referred to as self-lubrication, self-healing, and self-cleaning materials. In this chapter we discuss principles of self-organization at the interface and existing approaches to self-lubrication.

9.1 Principles of Self-Healing and Self-Lubricating Materials

One of the most interesting developments in materials science in the 2000s is the emergence of self-healing materials. These are materials that can repair minor or moderate damage similarly to self-healing in living nature. The common feature of most engineering materials is that they tend to deteriorate irreversibly with time due to wear and tear, fatigue, creep, brittle fracture and formation of cracks, corrosion, erosion, change of chemical structure, and so on. On the other hand, biological tissues and materials in living nature often have the ability to repair minor and moderate damage due to their ability to heal and regenerate.

Artificial self-healing materials, inspired by the self-healing ability in biological objects, are designed in such a way to include mechanisms that can at least partially repair damage, such as voids and cracks, and partially or completely restore macroscopic material properties. Typically, a healing agent (often a liquid) is stored in the matrix of the material and when a cavity (a crack or void) is formed, the healing agent fills the cavity and closes it due to a chemical reaction or a phase transition (e.g., solidification). The healing can be autonomous (without human intervention) or non-autonomous. The latter requires some external intervention, such as heating the material to trigger the repair process (Lumley 2007; Ghosh 2009; van der Swagg 2007; Nosonovsky and Rhatgi 2011).

From the thermodynamic point of view, healing is achieved by shifting the system away from thermodynamic equilibrium, which causes a restoring thermodynamic force (e.g., diffusion) to drive the system back to equilibrium.

The restoring force also drives the healing process that is characterized by a local decrease of entropy (Nosonovsky and Bhushan 2009, 2010a, 2010b; Nosonovsky et al. 2009). Shifting the system away from equilibrium can be achieved by placing it in a metastable state (e.g., creating an over-saturated solution), so that the rupture breaks the fragile meta-stable equilibrium and the system drives to the new most stable state. The metastability can also be achieved by heating that causes a phase transition (e.g., melting, the martensite–austenite transition in metals).

Several strategies of embedding the self-healing properties into engineering materials have been suggested so far (Figure 9.1). The most successful were the attempts to synthesize self-healing polymers because of their relatively large rates of diffusion and plasticity due to the presence of cross-molecular bonds. Self-healing polymers use thermosetting polymers, such as thermosetting epoxy that has the ability to cure (toughening or hardening by cross-linking of polymer chains). Epoxy is a polymer formed by a reaction of an epoxide resin with a polyamine hardener and it can serve as a healing agent that is stored within thin-walled, inert, brittle macro-capsules embedded into the matrix along with a catalyst or hardener (but separately from the latter). When a crack propagates, the capsules fracture, and the healing agent is released and propagates into the crack due to capillarity. Then the healing agent mixes with the catalyst in the matrix, which triggers the cross-linking reaction and hardening of the epoxy that seals the crack (Figure 9.1a).

A different approach involves thermoplastic polymers with various ways of incorporating the healing agent into the material. In this approach, heating is often required to initiate healing. Self-healing ceramic materials often use oxidative reactions because products of these reactions, including oxides, can be used to fill small cracks (van der Zwaag 2009).

It is much more difficult to heal metallic materials than polymers because metallic atoms are strongly bonded and have small volumes and low diffusion rates. Currently, there are three main directions in the development of self-healing metallic systems. First is the formation of precipitates at the defect sites that immobilize further growth until failure. Van der Zwaag (2009) and co-workers called this mechanism "damage prevention" because the idea is to prevent the formation of voids by the diffusion of the atoms to form a precipitate from an over-saturated but under-aged solid–solid solution (alloy). The driving mechanism for the diffusion is the excess surface energy of microscopic voids and cracks that serve as nucleation centers of the precipitate that plays the role of the healing agent (Figure 9.1b). As a result, the newly formed void is sealed by the deposit of atoms in the form of precipitates, before it grows, and thus minimizes creep and fatigue. Olson 2007, Manuel 2007, and co-workers used another approach: reinforcement of an alloy matrix with a micro-fiber or wires made of a shape-memory alloy (SMA), such as nitinol (NiTi). SMA wires have the ability to recover their original shape after some deformation has occurred if they are heated above

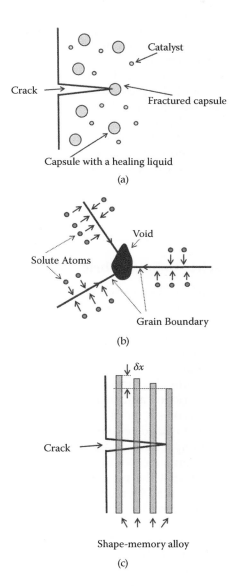

FIGURE 9.1
Self-healing by (a) encapsulation of a healing liquid (b) precipitation in an oversaturated alloy (c) embedding shape-memory alloy microwire. (Based on Nosonovsky and Rohatgi, 2012.)

the phase transformation temperature. If the composite undergoes crack formation, heating from the surface will activate the shape recovery effect of the SMA wires and close the cracks (Figure 9.1c).

The third approach is to use a healing agent (such as an alloy with a low melting temperature) embedded into a metallic solder matrix, similarly to the way it is done with polymers. However, encapsulation of a healing agent into a metallic material is a much more difficult task than in the case of polymers. The healing agent should be encapsulated in micro-capsules that serve as diffusion barriers and that fracture when a crack propagates.

Surface healing is another important field because the surface is the most vulnerable part of a material subjected to various modes of deterioration, including wear, corrosion, fretting fatigue, etc. Surface healing is opposite to wear and can be achieved by various means, including friction-induced in situ tribofilm formation.

9.2 Various Approaches to Self-Lubrication

The term "self-lubrication" has been used for more than two decades and refers to several methods and effects that reduce friction or wear. Among these methods are the deposition of self-lubricating coatings that either are hard (to reduce wear) or have low surface energy (to reduce adhesion and friction). In addition to coatings, self-lubrication can imply the development of metal-, polymer-, or ceramic-based composite self-lubricating materials, often with a matrix that provides structural integrity and reinforcement material that provides low friction and wear.

Nano-composites have become a focus of this research, as well as numerous attempts to include nano-sized reinforcement, carbon nano-tubes, and fullerene C_{60} molecules. Simple models assume that these large molecules and nano-sized particles serve as "rolling bearings" that reduce friction; however, it is obvious now that the mechanism can be more complicated and can involve self-organization. Dynamic self-organization is thought to be responsible for self-lubrication in atomic force microscopy experiments with atomic resolution.

A different approach involves a layer of lubricant that is being formed in situ during friction due to a chemical reaction. Such a reaction can be induced in situ by mechanical contact—for example, a copper protective layer formed at a metallic frictional interface due to the selective transfer of Cu ions from a copper-containing alloy (e.g., bronze) or from a lubricant. A protective layer can be formed also due to a chemical reaction of oxidation or a reaction with water vapor. For example, a self-lubricating layer of boric acid (H_3BO_3) is formed as a result of a reaction of water molecules with B_2O_3 coating. Another type of self-lubricating material involves a lubricant embedded into the matrix (e.g., inside micro-capsules that rupture during wear and release the lubricant).

Surface micro-texturing that provides holes and dimples that serve as reservoirs for a lubricant can be viewed as another method of providing self-lubrication. In addition, we should mention that self-lubrication is observed in many biological systems and that the term "self-lubrication" is used in geophysics, where it refers to anomaly low friction between tectonic plates that is observed during some earthquakes (Nosonovsky and Rohatgi 2011).

The design of coatings with hard and lubricious diamond-like carbon (DLC) surfaces requires a study of transitions between adhesive metal, load-supporting carbide, and wear-resistant DLC materials. Voevodin et al. (1997) investigated these transitions on a Ti-C system prepared by a hybrid of magnetron sputtering and pulsed laser deposition. Crystalline alpha-Ti, TiC, and amorphous DLC films were formed at 100°C substrate temperature by varying the film chemical composition. A gradual replacement of alpha-Ti with TiC and a two-phase region consisting of crystalline TiC and amorphous carbon (a-C) in transitions from Ti to TiC and from TiC to DLC were found. These transitions were reflected in mechanical properties investigated with nano-indentation that provided a hard coating with a low-friction surface, which also resisted brittle failure in tests with high-contact loads.

Neerinck et al. (1998) used pulsed laser deposition (PLD) to produce superhard (60–70 GPa) self-lubricating DLC with low friction and a low wear rate. They obtained thin (2–3 μm) DLC-based coatings for steel substrates, which could maintain friction coefficients of about 0.1 for several million cycles of unlubricated sliding at contact pressures above 1 GPa. Their scratch resistance exceeded that of conventional ceramic (TiN, TiC) coatings.

Vilar (1999) used laser cladding for the protection of materials against wear, corrosion, and oxidation, for the deposition of self-lubricating coatings and thermal barriers, and for the refurbishing of high-cost industrial components. Laser cladding is a hard-facing process that uses a high-powered laser beam to melt the coating material and a thin layer of the substrate to form a pore- and crack-free coating 50 μm–2 mm thick with low dilution that is perfectly bonded to the substrate. The process may be used for large area coverage by overlapping individual tracks, but it is the ability to protect smaller, localized areas that makes it unique.

Erdemir (1991), Erdemir, Fenske, and Eric (1990), Erdemir, Bindal, and Fenske (1996), and Erdeimir, Bindal, Zuiker, and Savrun (1996) investigated boric acid (H_3BO_3)-based applications and the self-lubrication mechanisms of boric acid films on boric oxide coatings prepared by vacuum evaporation. In particular, they measured the coefficients of friction of a steel ball sliding on a boric oxide–coated steel disk and a sapphire ball sliding on a boric oxide–coated alumina disk, which were 0.025–0.05 at steady state, depending on load and substrate material. This low friction was correlated with the formation of a lubricious boric acid film on boric oxide coatings exposed to open air. For the mechanism of self-lubrication, a layered triclinic crystal structure of boric acid was proposed. The atoms constituting each boric acid molecule were arrayed in closely packed and strongly bonded layers 0.318 nm apart and held together by weak forces, such as van der Waals. The authors hypothesized that, during sliding, these layers could align themselves parallel to the direction of relative motion and, once so aligned, could slide over one another with relative ease to provide low friction (Nosonovsky and Rohatgi 2011).

Boric oxide tends to react with water vapor present in air to form a boric acid protective coating:

$$B_2O_3 + 3H_2O \rightarrow 2H_3BO_3 \tag{9.1}$$

The protective coating, in turn, leads to reduced friction and wear. With its layered crystal structure, boric acid resembles those other solids known for their good lubrication capabilities (e.g., MoS_2, graphite, and hexagonal boron nitride; Erdemir, Bindal, and Fenske 1996; Erdeimire, Bindal, Zuiker, and Savrun 1996).

Many composite and nano-composite materials have been suggested for self-lubricating coatings. These include

- TiC/a-C:H nano-composite coatings (Pei, Galvan, and De Hosson 2005)
- Ti-B-N, Ti-B-N-C, and TiN/h-BN/TiB2 multi-layer coatings (Mollart et al. 1996)
- Various MMC materials (Kerr et al. 2000)
- Titanium nitride (TiN) coating (Akhadejdamrong et al. 2003)
- Aluminum/SiC/graphite hybrid composites with various amounts of graphite addition synthesized by semi-solid powder densification (SSPD; Guo and Tsao 2000)
- Plasma-sprayed cast iron splats on an aluminum alloy substrate (Morks et al. 2003)
- TZP-graphite self-lubricating ceramics (Liu and Xue 1996)
- CuO-doped yttria-stabilized tetragonal zirconia ceramics (Tocha et al. 2008)
- Carbon–carbon composites (Chen and Ju 1995)
- Nitride compounds (Zheng and Sun 2006)
- CrN-Ag self-lubricating hard coatings (Mulligan and Gall 2005)
- Super-hard self-lubricating Ti-Si-C-N nano-composite coatings (Ma et al. 2007)
- Micro-plasma oxidation on aluminum alloys in a solution of aluminate-graphite (Wu et al. 2008)

Zhang et al. (2008) investigated porous aluminum anodic oxide films fabricated by anodizing in phosphoric acid electrolyte containing organic acid. By controlling its micro-structure, a macro-porous and thick alumina template was obtained. Surface self-lubricating composites were prepared by taking ultrasonic impregnation in polytetrafluoroethylene (PTFE) latex and the relative subsequent heat treatment technology. Polcar, Evaristo, and Cavaleiro (2009) argued that transition metal dichalcogenides (TMDs) have been one of the best alternatives as low friction coatings for tribological applications, particularly in dry and vacuum environments; however, they have low load-bearing capacities. To increase the load-bearing capacity of these materials, alloying with C should be considered. They studied self-lubricating W-S-C and Mo-Se-C sputtered coatings and found self-lubricating behavior.

Skarvelis and Papadimitriou (2009) used a plasma transferred arc (PTA) technique to produce composite coatings based on co-melting of MoS_2, TiC, and iron ingredients, in an attempt to obtain wear-resistant layers with self-lubricating properties. Graphite and glassy carbon composites were investigated by Hokao et al. (2000). Strnad (2009) developed self-lubricated MoS_2-doped Ti-Al-Cr-N coatings in a multi-layer structure (Nosonovsky and Rohatgi 2011).

Various ceramics are also used for a self-lubricating effect. Suh et al. (2008) studied self-lubricating behavior of structural ceramic balls (ZrO_2, Al_2O_3, and SiC) sliding against the ZrO_2 disk. Blau et al. (1999) investigated self-lubricating properties of ceramic–matrix graphite composites. Lugscheider, Barwulf, and Barimani (1999) studied self-lubricating properties of tungsten and vanadium oxides deposited by the MSIP-PVD process. Bae et al. (1996) studied self-lubricating TiN-MoS_2 composite coatings and Mulligan and Gall (2005) studied CrN-Ag self-lubricating hard coatings.

Powder metallurgy is another area of interest. Li and Xiong (2008) prepared nickel-based self-lubricating composites with graphite and molybdenum disulfide, and lubricants were prepared by a powder metallurgy (PM) method, PM composites (Dellacorte and Sliney 1991), and MoS_2 precursor films on aluminum (Skeldon, Wang, and Thompson 1997).

Polymers and polymer composites and nano-composites are also used for the self-lubricating effect. Li, Xiang, and Lei (2008) prepared polyoxymethylene (POM) composites filled with low-density polyethylene (LDPE) and rice husk flour (RHF) prepared by injection molding. Quintelier et al. (2009) used polymer composites to develop self-lubricating coatings. Blanchet and Peng (1998) used self-lubricating fluorinated ethylene propylene (FEP) and PTFE composites (Nosonovsky and Rohatgi 2011).

With the advent of new carbon-based nano-materials, such as fullerene and carbon nano-tubes (CNTs), new opportunities emerged for tribologists. Thus, other materials were found to demonstrate a self-lubricating effect:

- Fullerene C_{60} (Bhushan et al. 1993)
- Fullerene-like WS_2 nano-particles (Rapoport, Fleischer, and Tenne 2003)

- Ni-based CNT (Wang et al. 2003; Scharf, Neira, and Hwang 2009)
- CNT on Al_2O_3 (Tu et al. 2004)
- CNT-reinforced Al composites (Zhou et al. 2007)
- Mg composites (Umeda, Kondoh, and Imai 2009)
- Boric acid nano-tubes, nano-tips, and nano-rods (Li, Ruoff, and Chang 2003)
- Composite coatings of Co plus fullerene-like WS_2 nano-particles on stainless steel substrate (Friedman et al. 2007)
- Ni-based alloy matrix sub-micron WS_2 self-lubricant composite coatings (Wang et al. 2008)

Alexandridou et al. (1995) developed wear-resistant MMC composite coatings and oil-containing self-lubricating metallic coatings. The latter have been produced by electrolytic co-deposition of oil-containing micro-capsules from Watts nickel plating baths. For this purpose, oil-containing polyterephthalamide micro-capsules were synthesized based on the interfacial polymerization of an oil-soluble monomer (terephthaloyl dichloride) and a mixture of two water-soluble monomers (diethylenetriamine and 1,6-hexamethylenediamine). The influence of several synthesis parameters (e.g., type of encapsulated organic phase, monomer concentration(s), and concentration ratio of the two amine monomers) on the size distribution and morphology of the oil-containing polyamide micro-capsules, as well as on their electrolytic co-deposition behavior, has been discussed. The morphological characteristics of the micro-capsules were affected to a great extent by the functionality of the water-soluble amine monomer. The composition of the core material of the micro-capsules showed a marked influence on their stability upon aging in the Watts nickel plating bath. The level of co-deposition was influenced by the presence of additives in the nickel electrolyte and was strongly dependent on the polymerization conditions employed in the micro-capsule synthesis (Nosonovsky and Rohatgi 2011).

Sui et al. (2005) decided to combine super-hydrophobicity with self-lubrication. They synthesized a carbon coating on Ti_3SiC_2 with combined super-hydrophobic and self-lubricating properties by chlorination at 1000°C followed by modification of the $CF_3(CF_2)(5)CH_2CH_2SiCl_3$ film. The porous structure as well as the organic film on the carbon coating endowed the surface with a super-hydrophobic property. Because of chemical inertia of the carbon coating and the modifier, the super-hydrophobic surface was very stable under various environments. Carbon coating was a good solid lubricant and greatly reduced the friction coefficient of Ti_3SiC_2 sliding against Si_3N_4, which was important for Ti(3)SiC(2) used as engineering material.

Material capable of forming in situ tribofilms is another big class of self-lubricating materials. Al_2O_3/TiC ceramic composites with the addition of CaF_2 solid lubricants showed reduced friction and wear due to an self-lubricating tribofilm formed in situ between the ring-block sliding couple. Deng and Cao (2007) found that two types of tribofilms are formed on the wear surface, depending on the CaF_2 content. A dense tribofilm with a smooth surface associated with a small friction coefficient and a low wear rate was formed by the releasing and smearing of CaF_2 solid lubricants on the wear surface with 10 vol% CaF_2 content. This dense tribofilm acted as a solid lubricant film between the sliding couple and thus significantly reduced the friction coefficient and the wear rate. Breakdown of the tribofilm on the surface associated with a large wear rate was observed on samples with 15 vol% CaF_2 content. This is due to the large degradation of mechanical properties of the composite with higher CaF_2 contents (Nosonovsky and Rohatgi 2011).

Aizawa et al. (2005) studied the self-lubrication mechanism via in situ-formed lubricious oxide tribofilms. They noted that while TiN and TiC ceramic coating films are frequently used as protective coatings for dies and cutting tools, these films often suffer from severe adhesive wearing in dry forming and machining. Chlorine ion implantation assists lubricious oxide film to be formed in situ during wearing. At the presence of chlorine atoms inside TiN or TiC films, in situ formation of lubricous intermediate titanium oxides with TiO and Ti_nO_{2n-1} is sustained to preserve low frictional and wearing state. The self-lubrication process works well in dry machining in order to reduce the flank wear of cutting tools—even in a higher cutting speed range up to 500 m/min.

Alexeyev (1993) studied self-lubricating composite material as a two-phase system with the plastic deformation of self-lubricating composite materials that contain soft second-phase particles. The soft phase flows toward the sliding surface. Thus, the properties of both the hard matrix and the soft second-phase particles, as well as the shape and size of the particles, control the processes of deformation and flow of the soft phase. The results may be used to optimize the micro-structure of self-lubricating composites to obtain the best tribological performance.

Self-lubrication was found also in atomic friction. Livshits and Shluger (1997) presented a theoretical model and conducted molecular dynamics (MD) simulation of the interaction between a crystalline sample and an atomic force microscopy (AFM) tip nano-asperity, combined with a semi-empirical treatment of the mesoscopic van der Waals attraction between tip and surface. They demonstrated that the adsorbed cluster could adjust itself to conditions of scanning by exchanging atoms with the surface and changing its structure and argued that this dynamic "self-organization" of the surface material on the tip during scanning could be a general effect that may explain why periodic surface images are often obtained by using a variety of tips and large tip loads (Nosonovsky and Rohatgi 2011).

Another effect closely related to atomic scale self-lubrication is superlubricity, or the regime of motion in which friction vanishes or very nearly vanishes. Super-lubricity may occur when two crystalline surfaces slide over each other in dry incommensurate contact (Dienwiebel et al. 2004). Thus, the atoms in graphite are oriented in a hexagonal manner and form an atomic asperity-and-valley landscape that resembles an egg crate. When the two graphite surfaces are in registry (every 60°), the friction force is high. When the two surfaces are rotated out of registry, the friction is largely reduced. A state of ultra-low friction can also be achieved when a sharp tip slides over a flat surface and the applied load is below a certain threshold.

Thermolubricity is another atomic-scale phenomenon resulting from the coupling of the thermal effects with friction. The thermal excitations of atoms can assist sliding by overcoming energy barriers. Jinesh et al. (2008) argued that friction at low velocities and low surface corrugations is much lower than the weak logarithmic velocity dependence predicted by thermally activated kinetic models of atomic friction. Furthermore, friction is zero in the zero-velocity limit. The effect was also demonstrated experimentally.

Two areas other than materials science in which the term "self-lubrication" is used are geophysics and biology. In geophysics, scientists have suggested that self-lubricating rheological mechanisms are most capable of generating plate-like motion out of fluid flows. The basic paradigm of self-lubrication is nominally derived from the feedback between viscous heating and temperature-dependent viscosity. Bercovici (1998) proposed an idealized self-lubrication mechanism based on void (such as pore and micro-crack) generation and volatile (e.g., water) ingestion. The term "self-lubrication" is also used in certain biomedical applications. It has been argued by Bejan and Marden (2009) that the tendency of the system to reduce lubrication is a common feature of geophysical and biological systems reflecting the tendency for self-organization.

9.3 Summary

Many methods of self-lubrication and surface healing have emerged in the past decade. These methods can be viewed as manifestations of friction-induced self-organization. It is important to learn how to embed self-organizing features into the structure of smart and composite materials (Table 9.1). Currently, self-healing and self-lubricating materials are designed and synthesized with a trial-and-error approach. However, thermodynamic theories of self-organization can help in finding a more systematic approach. For this purpose, the relevant structure parameters of the composite materials should be identified and structure–property relationships should be established for self-organization, after which the structure should be optimized for the desired properties (e.g., low friction/wear).

TABLE 9.1

Tribological Phenomena That Can Be Interpreted as Self-Organization Effects in Tribosystems

Effect	Mechanism/ Driving Force	Condition to Initiate	Final Configuration
Stationary micro-topography distribution after running-in	Feedback due to coupling of friction and wear	Wear affects micro-topography until it reaches the stationary value	Minimum friction and wear at the stationary micro-topography
In situ tribofilm formation	Chemical reaction leads to film growth	Wear decreases with increasing film thickness	Minimum friction and wear at the stationary film thickness
Slip waves	Dynamic instability	Unstable sliding	Reduced friction
Self-lubrication	Embedded self-lubrication mechanism	Thermodynamic criteria	Reduced friction and wear
Surface healing	Embedded self-healing mechanism	Proper coupling of degradation and healing	Reduced wear

References

Aizawa, A., Mitsuo, A., Yamamoto, S., Sumitomo, T., and Muraishic S. 2005. Self-lubrication mechanism via the in situ formed lubricious oxide tribofilm. *Wear* 259:708–718

Akhadejdamrong, T., Aizawa, M., Yoshitake, M., et al. 2003. Self-lubrication mechanism of chlorine implanted TiN coatings. *Wear* 254:668–679.

Alexandridou, S., Kiparissides, C., Fransaer, J., et al. 1995. On the synthesis of oil-containing microcapsules and their electrolytic codeposition. *Surface Coating Technology* 71:267–276.

Alexeyev, N. 1993. Mechanics of friction in self-lubricating composite materials. *Wear* 166:41–48.

Bae, Y. W., Lee, W. Y., and Besmann, T. M. et al. 1996. Preparation and friction characteristics of self-lubricating TiN-MoS2 composite coatings. *Mater. Sci. Eng. A* 209:372–376.

Bejan, A., and Marden, J. H. 2009. The constructural unification of biological and geophysical design. *Physics of Life Reviews* 6.

Bercovici, D. 1998. Generation of plate tectonics from lithosphere-mantle flow and void-volatile self-lubrication. *Earth and Planetary Science Letters* 154:139–151.

Bhushan, B., Gupta, B. K., and Vancleef, G. W. et al. 1993. Fullerene (C-60) films for solid lubrication. *Tribology Transactions* 36:573–580.

Blanchet, T. A., and Peng, Y. L. 1998. Wear-resistant irradiated FEP unirradiated PTFE composites. *Wear* 214:186–191.

Blau, P. J., Dumont, B., Braski, D. N., et al. 1999. Reciprocating friction and wear behavior of a ceramic–matrix graphite composite for possible use in diesel engine valve guides. *Wear* 225:1338–1349.

Chen, J. D., and Ju, C. P. 1995. Low energy tribological behavior of carbon–carbon composites. *Carbon* 3:7–62.

Dellacorte, C., and Sliney, H. E. 1991. Tribological properties of PM212—A high temperature, self-lubricating, powder-metallurgy composite. *Lubrication Engineering* 47:298–303.

Deng, J. X., and Cao, T. K. 2007.Self-lubricating mechanisms via the in situ formed tribofilm of sintered ceramics with CaF2 additions when sliding against hardened steel. *International Journal of Refractory Metals & Hard Materials* 25:189–197.

Dienwiebel, M., Verhoeven, G. S., Namboodiri, P., Frenken, J. W. M., Heimberg, J. A., and Zandbergen, H. W. 2004. Superlubricity of graphite. *Physical Review Letters* 92:126101.

Erdemir, A. 1991. Tribological properties of boric-acid and boric-acid-forming surfaces. *Lubrication Engineering* 47:168–173.

Erdemir, A., Bindal, C., and Fenske, G. R. 1996. Formation of ultralow friction surface films on boron carbide. *Applied Physics Letters* 68:1637–1639.

Erdemir, A., Bindal, C., Zuiker, C., and Savrun, E. 1996. Tribology of naturally occurring boric acid films on boron carbides. *Surface and Coating Technology* 86–87:507–510.

Erdemir, A., Fenske, G. R., and Erik, R. A. 1990. A study of the formation and self-lubrication mechanisms of boric-acid films on boric oxide coatings. *Surface and Coating Technology* 43–44:588–596.

Friedman, H., Eidelman, O., Feldman, Y., et al. 2007. Fabrication of self-lubricating cobalt coatings on metal surfaces. *Nanotechnology* 18:115703.

Ghosh. S. K., ed. 2009. *Self-healing materials: Fundamentals, design strategies, and applications.* Berlin: Wiley VCH, GmbH.

Guo, M. L. T., and Tsao, C. Y. A. 2000. Tribological behavior of self-lubricating aluminum/SiC/graphite hybrid composites synthesized by the semi-solid powder-densification method. *Composites Science Technology* 60:65–74.

Hokao, M., Hironaka, S., Suda, Y., et al. 2000. Friction and wear properties of graphite/glassy carbon composites. *Wear* 237:54–62.

Kerr, C., Barker, D., Walsh, F., et al. 2000. The electrodeposition of composite coatings based on metal matrix-included particle deposits. *Transactions of Institute of Metal Finishing* 78:171–178.

Li, J. L., and Xiong, D. S. 2008. Tribological properties of nickel-based self-lubricating composite at elevated temperature and counterface material selection. *Wear* 265:533–539.

Li, K., Xiang, D. H., and Lei, X. Y. 2008. Green and self-lubricating polyoxymethylene composites filled with low-density polyethylene and rice husk flour. *Journal of Applied Polymer Science* 108:2778–2786.

Li, Y., Ruoff, R. S., and Chang, R. P. H. 2003. Boric acid nanotubes, nanotips, nanorods, microtubes, and microtips. *Chemical Materials* 15:3276–3285.

Liu, H. W., and Xue, Q. J. 1996. The tribological properties of TZP-graphite self-lubricating ceramics. *Wear* 198:143–149.

Livshits, A. I., and Shluger, A. L. 1997. Self-lubrication in scanning-force-microscope image formation on ionic surfaces. *Physical Reviews B* 56:12482–12489.

Lugscheider, E., Barwulf, S., and Barimani, C. 1999. Properties of tungsten and vanadium oxides deposited by MSIP-PVD process for self-lubricating applications. *Surface Coating Technology* 120:458–464.

Lumley, R. 2007. Advances in self-healing of metals. In *Self-healing materials: An alternative approach to 20 centuries of materials science*, vol. 100, ed. S. van der Zwaag, 219–254. Springer series in materials science. Dordrecht: Springer.

Ma, S. L., Ma, D. Y., Guo, Y., et al. 2007. Synthesis and characterization of super hard, self-lubricating Ti-Si-C-N nanocomposite coatings. *Acta Materialia* 55:6350–6355.

Manuel, M. V. 2007. Design of a biomimetic self-healing alloy composite. PhD thesis, Northwestern University, Evanston, IL.

Mollart, T. P., Haupt, J., Gilmore, R., et al. 1996. Tribological behavior of homogeneous Ti-B-N, Ti-B-N-C and TiN/h-BN/TiB2 multilayer coatings. *Surface Coating Technology* 86:231–236.

Morks, M. F., Tsunekawa, Y., Okumiya, M., et al. 2003. Microstructure of plasma-sprayed cast iron splats with different particle sizes. *Materials Transactions* 44:743–748.

Mulligan, C. P., and Gall, D. 2005. CrN-Ag self-lubricating hard coatings. *Surface Coating Technology* 200:1495–1500.

Neerinck, D., Persoone, P., Sercu, M., et al. 1998. Diamond-like nanocomposite coatings for low-wear and low-friction applications in humid environments. *Thin Solid Films* 317:402–404.

Nosonovsky, M., Amano, R., Lucci, J. M., and Rohatgi, P. K. 2009. Physical chemistry of self-organization and self-healing in metals. *Phys. Chem.-Chem. Phys.* 11:9530–9536.

Nosonovsky, M., and Bhushan, B. 2009. Thermodynamics of surface degradation, self-organization, and self-healing for biomimetic surfaces. *Philosophical Transactions of Royal Society A* 367:1607–1627.

———. 2010a. Introduction to green tribology: Principles, research areas, and challenges. *Philosophical Transactions of Royal Society A* 368:4677–4694.

———. 2010b. Surface self-organization: From wear to self-healing in biological and technical surfaces. *Applied Surface Science* 256:3982–3987.

Nosonovsky, M., and Rohatgi, P. K. 2011. *Biomimetics in materials science*. New York: Springer.

Olson, G. B. 1997. Computational design of hierarchically structured materials. *Science* 277:1237–1242.

Pei, Y. T., Galvan, D., and De Hosson, J. T. M. 2005. Nanostructure and properties of TiC/a-C: H composite coatings *Acta Materialia* 53:505–4521.

Polcar, T., Evaristo, M., and Cavaleiro, A. 2009. Comparative study of the tribological behavior of self-lubricating W-S-C and Mo-Se-C sputtered coatings. *Wear* 266:388–392.

Quintelier, J., Samyn, P., De Doncker, L., et al. 2009. Self-lubricating and self-protecting properties of polymer composites for wear and friction applications. *Polymer Composites* 30:932–940.

Rapoport, L., Fleischer, N., and Tenne, R. 2003. Fullerene-like WS2 nanoparticles: Superior lubricants for harsh conditions *Advanced Materials* 15:I651–655.

Scharf, T. W., Neira, A., and Hwang, J. Y. 2009. Self-lubricating carbon nanotube reinforced nickel matrix composites. *Journal of Applied Physics* 106:013508.

Skarvelis, P., and Papadimitriou, G. D. 2009. Plasma transferred arc composite coatings with self lubricating properties, based on Fe and Ti sulfides: Microstructure and tribological behavior. *Surface Coating Technology* 203:1384–1394.

Skeldon, P., Wang, H. W., and Thompson, G. E. 1997. Formation and characterization of self-lubricating MoS2 precursor films on anodized aluminum. *Wear* 206:187–196.

Strnad, G., Biro, D., Bolos, V., et al. 2009. Researches on nanocomposite self-lubricating coatings. *Metallurgia International* 14:121–124.

Suh, M. S., Chae, Y. H., and Kim, S. S. 2008. Friction and wear behavior of structural ceramics sliding against zirconia. *Wear* 264:800–806.

Sun, M., Luo, C., Xu, L., Ji, H., Ouyang, Q., Yu, D., and Chen, Y. 2005. Artificial lotus leaf by nanocasting. *Langmuir* 21:8978–8981.

Tocha, E., Pasaribu, H. R., Schipper, D. J., et al. 2008. Low friction in CuO-doped yttria-stabilized tetragonal zirconia ceramics: A complementary macro- and nanotribology study. *Journal of American Ceramic Society* 91:1646–1652.

Tu, J. P., Jiang, C. X., Guo, S. Y., et al. 2004. Micro-friction characteristics of aligned carbon nanotube film on an anodic aluminum oxide template. *Materials Letters* 58:1646–1649.

Umeda, J., Kondoh, K., and Imai, H. 2009. Friction and wear behavior of sintered magnesium composite reinforced with CNT-Mg2Si/MgO. *Materials Science & Engineering A* 504:157–162.

van der Zwaag, S., ed. 2007. *Self-healing materials—An alternative approach to 20 centuries of materials science*. Dordrecht, the Netherlands: Springer.

van der Zwaag, S. 2009. Self-healing behaviour in man-made engineering materials: Bioinspired but taking into account their intrinsic character. *Phil. Trans. R. Soc. A* 367:1689–1704.

Vilar, R. 1999. Laser cladding. *Journal of Laser Applications* 11:64–79.

Voevodin, A. A., Capano, M. A., Laube, S. J. P., Donley, M. S., and Zabinski, J. S. 1997. Design of a Ti/TiC/DLC functionally gradient coating based on studies of structural transitions in Ti-C thin films. *Thin Solid Films* 298:107–115.

Wang, A. H., Zhang, X. L., Zhang, X. F., et al. 2008. Ni-based alloy/submicron WS2 self-lubricating composite coating synthesized by Nd: YAG laser cladding *Materials Science Engineering A* 475:312–318.

Wang, L. Y., Tu, J. P., Chen, W. X., et al. 2003. Friction and wear behavior of electroless Ni-based CNT composite coatings. *Wear* 254:1289–1293.

Wu, X. H., Qin, W., Guo, Y., et al. 2008. Self-lubricative coating grown by micro-plasma oxidation on aluminum alloys in the solution of aluminate–graphite. *Applied Surface Science* 254: 6395–6399.

Zhang, W. J., Zhang, D., Le, Y. K., et al. 2008. Fabrication of surface self-lubricating composites of aluminum alloy. *Applied Surface Science* 255 2671–2674.

Zheng, W. T., and Sun, C. Q. 2006. Electronic process of nitriding: Mechanism and applications. *Progress in Solid State Chemistry* 34:1–20.

Zhou, S. M., Zhang, X. B., Ding, Z. P., et al. 2007. Fabrication and tribological properties of carbon nanotubes reinforced Al composites prepared by pressureless infiltration technique. *Composites A—Applied Science & Manufacturing* 38:301–306.

10

Outlook

As we discussed in the introductory chapter of this book, friction played an important role in ancient and medieval physics, somewhat similar to the force of inertia. These two forces—inertia and friction—cannot exist without each other. To put it simply, if friction existed without inertia, nothing could move, whereas if inertia existed without friction, nothing could stop. However, Aristotle's *Physics* lacked a correct understanding of both friction and inertia. Aristotle believed that a body keeps moving only when a force is applied and that velocity is proportional to the applied force. He also believed that straightforward and circular motion are two unrelated types of motion, as well as terrestrial and celestial motion, and that motion and rest are two unrelated states, leading to numerous contradictions in his concepts.

Further progress of mechanics helped to resolve these contradictions. Isaac Newton showed that terrestrial and celestial motions are the same. (A planet moves in accordance with the same laws of motion as an apple does.) He also showed, by discovering infinitesimal calculus simultaneously with Leibnitz, that the state of rest was nothing more than a special case of motion, rather than a separate essence. Euler showed that rotational motion is similar to translational motion. For the inertia force, it took centuries after Aristotle until the concept of inertial motion was introduced and correctly understood by Galileo and then incorporated into Newton's laws of motion as a part of the laws of nature. Euler's laws of motion included essentially two laws: for translational and rotational motion. Two laws of thermodynamics that are essential to many problems of continuum mechanics should be added to these, making thus the number of fundamental laws of mechanics equal to four.

The situation with friction was different. While a lot of experimental data were accumulated and led to the empirical Amontons–Coulomb laws of friction, friction force was never considered a fundamental force of nature. It was introduced into the equations of motion in an artificial ad hoc manner and friction was viewed as a complex phenomenon involving various unrelated mechanisms (adhesion, deformation, fracture, etc.). Despite that, friction possesses a large degree of universality and generality. It is amazing that the same friction laws apply to very diverse situations: different mechanisms of friction, different materials and classes of materials, and loads ranging from nanonewtons to millions of tons (thus covering the range of about 20 orders of magnitude).

The development of thermodynamic methods of analysis in recent decades makes it possible to view friction as a coherent phenomenon covered by the same set of general rules and governed by the second law of thermodynamics. First, it was realized that friction is not always compatible with the rest of the laws of mechanics, leading to the so-called Painleve paradoxes of non-unique or non-existent solutions. This is a result of over-simplification of a mechanical system, which is assumed to be rigid (non-deformable) and covered by a simple Coulomb friction law. If elastic deformation is introduced, most paradoxes can be resolved. However, in that case a new problem emerges: dynamic instabilities of the solutions. These instabilities lead to friction-induced self-excited vibrations and can result in the formation of newly organized patterns (such as periodic stick–slip). It was therefore realized that friction not only can stabilize a mechanical system and lead to dissipation and deterioration, but also, counter-intuitively, it can be a cause of instabilities and lead to the formation of new "secondary" structures at the frictional interface.

It was further understood that mechanical friction-induced vibrations and self-organizing processes can be viewed as a special case of more general thermodynamic effects. An entropic stability criterion (Equation 3.37) can be formulated for the rate of entropy production in the form of $\delta^2 \dot{S} > 0$. Depending on the processes involved, the rate of entropy production can include either pure mechanical terms or those involving heat and mass transfer, chemical reactions, electromagnetic interactions, and so on. The procedure of stability analysis for a particular system, as described in Chapter 3, should consist in establishing relationships between small fluctuations (variations) of parameters relevant to a particular system. Instability can indicate a possibility for the onset of self-organizing behavior.

Another important observation about friction is that, in most cases, a system with friction tends to evolve with time into a state of minimum (or, at least, reduced) dissipation. This is manifested on several levels: Friction is usually reduced during the run-in and shake-down transient processes, friction tends to increase with the age of contact and decrease with increasing velocity, and friction usually grows when something is changed in the system and then tends gradually to decrease. This is parallel to the non-equilibrium thermodynamic principle of minimum entropy production, which is valid in many situations (although a rigorous proof or definition of conditions in which it is valid remains controversial).

Friction is also related to the critical phenomena associated with the phase transitions of the second order. Two possible regimes of friction—stick and slip—can be viewed as two phase states. The onset of sliding is associated with the stick–slip transition. Such phase transitions are characterized by universal critical behavior, which, at least partially, can be responsible for the universal properties of frictional phenomena, which are largely independent of a particular mechanism of friction. The phase transitions are similar to destabilization to a certain extent; in both cases, there is a critical value of a control parameter near which critical behavior is observed. Thus, the potential $U(x) = ax^2 + bx^4$ with $b > 0$ has one minimum $x = 0$ for $a > 0$, one maximum at $x = 0$, and two minima $x = \pm\sqrt{-a/b}$ at $a < 0$, while $a = 0$ is a critical state with a triple root at $x = 0$. Such potential is used for the study of both destabilization and phase transitions of the second order.

Friction is a result of imperfections and various non-idealities in the material. In an ideal world of absolutely rigid, non-deformable bodies and conservative interactions, there would be no energy dissipation and, therefore, no friction. However, various defects, heterogeneities, asperities, and non-ideality of the material result in the statistical tendency of energy to dissipate during sliding, as reflected by the second law of thermodynamics. When this irreversible energy dissipation is spent for breaking the bonds between the particles of the material, this leads to the formation of debris and material removal and is observed as wear. When energy is spent for heating without visible material damage, this is observed as friction. In principle, the rate of dissipation should be related to the density of defects in the materials. A rigorous quantitative theory of how defect density affects dissipation on the basis of statistical mechanics remains to be built. However, the use of Shannon entropy as a surface characteristic is a first step in that direction.

Another important feature of all friction mechanisms is that they involve interactions along the interface, which are typically small in comparison with bulk interactions. For example, the contact occurs only on the tops of the asperities and thus the real area of contact constitutes only a small fraction of the total area of contact. Furthermore, the adhesion force through the frictional interface is usually weak in comparison with the strong bonds in the bulk of the body. As a result, a small non-dimensional parameter can be identified in frictional systems, leading to linear behavior. Furthermore, the frictional interface is very anisotropic in the sense that a small tangential force results in significant displacement in comparison with a normal force.

Due to the presence of the small parameter, one can use the Onsager phenomenological linear relationships (which are normally valid only in the quasi-linear case) to derive the viscous-type relationship between the velocity and the friction force. By further applying the asymptotic decomposition, we obtain the Amontons–Coulomb friction law. The positive value of friction is the consequence of the second law of thermodynamics; thus, conceptually one can relate the friction law to the second law of thermodynamics rather than consider it a phenomenological rule.

The area of friction-induced mechanical vibrations has been well established and studied. The thermodynamic approach provides an enhancement of this field by including heat propagation, mass transfer (due to wear or due to fraction separation in a composite material), tribochemical, triboelectrical, and other processes into the same conceptual approach of stable versus unstable processes at the sliding interface. This leads to the understanding of such phenomena as thermolubricity, electrolubricity, self-lubrication, in situ formed tribofilms, and third-body mechanisms of friction.

On a more practical side, we studied a number of frictional phenomena related to friction-induced instabilities, vibrations, waves, and self-organized patterns. Sliding of pure elastic half-spaces with a constant coefficient of friction may result in dynamic Adams–Martins instabilities. A similar but related type of instability occurs when the coefficient of friction decreases with increasing sliding velocity and when heat generation is sufficient to expand the material (thermoelastic instabilities). These frictional instabilities can lead to various, usually undesirable, effects, such as break squeal, belt vibration, or irritating door hinge noise. They are also used for sound generation in many musical instruments as well as in living nature.

Simple mechanical instabilities can result in a stick–slip motion with a certain pattern, which can result in "micro-slip" waves propagating along the frictional interface and be observed as friction reduction. The pattern-formation effects are especially significant and diverse when the processes of heat and mass transfer, chemical reactions, wear, electrical conduction, and others are involved. These processes can lead to the formation of in situ protective tribofilms and other effects that are helpful to control friction and wear. It is important to understand structure–property relationships involved in these processes, in order to be able to embed these mechanisms into new, smart self-lubricating materials. The recent progress in self-lubricating and self-healing materials makes these developments very encouraging.

Index

For Product Safety Concerns and Information please contact our EU
representative GPSR@taylorandfrancis.com Taylor & Francis Verlag GmbH,
Kaufingerstraße 24, 80331 München, Germany

Printed and bound by CPI Group (UK) Ltd, Croydon, CR0 4YY
01/05/2025
01858572-0001